LARGE ELASTIC
DEFORMATIONS

069 8463

THE UNIVERSITY OF ASTON IN BIRMINGHAM LIBRARY

1543311-01-03

531.3823 GRE

138584

LARGE ELASTIC DEFORMATIONS. 2ND ED
[1970] 0198533349

GREEN A E

ASTON UNIVERSITY LIBRARY

When on loan, this book is due for return on the last date stamped below.

17 DEC 1971	-8. JUL. 1977	29 OCT 1986 ASTON U.L.
24 MAR 1972	27 JUN 1980	
30 JUN 1972		
15 DEC 1972	-3 JUL 198	
23 MAR 1973	28 APR 1982 ASTON U.L.	
14 DEC 1973	18 JUN 1982 ASTON U.L.	
28 JUN 1974	29 MAY 1984 ASTON U.L.	
13 DEC 1974		
-4. OCT. 1976		
17 DEC 1976		
-1. APR. 1977		

ASTON UNIVERSITY
LIBRARY SERVICES

WITHDRAWN
FROM STOCK

LARGE ELASTIC DEFORMATIONS

BY

A. E. GREEN

SEDLEIAN PROFESSOR OF NATURAL PHILOSOPHY
UNIVERSITY OF OXFORD

AND

J. E. ADKINS

LATE PROFESSOR OF THEORETICAL MECHANICS
UNIVERSITY OF NOTTINGHAM

SECOND EDITION

REVISED BY

A. E. GREEN

CLARENDON PRESS · OXFORD

1970

Oxford University Press, Ely House, London W. 1

GLASGOW NEW YORK TORONTO MELBOURNE WELLINGTON
CAPE TOWN SALISBURY IBADAN NAIROBI DAR ES SALAAM LUSAKA ADDIS ABABA
BOMBAY CALCUTTA MADRAS KARACHI LAHORE DACCA
KUALA LUMPUR SINGAPORE HONG KONG TOKYO

© OXFORD UNIVERSITY PRESS 1960, 1970

FIRST EDITION 1960
SECOND EDITION 1970

PRINTED IN NORTHERN IRELAND
AT THE UNIVERSITIES PRESS, BELFAST

PREFACE TO SECOND EDITION

The first edition contained a final Chapter XI giving an account of some developments in non-linear continuum mechanics that were a natural extension of those of finite elasticity when the assumption of ideal elasticity was relaxed. Rapid advances have now been made in this field so that either a much larger book is needed for its exposition or the chapter must be deleted. A similar situation holds for Chapter IX on the theory of stability. Partly owing to the death of one of the authors a major extension of the book is not possible at present, so that both Chapters IX and XI of the first edition have been omitted. The rest of the book is mainly concerned with elastostatics and is unaltered apart from some corrections and additions in Chapter 2. Some changes have, however, been made in the chapter on thermodynamics.

A. E. G.

PREFACE TO THE FIRST EDITION

THE rapid development of the theory of large elastic deformations during the past decade was motivated in the first instance by the increasing importance of rubber in industrial applications and the need to get a clear picture of its mechanical properties. For this reason, the early work in the field was concentrated largely on ideally elastic, isotropic, incompressible materials with vulcanized rubber as a typical example of a substance of this kind. The initial success in solving a number of simple problems led to corresponding investigations for compressible and aeolotropic bodies, while the difficulty of solving all but the simplest problems in complete generality has led to the development of approximation procedures.

The general theory of elasticity for finite deformations has been given by Green and Zerna in the book *Theoretical Elasticity* (subsequently referred to as $T.E.$) published by the Clarendon Press (1954). This theory is presented in compact form with the aid of tensor notation and the results are applied to solve a number of special problems, mainly for isotropic incompressible materials. In the present book attention is concentrated on subsequent developments. A summary of the essential basic formulae of the finite theory is given in Chapter I, again using the tensor notation of $T.E.$; for proofs of these formulae and an exposition of the elementary tensor analysis required the reader is referred to the earlier book. Chapter I then proceeds to an examination of the form of the strain energy function for the basic crystal classes together with the development of the stress-strain relations for orthotropic and transversely isotropic materials. Curvilinear aeolotropy and materials subject to constraints are also examined.

Chapter II contains some of the exact solutions of the finite theory, mainly for aeolotropic bodies. The earlier part of this chapter is concerned with cylindrically symmetrical problems and the flexure problem in which the results apply for a general form of strain energy. In later sections solutions are derived using the restricted Mooney form of strain energy for rubber-like materials.

In Chapter III the theory of plane strain is developed using tensor notation initially and specializing subsequently to give a complex variable formulation. A number of special problems are examined using this theory. Chapter IV deals with plane stress and the membrane theory of thin shells. In applications attention is confined to axially symmetrical problems which involve only one independent variable.

For some of these problems, the membrane equations can be solved analytically; for others, simple numerical methods of integration are available.

A method of successive approximation is developed in Chapter V and illustrated by simple examples. Chapter VI deals with the application of this approximation method to two-dimensional problems, attention being focused largely upon plane strain. The equations derived resemble those of the classical infinitesimal theory, and a complex variable formulation permits the use of the powerful techniques evolved for the classical theory by Muskhelishvili.

The reinforcement of elastic materials by systems of thin flexible inextensible cords is considered in Chapter VII, the main problems examined being either two-dimensional or those possessing cylindrical symmetry. Such problems arise in many industrial applications of rubber. Chapters VIII and IX are concerned with the theories of thermoelasticity and elastic stability respectively for finite deformations. A description of some of the more important physical experiments which have been carried out on vulcanized rubber is contained in Chapter X. These experiments illustrate how a completely general theory, such as that developed in the earlier chapters, may be applied to evaluate the mechanical properties of real materials.

In the last chapter of the book an account is given of some of the recent developments in non-linear continuum mechanics. The theory may be regarded as a natural extension of that of finite elasticity when the assumption of ideal elasticity is relaxed. Owing to the rapid advances now being made in this field attention is confined to a consideration of the kinematics of deformation and some of the simpler forms of the stress deformation relations.

Many of the developments described in this book have been due to the initiative of Professor R. S. Rivlin, and the debt which the subject owes him is apparent from the references in the text. The authors record with pleasure their own indebtedness to him for many stimulating discussions and contacts during the past few years. Acknowledgement should also be made of the encouragement and financial support given to a number of workers in the field by the British Rubber Producer's Research Association with whom Professor Rivlin and one of the present authors (J. E. A.) were at one time associated.

It is a pleasure to thank Dr. L. R. G. Treloar for helpful comments on an earlier draft of Chapter X and for permission to reproduce Figs. 10.4 and 10.6 from his book, Miss M. J. Haining and Miss A. P.

Dick for typing the manuscript, and the Press for their care and attention in getting the book into print.

Thanks are due to the Royal Society for permission to reproduce Figs. 4.2–4.6, 10.3, 10.5, and 10.7–10.10 from their *Philosophical Transactions*.

A. E. G.
J. E. A.

CONTENTS

I. THE STRAIN ENERGY FUNCTION — 1
  1.1. Summary of notation and formulae — 1
  1.2. Theorems on invariants — 7
  1.3. Strain energy — 8
  1.4. Crystal classes — 10
  1.5. Form-invariance — 11
  1.6. Symmetry transformations — 11
  1.7. The triclinic system — 13
  1.8. The monoclinic system — 14
  1.9. The rhombic system — 15
  1.10. The tetragonal system — 15
  1.11. The cubic system — 17
  1.12. The hexagonal system — 19
  1.13. Transverse isotropy — 24
  1.14. Isotropy — 26
  1.15. Stress–strain relations — 27
  1.16. Curvilinear aeolotropy — 29
  1.17. Geometrical constraints — 33
  1.18. Stress–strain relations for materials subject to constraints — 34

II. GENERAL SOLUTIONS OF PROBLEMS — 37
  2.1. Cylindrically symmetrical problems: incompressible transverse isotropy — 39
  2.2. Extension and torsion of an incompressible solid circular cylinder — 43
  2.3. Extension, inflation, and torsion of an incompressible circular cylindrical tube — 44
  2.4. Incompressible circular tube turned inside out — 47
  2.5. Flexure of an initially curved incompressible cuboid — 47
  2.6. Straightening, stretching, and shearing of an initially-curved cuboid — 49
  2.7. Flexure of a cuboid of incompressible transversely isotropic material — 50
  2.8. Homogeneous strain of an aeolotropic cuboid — 55
  2.9. Flexure of an incompressible aeolotropic cuboid — 57
  2.10. Cylindrically symmetrical problems: incompressible cylindrical aeolotropy — 59
  2.11. Combined flexure and homogeneous strain of an incompressible aeolotropic cuboid — 62
  2.12. Cylindrical symmetry: compressible transversely isotropic material — 65
  2.13. Extension and torsion of a solid compressible circular cylinder — 67
  2.14. Extension, inflation, and torsion of a compressible circular cylindrical tube — 68

- 2.15. Flexure of a cuboid of compressible aeolotropic material ... 69
- 2.16. Cylindrically symmetrical problems; compressible cylindrically aeolotropic bodies ... 72
- 2.17. Combined flexure and homogeneous strain of a compressible aeolotropic cuboid ... 74
- 2.18. Cylindrically symmetrical problems: parametric forms ... 75
- 2.19. Combined flexure and homogeneous strain of an aeolotropic cuboid: parametric forms ... 82
- 2.20. Inversion and expansion of a spherical shell ... 84
- 2.21. Generalizations of the shear problem ... 87
- 2.22. Generalized shear: complex variable ... 91
- 2.23. Generalized shear: conformal transformations ... 93
- 2.24. Cylindrical annuli ... 96

III. FINITE PLANE STRAIN ... 101
- 3.1. Plane strain superposed on uniform finite extension ... 102
- 3.2. Airy's stress function ... 104
- 3.3. Stress–strain relations: equations of equilibrium ... 107
- 3.4. Complex variable formulation: geometrical relations ... 112
- 3.5. Complex variable: stresses; equations of equilibrium ... 116
- 3.6. Incompressible materials: reciprocal equations ... 120
- 3.7. Flexure of a cuboid ... 123
- 3.8. Generalized shear ... 127
- 3.9. Further special solutions for a Mooney material ... 129

IV. THEORY OF ELASTIC MEMBRANES ... 133
- 4.1. Stress resultants and loads: plane sheet ... 134
- 4.2. Equations of equilibrium: Airy's stress function ... 136
- 4.3. Values of strains and stresses at middle surface ... 138
- 4.4. Stress resultants: equations of equilibrium ... 141
- 4.5. Geometrical relations for a deformed shell ... 143
- 4.6. Stress resultants and stress couples ... 144
- 4.7. Surface conditions ... 145
- 4.8. Equations of equilibrium ... 147
- 4.9. Stress resultants ... 147
- 4.10. Radial deformation of a plane sheet containing a circular hole or inclusion ... 152
- 4.11. Axially symmetrical problems ... 156
- 4.12. Inflation of a circular plane sheet ... 161
- 4.13. Inflation of a spherical shell ... 169
- 4.14. Deformation of surface of revolution into one of similar shape ... 171

V. THEORY OF SUCCESSIVE APPROXIMATIONS ... 174
- 5.1. Uniqueness of series solution ... 175

5.2.	Expansion of displacement, strain, and invariants	178
5.3.	Expansion of stress: compressible body	180
5.4.	Expansion of stress: incompressible body	182
5.5.	Equations of equilibrium: surface stresses	184
5.6.	Hydrostatic pressure	185
5.7.	Simple extension	186
5.8.	Simple shear	187
5.9.	Deformation of a cylinder	188
5.10.	Torsion of a compressible cylinder: classical theory	191
5.11.	Torsion: second order	192
5.12.	Simply-connected cross-section	198

VI. APPROXIMATE SOLUTION OF TWO-DIMENSIONAL PROBLEMS 200

6.1.	Plane strain: isotropic incompressible bodies	201
6.2.	Coordinates in the undeformed body	205
6.3.	Force and couple resultants	206
6.4.	Complex potentials for an infinite body	209
6.5.	General theory for two-dimensional problems	212
6.6.	Simultaneous solution of first and second boundary value problems	216
6.7.	Infinite region containing a circular boundary	219
6.8.	Incompressible plane strain: infinite body containing a circular hole	222
6.9.	Incompressible plane strain: infinite body containing a circular rigid inclusion	226

VII. REINFORCEMENT BY INEXTENSIBLE CORDS 228

7.1.	Physical assumptions	228
7.2.	Uniform stretching of a plane reinforced sheet	229
7.3.	Simple extension of a reinforced incompressible isotropic sheet	232
7.4.	Plane deformation of a thin reinforced sheet	234
7.5.	Plane deformation of a reinforced sheet: stress resultants	237
7.6.	Plane deformation of a reinforced sheet: Airy's stress function	238
7.7.	Plane deformation of a sheet reinforced with systems of parallel straight cords	240
7.8.	Solutions of constraint equations for a sheet reinforced with initially straight cords	243
7.9.	Plane deformation of a network of inextensible cords	248
7.10.	Cylindrically symmetrical deformations: constraint conditions	252
7.11.	Cylindrically symmetrical problems: stress resultants in a layer of cords	254
7.12.	Cylindrical symmetry: forces on the composite body	255
7.13.	Flexure of an initially curved reinforced block	259
7.14.	Inflation and extension of a cylindrical tube	261
7.15.	Combined flexure and homogeneous strain of a reinforced cuboid	264

7.16.	Flexure of a reinforced cuboid	266
7.17.	Simple extension and flexure of a thin reinforced isotropic sheet	267
7.18.	Reinforced bodies and materials subject to constraints	271

VIII. THERMOELASTICITY 273

8.1.	Energy equation	273
8.2.	Entropy production inequality	275
8.3.	Equations for an elastic body	275
8.4.	Incompressible bodies	278
8.5.	Heat conduction vector for isotropic bodies	280
8.6.	Heat conduction vector for isotropic bodies: alternative form	283
8.7.	Classical theory for isotropic bodies	285

IX. EXPERIMENTAL APPLICATIONS 288

9.1.	Nature of the problem	288
9.2.	Preliminary investigations	290
9.3.	The functional form of W	293
9.4.	Experiments on rubber	296
9.5.	Pure homogeneous strain of a thin rubber sheet	297
9.6.	Pure shear	300
9.7.	Simple elongation	301
9.8.	Uniform two-dimensional extension	304
9.9.	Experiments on torsion	307
9.10.	Torsion of a cylindrical rod	311
9.11.	Inflation, extension, and torsion of a cylindrical tube	313
9.12.	Further experiments	314

APPENDIX. REDUCTION OF A MATRIX POLYNOMIAL OF TWO MATRICES 315

AUTHOR INDEX 319

SUBJECT INDEX 321

I
THE STRAIN ENERGY FUNCTION

An historical and critical account of large elastic deformation theory and applications up to about 1952 has been given by Truesdell[†] and this has been followed by a review of progress in non-linear elasticity up to 1955 by Doyle and Ericksen.[‡] Both these reviews contain extensive references which will not be repeated in full here. Additional references are given in monographs by Truesdell and Toupin and by Truesdell and Noll.[§] In the present book we follow, as far as possible, the notations used by Green and Zerna[||] in *Theoretical Elasticity*.

A summary of basic notations and formulae is given in § 1.1 and the reader is referred to *T.E.* for detailed derivation of the formulae. Most of the rest of Chapter I is devoted to a discussion of the form of the elastic potential or strain energy for large elastic deformations for the crystal classes. This work is based on a paper by Smith and Rivlin[††] who also included electrical effects,[‡‡] but these will not be considered here. Since the discussion depends on known theorems in classical invariant theory these theorems are summarized in § 1.2.

1.1. Summary of notation and formulae

Throughout the book Latin indices take the values 1, 2, 3 unless stated otherwise. We consider a body of perfectly elastic material to undergo a deformation in which a point P_0 initially at x_i referred to a fixed rectangular cartesian coordinate system moves to y_i in the same coordinate system.[§§] We also define a general curvilinear coordinate

---

[†] C. Truesdell, *J. rat. Mech. Analysis* **1** (1952), 125; **2** (1953), 593.

[‡] T. C. Doyle and J. L. Ericksen, *Adv. appl. Mech.* **4** (1956), 53. See also F. R. Eirich, *Rheology* (New York, 1956), p. 351; R. S. Rivlin, Brown University Report No. C11–43.

[§] C. Truesdell and R. A. Toupin, The classical field thesis, *Handbuch der Physik* (Berlin–Gottingen–Heidelberg 1960). C. Truesdell and W. Noll, The non-linear field theories of mechanics, *Handbuch der Physik* (Berlin–Heidelberg–New York 1965).

[||] A. E. Green and W. Zerna, *Theoretical Elasticity* (edn. 2. Clarendon Press, Oxford, 1968), referred to subsequently as *T.E.*

[††] G. F. Smith and R. S. Rivlin, *Trans. Am. math. Soc.* **88** (1958), 175.

[‡‡] Brown University Reports (1956) DA-3487/4, 9.

[§§] The position of the index on coordinates x_i, y_i and θ^i is immaterial and it is convenient to use either upper or lower indices. The indices will be omitted if, for example, only a general reference to the coordinates θ is required. The differential and partial derivative involving general curvilinear coordinates θ will always be denoted by $d\theta^i$ and $\partial\theta^i$ respectively since $d\theta_i$ has a different meaning and is not a differential. For rectangular coordinates, however, we use either dx_i, dx^i for differentials and ∂x_i, ∂x^i for partial derivatives, since $dx_i = dx^i$.

system θ^i ($= \theta_i$) so that
$$x_i = x_i(\theta_1, \theta_2, \theta_3), \tag{1.1.1}$$

where $x_i(\theta_1, \theta_2, \theta_3)$ is a single-valued function possessing a unique inverse and continuous derivatives up to any required order, except possibly at singular points, lines, or surfaces. It follows that $|\partial x^i/\partial \theta^r|$ is non-zero and we assume that

$$\left|\frac{\partial x^i}{\partial \theta^r}\right| > 0 \tag{1.1.2}$$

everywhere. The deformation of the body is then defined by

$$y_i = y_i(x_1, x_2, x_3, t), \tag{1.1.3}$$

or by
$$y_i = y_i(\theta_1, \theta_2, \theta_3, t), \tag{1.1.4}$$

the functional forms in (1.1.3) and (1.1.4) being different in general. The coordinates θ (and x) are called convected or material coordinates. The functions $y_i(\theta_1, \theta_2, \theta_3, t)$ are assumed to be single-valued and continuously differentiable with respect to θ and t as many times as required, except possibly at singular points, lines, and surfaces, and for each value of t they have unique inverses. Also, if the deformation is to be possible in a real material, we have

$$\left|\frac{\partial y^i}{\partial \theta^r}\right| > 0. \tag{1.1.5}$$

The position vector of the point initially at x_i is $\mathbf{r}$, where

$$\mathbf{r} = \mathbf{r}(\theta_1, \theta_2, \theta_3). \tag{1.1.6}$$

The corresponding position vector of the point y_i is $\mathbf{R}$, where

$$\mathbf{R} = \mathbf{R}(\theta_1, \theta_2, \theta_3, t). \tag{1.1.7}$$

Base vectors and metric tensors for the coordinate system θ may be defined in both the undeformed and deformed bodies. Thus

$$\left.\begin{aligned}
\mathbf{g}_i &= \frac{\partial \mathbf{r}}{\partial \theta^i}, \quad \mathbf{G}_i = \frac{\partial \mathbf{R}}{\partial \theta^i}, \quad g_{ij} = \mathbf{g}_i \cdot \mathbf{g}_j = \frac{\partial x^r}{\partial \theta^i}\frac{\partial x^r}{\partial \theta^j} \\
G_{ij} &= \mathbf{G}_i \cdot \mathbf{G}_j = \frac{\partial y^r}{\partial \theta^i}\frac{\partial y^r}{\partial \theta^j}, \quad g^{ir}g_{rj} = \delta^i_j, \quad G^{ir}G_{rj} = \delta^i_j \\
\mathbf{g}^i &= g^{ir}\mathbf{g}_r, \quad \mathbf{G}^i = G^{ir}\mathbf{G}_r
\end{aligned}\right\} \tag{1.1.8}$$

If
$$g = |g_{ij}|, \quad G = |G_{ij}|, \tag{1.1.9}$$

1.1 THE STRAIN ENERGY FUNCTION

we see that

$$\sqrt{g} = \left|\frac{\partial x^i}{\partial \theta^j}\right|, \quad \sqrt{G} = \left|\frac{\partial y^i}{\partial \theta^j}\right|, \quad \sqrt{\left(\frac{G}{g}\right)} = \left|\frac{\partial y^i}{\partial x^j}\right| > 0, \quad (1.1.10)$$

positive roots being chosen in view of (1.1.2) and (1.1.5).

The displacement vector $\mathbf{v}(\theta_1, \theta_2, \theta_3, t)$ may be expressed in the alternative forms

$$\mathbf{v} = v^i \mathbf{g}_i = v_j \mathbf{g}^j, \quad \mathbf{v} = V^i \mathbf{G}_i = V_j \mathbf{G}^j. \quad (1.1.11)$$

The strain tensor is defined to be

$$\gamma_{ij} = \tfrac{1}{2}(G_{ij} - g_{ij}), \quad (1.1.12)$$

and in terms of the displacement vector this becomes

$$\begin{aligned}\gamma_{ij} &= \tfrac{1}{2}(\mathbf{g}_i \cdot \mathbf{v}_{,j} + \mathbf{g}_j \cdot \mathbf{v}_{,i} + \mathbf{v}_{,i} \cdot \mathbf{v}_{,j}) \\ &= \tfrac{1}{2}(v_i|_j + v_j|_i + v^r|_i v_r|_j)\end{aligned}, \quad (1.1.13)$$

$$\begin{aligned}\gamma_{ij} &= \tfrac{1}{2}(\mathbf{G}_i \cdot \mathbf{v}_{,j} + \mathbf{G}_j \cdot \mathbf{v}_{,i} - \mathbf{v}_{,i} \cdot \mathbf{v}_{,j}) \\ &= \tfrac{1}{2}(V_i\|_j + V_j\|_i - V^r\|_i V_r\|_j)\end{aligned}. \quad (1.1.14)$$

In these formulae a comma denotes partial differentiation with respect to θ, a single line denotes covariant differentiation with respect to θ using the metric tensors g_{ij}, g^{ij}, and a double line denotes covariant differentiation with respect to θ using the metric tensors G_{ij}, G^{ij}.

When the coordinates θ are taken to coincide with the rectangular cartesian coordinates x which define points in the undeformed body we denote γ_{ij} by e_{ij}, so that

$$e_{rs} = \frac{\partial \theta^i}{\partial x^r} \frac{\partial \theta^j}{\partial x^s} \gamma_{ij} = \frac{1}{2}\left(\frac{\partial y^m}{\partial x^r} \frac{\partial y^m}{\partial x^s} - \delta_{rs}\right), \quad (1.1.15)$$

and, from (1.1.13),

$$e_{ij} = \frac{1}{2}\left(\frac{\partial u_i}{\partial x_j} + \frac{\partial u_j}{\partial x_i} + \frac{\partial u_r}{\partial x_i}\frac{\partial u_r}{\partial x_j}\right), \quad (1.1.16)$$

where u_r are the components of displacement referred to axes x_r in the undeformed body. We observe that e_{rs}, regarded as a function of x_i, is a cartesian tensor under all changes of rectangular axes x_i. If the components of displacement are referred to rectangular cartesian axes y_i in the deformed body, and are denoted† by U_i, then putting

---

† Since we have the same rectangular frame of reference for the undeformed and deformed body $U_i = u_i$, but it is convenient to retain the notational difference.

$\theta = y$ in (1.1.14), this particular value of γ_{ij} becomes

$$\frac{1}{2}\left(\frac{\partial U_i}{\partial y_j} + \frac{\partial U_j}{\partial y_i} - \frac{\partial U_r}{\partial y_i}\frac{\partial U_r}{\partial y_j}\right). \tag{1.1.17}$$

Strain invariants I_1, I_2, I_3 are given by

$$\left.\begin{aligned} I_1 &= 3+2\gamma_r^r = g^{rs}G_{rs} = 3+2e_{rr} \\ I_2 &= 3+4\gamma_r^r+2(\gamma_r^r\gamma_s^s-\gamma_s^r\gamma_r^s) = g_{rs}G^{rs}I_3 \\ &= 3+4e_{rr}+2(e_{rr}e_{ss}-e_{rs}e_{rs}) \\ I_3 &= |\delta_s^r+2\gamma_s^r| = G/g = |\delta_{rs}+2e_{rs}| \end{aligned}\right\}, \tag{1.1.18}$$

where
$$\gamma_s^r = g^{rm}\gamma_{ms}. \tag{1.1.19}$$

We shall also use strain invariants J_1, J_2, J_3 defined in terms of I_1, I_2, I_3 by the equations

$$\left.\begin{aligned} J_1 &= I_1-3 \\ J_2 &= I_2-2I_1+3 \\ J_3 &= I_3-I_2+I_1-1 \end{aligned}\right\}. \tag{1.1.20}$$

The state of stress at a point in the deformed body, referred to curvilinear coordinates θ, is determined by the symmetric stress tensor τ^{ij}, measured *per unit area of the deformed body*. The stress vectors $\mathbf{t}_i$ associated with the θ^i-surfaces in the deformed body are then given by

$$\mathbf{t}_i\sqrt{G^{ii}} = \tau^{ij}\mathbf{G}_j = \tau_j^i\mathbf{G}^j, \qquad \tau^{ij} = \tau^{ji}. \tag{1.1.21}$$

Indices appearing more than twice are not summed unless explicitly indicated. Alternatively,

$$\mathbf{t}_i = \sum_{j=1}^{3}\sigma_{ij}\frac{\mathbf{G}_j}{\sqrt{G_{jj}}}, \tag{1.1.22}$$

where
$$\sigma_{ij} = \sqrt{(G_{jj}/G^{ii})}\tau^{ij} \tag{1.1.23}$$

are physical components of stress, referred to oblique unit vectors. The stress vector $\mathbf{t}$ associated with a surface in the deformed body, whose unit normal is $\mathbf{n}$, is given by

$$\mathbf{t} = \tau^{ij}n_i\mathbf{G}_j = \tau_j^i n_i\mathbf{G}^j = \frac{n_i\mathbf{T}_i}{\sqrt{G}}, \tag{1.1.24}$$

where
$$\mathbf{n} = n_i\mathbf{G}^i = n^i\mathbf{G}_i, \tag{1.1.25}$$

and
$$\mathbf{T}_i = \mathbf{t}_i\sqrt{(GG^{ii})} = \sqrt{(G)}\tau^{ij}\mathbf{G}_j = \sqrt{(G)}\tau_j^i\mathbf{G}^j. \tag{1.1.26}$$

The equations of motion have the alternative forms

$$\mathbf{T}_{i,i}+\rho\mathbf{F}\sqrt{G} = \rho\mathbf{f}\sqrt{G}, \tag{1.1.27}$$

or
$$\tau^{ij}\|_i + \rho F^j = \rho f^j, \tag{1.1.28}$$

or
$$\tau^i_j\|_i + \rho F_j = \rho f_j, \tag{1.1.29}$$

where
$$\mathbf{F} = F^i\mathbf{G}_i = F_j\mathbf{G}^j, \qquad \mathbf{f} = f^i\mathbf{G}_i = f_j\mathbf{G}^j, \tag{1.1.30}$$

and $\mathbf{F}$, $\mathbf{f}$ are respectively the body-force vector, per unit mass, and the acceleration vector. Also, ρ is the density of the deformed body, and

$$\rho\sqrt{G} = \rho_0\sqrt{g}, \tag{1.1.31}$$

where ρ_0 is the density of the undeformed body.

When the body is perfectly elastic an elastic potential or strain energy function W exists, measured per unit volume of the undeformed body, which depends on the strain components γ_{ij} and has the property

$$\dot{W} = \frac{\mathbf{T}_i \cdot \dot{\mathbf{v}}_{,i}}{\sqrt{g}} = \sqrt{(I_3)}\tau^{ij}\mathbf{G}_j \cdot \dot{\mathbf{v}}_{,i} = \sqrt{(I_3)}\tau^{ij}\dot{\gamma}_{ij}, \tag{1.1.32}$$

where a superposed dot denotes partial differentiation with respect to t holding θ fixed. Since $\dot{\gamma}_{ij} = \dot{\gamma}_{ji}$ we have, for a compressible body,

$$\tau^{ij} = \frac{1}{2\sqrt{I_3}}\left(\frac{\partial W}{\partial \gamma_{ij}} + \frac{\partial W}{\partial \gamma_{ji}}\right). \tag{1.1.33}$$

When the body is incompressible,

$$I_3 = 1, \qquad G^{ij}\dot{\gamma}_{ij} = 0, \tag{1.1.34}$$

so that (1.1.33) is replaced by

$$\tau^{ij} = pG^{ij} + \frac{1}{2}\left(\frac{\partial W}{\partial \gamma_{ij}} + \frac{\partial W}{\partial \gamma_{ji}}\right), \tag{1.1.35}$$

where p is a scalar function of the coordinates θ.

For some purposes it is convenient to define the state of stress in the deformed body by a symmetric stress tensor s^{ij} measured *per unit area of the undeformed body*. The analysis of stress using s^{ij} is similar to that given in $T.E.$ using τ^{ij} so only the main results are quoted here. Stress vectors $_0\mathbf{t}_i$ measured per unit area of the undeformed body but associated with the θ^i-surfaces in the deformed body are given by

$$_0\mathbf{t}_i\sqrt{g^{ii}} = s^{ij}\mathbf{G}_j = s^{ir}(\delta^j_r + v^j|_r)\mathbf{g}_j, \qquad s^{ij} = s^{ji}. \tag{1.1.36}$$

The stress vector $_0\mathbf{t}$, per unit area of the undeformed body, associated with a surface in the deformed body, whose unit normal in its undeformed position is $_0\mathbf{n}$, is

$$_0\mathbf{t} = s^{ij}{}_0n_i\mathbf{G}_j = s^i_j{}_0n_i\mathbf{G}^j, \tag{1.1.37}$$

where
$$_0\mathbf{n} = {_0n_i}\mathbf{g}^i = {_0n^i}\mathbf{g}_i. \tag{1.1.38}$$

Also,
$$_0\mathbf{t} = \frac{_0n_i \mathbf{T}_i}{\sqrt{g}}, \quad \mathbf{T}_i = \sqrt{(gg^{ii})}{_0\mathbf{t}_i} = \sqrt{(g)}s^{ij}\mathbf{G}_j, \tag{1.1.39}$$

and the vector equation of motion becomes
$$\mathbf{T}_{i,i} + \rho_0 \mathbf{F}\sqrt{g} = \rho_0 \mathbf{f}\sqrt{g}. \tag{1.1.40}$$

Alternatively
$$[s^{ir}(\delta_r^j + v^j|_r)]|_i + \rho_0({_0F^j}) = \rho_0({_0f^j}), \tag{1.1.41}$$

where
$$\mathbf{F} = {_0F^j}\mathbf{g}_j, \quad \mathbf{f} = {_0f^j}\mathbf{g}_j. \tag{1.1.42}$$

Finally,
$$s^{ij} = \frac{1}{2}\left(\frac{\partial W}{\partial \gamma_{ij}} + \frac{\partial W}{\partial \gamma_{ji}}\right) = \tau^{ij}\sqrt{I_3} \tag{1.1.43}$$

for a compressible body, and
$$s^{ij} = pG^{ij} + \frac{1}{2}\left(\frac{\partial W}{\partial \gamma_{ij}} + \frac{\partial W}{\partial \gamma_{ji}}\right) \tag{1.1.44}$$

for an incompressible body.

The stress vectors $\mathbf{t}_i$, ${_0\mathbf{t}_i}$ in (1.1.21) and (1.1.36) have been expressed in terms of components referred to base vectors $\mathbf{G}_j$ in the deformed body. These vectors could, of course, be expressed in terms of components referred to base vectors $\mathbf{g}_j$ in the undeformed body and we add these forms here for completeness. Thus

$$\left.\begin{array}{l}_0\mathbf{t}_i\sqrt{g^{ii}} = t^{ij}\mathbf{g}_j, \quad {_0\mathbf{t}} = {_0n_i}\, t^{ij}\mathbf{g}_j \\ \mathbf{t}_i\sqrt{G^{ii}} = \pi^{ij}\mathbf{g}_j, \quad \mathbf{t} = \pi^{ij}n_i\mathbf{g}_j \\ \mathbf{T}_i = \sqrt{(g)}t^{ij}\mathbf{g}_j = \sqrt{(G)}\pi^{ij}\mathbf{g}_j\end{array}\right\}, \tag{1.1.45}$$

where
$$t^{ij} = s^{ir}(\delta_r^j + v^j|_r), \quad \pi^{ij} = \tau^{ir}(\delta_r^j + v^j|_r). \tag{1.1.46}$$

The equations of motion in terms of t^{ij} are
$$t^{ij}|_i + \rho_0({_0F^j}) = \rho_0({_0f^j}), \tag{1.1.47}$$

where t^{ij} is not symmetric but satisfies the equation
$$t^{im}\mathbf{g}_m \cdot \mathbf{G}^j = t^{jm}\mathbf{g}_m \cdot \mathbf{G}^i. \tag{1.1.48}$$

Further, from (1.1.32) and (1.1.45),
$$\dot{W} = t^{ij}\dot{v}_j|_i = \sqrt{(I_3)}\pi^{ij}\dot{v}_j|_i, \tag{1.1.49}$$

so that
$$\sqrt{(I_3)}\pi^{ij} = t^{ij} = \frac{\partial W}{\partial v_j|_i}. \tag{1.1.50}$$

For ease of reference we record the most important stress relations in Table I. The last entries in this table are alternative forms of equations of equilibrium when body forces are zero.

TABLE I

$$\mathbf{t}_i\sqrt{G^{ii}} = \tau^{ij}\mathbf{G}_j = \pi^{ij}\mathbf{g}_j, \qquad \tau^{ij} = \tau^{ji}$$

$$_0\mathbf{t}_i\sqrt{g^{ii}} = s^{ij}\mathbf{G}_j = t^{ij}\mathbf{g}_j, \qquad s^{ij} = s^{ji}$$

$$\mathbf{T}_i = \sqrt{(G)}\tau^{ij}\mathbf{G}_j = \sqrt{(g)}s^{ij}\mathbf{G}_j = \sqrt{(G)}\pi^{ij}\mathbf{g}_j = \sqrt{(g)}t^{ij}\mathbf{g}_j$$

$$\mathbf{t} = \tau^{ij}n_i\mathbf{G}_j = \pi^{ij}n_i\mathbf{g}_j = \frac{n_i\mathbf{T}_i}{\sqrt{G}}$$

$$_0\mathbf{t} = s^{ij}{}_0n_i\mathbf{G}_j = t^{ij}{}_0n_i\mathbf{g}_j = \frac{_0n_i\mathbf{T}_i}{\sqrt{g}}$$

$$\tau^{ij}\sqrt{I_3} = s^{ij} = \frac{1}{2}\left(\frac{\partial W}{\partial \gamma_{ij}} + \frac{\partial W}{\partial \gamma_{ji}}\right), \qquad \pi^{ij}\sqrt{I_3} = t^{ij} = \frac{\partial W}{\partial v_j|_i}$$

$$t^{ij} = s^{ir}(\delta_r^j + v^j|_r), \qquad \pi^{ij} = \tau^{ir}(\delta_r^j + v^j|_r)$$

$$\mathbf{T}_{i,i} = 0, \qquad \tau^{ij}\|_i = 0, \qquad t^{ij}|_i = 0$$

1.2. Theorems on invariants†

THEOREM 1. *A polynomial basis for a polynomial in the n vectors* $\alpha^{(r)} = (\alpha_1^{(r)}, \alpha_2^{(r)}, ..., \alpha_n^{(r)}) (r = 1, 2, ..., n)$ *in n-dimensional space, which is invariant under all proper orthogonal transformations, is formed by the scalar products*

$$\alpha_i^{(r)}\alpha_i^{(s)}, \qquad (1.2.1)$$

and the *determinant*
$$|\alpha_s^{(r)}|. \qquad (1.2.2)$$

All indices and suffixes here take values 1, 2,..., *n*.

THEOREM 2. *A polynomial basis for polynomials which are symmetric in the two sets of variables* $(y_1, y_2, ..., y_n)$ *and* $(z_1, z_2, ..., z_n)$ *is formed from the quantities*

$$\left.\begin{array}{ll} K_j = \tfrac{1}{2}(y_j+z_j) & (j = 1, 2, ..., n) \\ K_{jk} = \tfrac{1}{2}(y_jz_k+y_kz_j) & (j, k = 1, 2, ..., n) \end{array}\right\}. \qquad (1.2.3)$$

THEOREM 3. *A polynomial basis for polynomials which are symmetric in the three pairs of variables* (y_1, z_1), (y_2, z_2), *and* (y_3, z_3) *is*

† See, for example, H. Weyl, *The Classical Groups* (Princeton University Press, 1946), pp. 36, 53, 276.

formed by the quantities

$$\left.\begin{aligned}
&L_1 = y_1+y_2+y_3, \quad L_2 = y_2y_3+y_3y_1+y_1y_2, \quad L_3 = y_1y_2y_3 \\
&L_4 = z_1+z_2+z_3, \quad L_5 = z_2z_3+z_3z_1+z_1z_2, \quad L_6 = z_1z_2z_3 \\
&L_7 = y_2z_3+y_3z_2+y_3z_1+y_1z_3+y_1z_2+y_2z_1 \\
&L_8 = y_1z_2z_3+y_2z_3z_1+y_3z_1z_2 \\
&L_9 = z_1y_2y_3+z_2y_3y_1+z_3y_1y_2
\end{aligned}\right\}. \quad (1.2.4)$$

THEOREM 4. *A polynomial basis for polynomials in the variables $y_1, y_2, y_3, z_1, z_2, z_3$ which are form-invariant under cyclic rotation of the subscripts 1, 2, 3 is formed by the quantities*

$$\left.\begin{aligned}
&M_1 = y_1+y_2+y_3, & &M_2 = y_2y_3+y_3y_1+y_1y_2 \\
&M_3 = y_1y_2y_3, & &M_4 = z_1+z_2+z_3 \\
&M_5 = z_2z_3+z_3z_1+z_1z_2, & &M_6 = z_1z_2z_3 \\
&M_7 = y_2z_3+y_3z_1+y_1z_2, & &M_8 = z_2y_3+z_3y_1+z_1y_2 \\
&M_9 = y_3y_2^2+y_1y_3^2+y_2y_1^2, & &M_{10} = z_3z_2^2+z_1z_3^2+z_2z_1^2 \\
&M_{11} = y_1z_2z_3+y_2z_3z_1+y_3z_1z_2, & &M_{12} = z_1y_2y_3+z_2y_3y_1+z_3y_1y_2 \\
&M_{13} = y_1y_2z_2+y_2y_3z_3+y_3y_1z_1, & &M_{14} = z_1z_2y_2+z_2z_3y_3+z_3z_1y_1
\end{aligned}\right\}. \quad (1.2.5)$$

THEOREM 5. *A polynomial basis for a polynomial in the variables $y_1, y_2, \ldots, z_n, N_1, N_2, \ldots, N_k$, which is form-invariant under a group of transformations under which group $N_1, N_2, \ldots, N_k$ are invariant is formed by adjoining to the quantities $N_1, N_2, \ldots, N_k$ the polynomial basis for polynomials in the variables $y_1, \ldots, z_n$ which are form-invariant under the given group of transformations.*

1.3. Strain energy

In *T.E.* the strain energy W was assumed to be a scalar function of the strain components γ_{ij}, the metric tensor g_{ij}, and tensors which represent the physical properties of the undeformed body. Attention was mainly directed to homogeneous isotropic bodies when the deformations were large. For classical infinitesimal elasticity theory the form of W was examined for bodies which are initially homogeneous and which have certain kinds of elastic symmetry. It was pointed out by Green and Wilkes[†] that some of this discussion is valid for large deformation theory. A very comprehensive discussion of the forms assumed by W for various crystal systems has been given by Smith

---

[†] A. E. Green and E. W. Wilkes, *J. rat. Mech. Analysis* **3** (1954), 713.

and Rivlin.† The work of the rest of this chapter is based on that of Smith and Rivlin and the starting-point for our discussion differs from that used in *T.E.*

We suppose that the deformation of the body is described in rectangular cartesian coordinates by (1.1.3) and we assume that W is expressible as a function of the nine displacement gradients $\partial y_i/\partial x_j$, i.e.,
$$W = W\left(\frac{\partial y_i}{\partial x_j}\right). \tag{1.3.1}$$

We also assume that the initial body is homogeneous so that any material coefficients in (1.3.1) are constants. More general kinds of bodies are considered in § 1.16. We could also assume that W depends on the three components of displacement $y_i - x_i$ but these may be omitted since W must be unaltered by arbitrary rigid-body translations. Finally, W also depends on the temperature‡ T, a scalar function of position and time, but this is not included explicitly in this chapter as it does not affect the discussion.

We now assume that the deformed body is subject to a rigid-body rotation, in which the point y_i moves to $\bar{y}_i$, where
$$\bar{y}_i = a_{ij} y_j, \tag{1.3.2}$$
and
$$a_{ik} a_{jk} = a_{ki} a_{kj} = \delta_{ij}, \quad |a_{ij}| = 1. \tag{1.3.3}$$

From (1.3.1) the strain energy W, per unit initial volume, after the rigid-body rotation is given by
$$W\left(\frac{\partial \bar{y}_i}{\partial x_j}\right). \tag{1.3.4}$$

A rigid-body rotation cannot alter the energy stored per unit initial volume so that, from (1.3.1), (1.3.2), and (1.3.4),
$$W\left(\frac{\partial y_i}{\partial x_j}\right) = W\left(a_{ir}\frac{\partial y_r}{\partial x_j}\right) \tag{1.3.5}$$

for all tensor functions a_{ir} of time satisfying the relations (1.3.3). In view of the last condition in (1.1.10) we may put
$$\frac{\partial y_r}{\partial x_j} = R_{rk} M_{kj}$$

† G. F. Smith and R. S. Rivlin, loc. cit., p. 1.
‡ Or alternatively, on the entropy S; see Chapter VIII.

where M_{kj} is a positive definite symmetric tensor and R_{rk} is a proper orthogonal tensor. Also

$$M_{ki}M_{kj} = \frac{\partial y_k}{\partial x_i}\frac{\partial y_k}{\partial x_j} = 2e_{ij}+\delta_{ij}. \tag{1.3.6}$$

The relation (1.3.5) holds for a particular value of x_i so we may choose the special value R_{ri} for a_{ir}, so that

$$W\left(\frac{\partial y_i}{\partial x_j}\right) = W(M_{ij}). \tag{1.3.7}$$

Since M_{ij} is a positive definite symmetric tensor and (1.3.6) holds, we may replace W by the different function

$$W = W(e_{ij}). \tag{1.3.8}$$

It may be verified that the expression (1.3.8) for W satisfies (1.3.5) for all proper orthogonal tensor functions a_{ir} of time. In the rest of this chapter we assume that W is a polynomial in $e_{11}, e_{22}, e_{33}, e_{12}, e_{23}, e_{31}$ but much of the discussion can be extended to single valued functions of these variables with the help of some results of A. C. Pipkin and A. S. Wineman.[†]

1.4. Crystal classes

We consider special types of aeolotropic materials known as crystal classes for which there exist, in the undeformed state, three preferred directions which may be defined by the unit vectors $\mathbf{h}_i$. Associated with each crystal class is a group of symmetry transformations which characterizes the symmetry properties of the undeformed material. The crystal, when subjected to one of these transformations, is carried into a configuration which is indistinguishable from the original configuration. In other words, associated with each symmetry class we have three preferred directions intrinsically defined in the material and a set of transformations which essentially gives the equivalent positions of the vectors $\mathbf{h}_i$. We shall require that the strain energy function W for a certain crystal be form-invariant under the group of transformations associated with that crystal class. Thus, the symmetry properties of the material impose restrictions upon the manner in which W depends on the quantities e_{ij}. For a particular crystal class W may be represented as a polynomial in the quantities e_{ij} which is form-invariant under the group of transformations associated with that

[†] A. C. Pipkin and A. S. Wineman, *Arch ration. Mech. Analysis* 12 (1963), 420; 17 (1964), 184.

crystal class. We must therefore find a set of invariants of e_{ij} which forms a polynomial basis for the group of transformations describing the symmetry properties of the crystal class under consideration. The strain energy function W is then expressible as a polynomial in these invariants.

1.5. Form-invariance

We now suppose that the symmetry transformation $\mathbf{B}$ ($= \|b_{ij}\|$) transforms the rectangular cartesian coordinate system x_i into the rectangular cartesian coordinate system x'_i, so that

$$\left.\begin{array}{c} x'_i = b_{ij}x_j, \qquad y'_i = b_{ij}y_j \\ b_{ir}b_{jr} = b_{ri}b_{rj} = \delta_{ij} \end{array}\right\}. \qquad (1.5.1)$$

Since the strain components e_{ij}, given by (1.1.15), are components of a cartesian tensor for all changes of rectangular cartesian coordinates of the form (1.5.1), it follows that the corresponding strain components e'_{ij}, referred to x'_i axes, are given by

$$e'_{ij} = \frac{1}{2}\left(\frac{\partial y'_r}{\partial x'_i}\frac{\partial y'_r}{\partial x'_j} - \delta_{ij}\right) = b_{ir}b_{js}e_{rs}. \qquad (1.5.2)$$

The condition that W be form-invariant under the transformation $\mathbf{B}$ may be expressed as

$$W(e'_{ij}) \equiv W(e_{ij}), \qquad (1.5.3)$$

and with (1.5.2) this becomes

$$W(b_{ir}b_{js}e_{rs}) \equiv W(e_{ij}). \qquad (1.5.4)$$

In the next section we list the transformations which are sufficient to describe the symmetry properties of all the crystal classes.†

1.6. Symmetry transformations

Here we list the required transformations together with the corresponding transformations of the strain components e_{ij} as given by (1.5.2).

† We follow G. F. Smith and R. S. Rivlin, loc. cit., p. 1, who used nomenclature for the crystal classes given by J. D. Dana and C. S. Hurlbut, *Dana's Manual of Mineralogy* (New York, 1952). The transformations associated with the crystal classes are obtained from E. S. Dana and W. E. Ford, *Dana's Textbook of Mineralogy* (New York, 1932).

$\mathbf{I}$: $\quad x_1' = x_1, \quad x_2' = x_2, \quad x_3' = x_3; \quad e_{ij}' = e_{ij}.$ \hfill (1.6.1)

$\mathbf{C}$: $\quad x_1' = -x_1, \quad x_2' = -x_2, \quad x_3' = -x_3; \quad e_{ij}' = e_{ij}.$ \hfill (1.6.2)

$\mathbf{R_1}$: $\left. \begin{array}{l} x_1' = -x_1, \quad x_2' = x_2, \quad x_3' = x_3 \\ e_{11}' = e_{11}, \quad e_{22}' = e_{22}, \quad e_{33}' = e_{33} \\ e_{12}' = -e_{12}, \quad e_{23}' = e_{23}, \quad e_{31}' = -e_{31} \end{array} \right\}.$ \hfill (1.6.3)

$\mathbf{R_2}$: $\left. \begin{array}{l} x_1' = x_1, \quad x_2' = -x_2, \quad x_3' = x_3 \\ e_{11}' = e_{11}, \quad e_{22}' = e_{22}, \quad e_{33}' = e_{33} \\ e_{12}' = -e_{12}, \quad e_{23}' = -e_{23}, \quad e_{31}' = e_{31} \end{array} \right\}.$ \hfill (1.6.4)

$\mathbf{R_3}$: $\left. \begin{array}{l} x_1' = x_1, \quad x_2' = x_2, \quad x_3' = -x_3 \\ e_{11}' = e_{11}, \quad e_{22}' = e_{22}, \quad e_{33}' = e_{33} \\ e_{12}' = e_{12}, \quad e_{23}' = -e_{23}, \quad e_{31}' = -e_{31} \end{array} \right\}.$ \hfill (1.6.5)

$\mathbf{D_1}$: $\left. \begin{array}{l} x_1' = x_1, \quad x_2' = -x_2, \quad x_3' = -x_3 \\ e_{11}' = e_{11}, \quad e_{22}' = e_{22}, \quad e_{33}' = e_{33} \\ e_{12}' = -e_{12}, \quad e_{23}' = e_{23}, \quad e_{31}' = -e_{31} \end{array} \right\}.$ \hfill (1.6.6)

$\mathbf{D_2}$: $\left. \begin{array}{l} x_1' = -x_1, \quad x_2' = x_2, \quad x_3' = -x_3 \\ e_{11}' = e_{11}, \quad e_{22}' = e_{22}, \quad e_{33}' = e_{33} \\ e_{12}' = -e_{12}, \quad e_{23}' = -e_{23}, \quad e_{31}' = e_{31} \end{array} \right\}.$ \hfill (1.6.7)

$\mathbf{D_3}$: $\left. \begin{array}{l} x_1' = -x_1, \quad x_2' = -x_2, \quad x_3' = x_3 \\ e_{11}' = e_{11}, \quad e_{22}' = e_{22}, \quad e_{33}' = e_{33} \\ e_{12}' = e_{12}, \quad e_{23}' = -e_{23}, \quad e_{31}' = -e_{31} \end{array} \right\}.$ \hfill (1.6.8)

$\mathbf{T_1}$: $\left. \begin{array}{l} x_1' = x_1, \quad x_2' = x_3, \quad x_3' = x_2 \\ e_{11}' = e_{11}, \quad e_{22}' = e_{33}, \quad e_{33}' = e_{22} \\ e_{12}' = e_{13}, \quad e_{23}' = e_{23}, \quad e_{31}' = e_{12} \end{array} \right\}.$ \hfill (1.6.9)

$\mathbf{T_2}$: $\left. \begin{array}{l} x_1' = x_3, \quad x_2' = x_2, \quad x_3' = x_1 \\ e_{11}' = e_{33}, \quad e_{22}' = e_{22}, \quad e_{33}' = e_{11} \\ e_{12}' = e_{23}, \quad e_{23}' = e_{12}, \quad e_{31}' = e_{31} \end{array} \right\}.$ \hfill (1.6.10)

$\mathbf{T_3}$: $\left. \begin{array}{l} x_1' = x_2, \quad x_2' = x_1, \quad x_3' = x_3 \\ e_{11}' = e_{22}, \quad e_{22}' = e_{11}, \quad e_{33}' = e_{33} \\ e_{12}' = e_{12}, \quad e_{23}' = e_{31}, \quad e_{31}' = e_{23} \end{array} \right\}.$ \hfill (1.6.11)

$\mathbf{M_1}$: $\left. \begin{array}{l} x_1' = x_2, \quad x_2' = x_3, \quad x_3' = x_1 \\ e_{11}' = e_{22}, \quad e_{22}' = e_{33}, \quad e_{33}' = e_{11} \\ e_{12}' = e_{23}, \quad e_{23}' = e_{31}, \quad e_{31}' = e_{12} \end{array} \right\}.$ \hfill (1.6.12)

$$\mathbf{M}_2: \quad \begin{aligned} x_1' &= x_3, & x_2' &= x_1, & x_3' &= x_2 \\ e_{11}' &= e_{33}, & e_{22}' &= e_{11}, & e_{33}' &= e_{22} \\ e_{12}' &= e_{31}, & e_{23}' &= e_{12}, & e_{31}' &= e_{23} \end{aligned} \Bigg\} \qquad (1.6.13)$$

$$\mathbf{S}_1: \quad \begin{aligned} x_1' + ix_2' &= (x_1 + ix_2)e^{-2i\pi/3}, \quad x_3' = x_3 \\ e_{11}' + e_{22}' &= e_{11} + e_{22}, \quad e_{33}' = e_{33} \\ e_{11}' - e_{22}' + 2ie_{12}' &= (e_{11} - e_{22} + 2ie_{12})e^{-4i\pi/3} \\ e_{13}' + ie_{23}' &= (e_{13} + ie_{23})e^{-2i\pi/3} \end{aligned} \Bigg\} \qquad (1.6.14)$$

$$\mathbf{S}_2: \quad \begin{aligned} x_1' + ix_2' &= (x_1 + ix_2)e^{2i\pi/3}, \quad x_3' = x_3 \\ e_{11}' + e_{22}' &= e_{11} + e_{22}, \quad e_{33}' = e_{33} \\ e_{11}' - e_{22}' + 2ie_{12}' &= (e_{11} - e_{22} + 2ie_{12})e^{4i\pi/3} \\ e_{13}' + ie_{23}' &= (e_{13} + ie_{23})e^{2i\pi/3} \end{aligned} \Bigg\} \qquad (1.6.15)$$

The transformations $\mathbf{S}_1$, $\mathbf{S}_2$ have been abbreviated with the use of complex variable notation. In addition, we shall need transformations which can be represented by a product. For example, $\mathbf{CT}_1$ denotes the transformation

$$\mathbf{CT}_1: \quad x_1' = -x_1, \quad x_2' = -x_3, \quad x_3' = -x_2,$$

and $\mathbf{CS}_1$ denotes

$$\mathbf{CS}_1: \quad x_1' + ix_2' = -(x_1 + ix_2)e^{-2i\pi/3}, \quad x_3' = -x_3.$$

The transformation $\mathbf{I}$ is the identity transformation and $\mathbf{C}$ is central inversion. $\mathbf{R}_1$, $\mathbf{R}_2$, $\mathbf{R}_3$ are reflections in planes normal to the x_1-, x_2-, x_3-axes respectively. $\mathbf{D}_1$, $\mathbf{D}_2$, $\mathbf{D}_3$ are rotations through 180° of the coordinate system about the x_1-, x_2-, x_3-axes respectively. The transformation $\mathbf{T}_1$ is a reflection in the plane through the axis x_1 bisecting the angle between the x_2- and x_3-axes; $\mathbf{T}_2$ is a reflection in the plane through the axis x_2 bisecting the angle between the x_3- and x_1-axes; $\mathbf{T}_3$ is a reflection in the plane through the axis x_3 bisecting the angle between the x_1- and x_2-axes. The transformations $\mathbf{M}_1$, $\mathbf{M}_2$ are rotations of the axes through 120° and 240° respectively about a line passing through the origin and the point 1, 1, 1. The transformations $\mathbf{S}_1$, $\mathbf{S}_2$ are rotations of the axes through 120° and 240° respectively about the x_3-axis.

1.7. The triclinic system

For a material having triclinic symmetry there is no restriction on the orientation of the unit vectors $\mathbf{h}_i$ defining the preferred directions. We may therefore choose any rectangular cartesian coordinate system as a reference system.

The triclinic system contains two crystal classes, which, together with the symmetry transformations characterizing them, are:

$$\text{pedial} \quad \mathbf{I},$$
$$\text{pinacoidal} \quad \mathbf{I}, \mathbf{C}.$$

From (1.6.1), (1.6.2) we see that e_{ij} are unaltered by the transformations $\mathbf{I}$ or $\mathbf{C}$. Hence, from (1.5.3), no restriction is placed on the form of W so that

$$W = W(e_{11}, e_{22}, e_{33}, e_{12}, e_{23}, e_{31}), \qquad (1.7.1)$$

where W is a polynomial in the arguments shown.

1.8. The monoclinic system

For a material with monoclinic symmetry, the unit vectors $\mathbf{h}_2$, $\mathbf{h}_3$ are not at right angles and the unit vector $\mathbf{h}_1$ is perpendicular to the plane defined by $\mathbf{h}_2$ and $\mathbf{h}_3$. We take as our reference system a rectangular cartesian coordinate system x_i, whose x_1-axis is parallel to $\mathbf{h}_1$. The axes x_2, x_3 may be in arbitrary perpendicular directions in the $\mathbf{h}_2$, $\mathbf{h}_3$ plane.

The monoclinic system contains three classes which, together with the symmetry transformations defining them, are:

$$\text{domatic} \quad \mathbf{I}, \mathbf{R}_1.$$
$$\text{sphenoidal} \quad \mathbf{I}, \mathbf{D}_1,$$
$$\text{prismatic} \quad \mathbf{I}, \mathbf{C}, \mathbf{R}_1, \mathbf{D}_1.$$

From § 1.6 and (1.5.3) it is seen that, for each of the three crystal classes of the monoclinic system, the limitation imposed on the form of W is

$$W(e_{11}, e_{22}, e_{33}, e_{12}, e_{23}, e_{31}) = W(e_{11}, e_{22}, e_{33}, -e_{12}, e_{23}, -e_{31}). \qquad (1.8.1)$$

In Theorems 2 and 5 we take

$$(y_1, y_2) = (e_{12}, e_{31}), \qquad (z_1, z_2) = (-e_{12}, -e_{31}),$$
$$(N_1, N_2, N_3, N_4) = (e_{11}, e_{22}, e_{33}, e_{23}),$$

so that W must be expressible as a polynomial in the seven quantities e_{11}, e_{22}, e_{33}, e_{23}, e_{31}^2, e_{12}^2, $e_{31}e_{12}$. Thus†

$$W = W(e_{11}, e_{22}, e_{33}, e_{23}, e_{31}^2, e_{12}^2, e_{31}e_{12}), \qquad (1.8.2)$$

where W is a polynomial.

† The functional form in (1.8.2) is not, of course, the same as that in (1.8.1), but it is convenient to use W in both cases. A similar situation occurs throughout the rest of this chapter without further comment.

1.9. The rhombic system

For a material having rhombic symmetry, the unit vectors $\mathbf{h}_i$ are mutually perpendicular. We take as our reference system a rectangular cartesian coordinate system x_i, the axes of which are parallel to the vectors $\mathbf{h}_i$.

The rhombic system contains three classes which, together with the symmetry transformations characterizing them, are:

rhombic-pyramidal $\mathbf{I}, \mathbf{R}_2, \mathbf{R}_3, \mathbf{D}_1$,
rhombic-disphenoidal $\mathbf{I}, \mathbf{D}_1, \mathbf{D}_2, \mathbf{D}_3$,
rhombic-dipyramidal $\mathbf{I}, \mathbf{C}, \mathbf{R}_1, \mathbf{R}_2, \mathbf{R}_3$,
$\mathbf{D}_1, \mathbf{D}_2, \mathbf{D}_3$.

From §§ 1.6, 1.8 we see that, for each of the three crystal classes of the rhombic system, the extra limitation imposed on the form of W as given by (1.8.2) is

$$W(e_{11}, e_{22}, e_{33}, e_{23}, e_{31}^2, e_{12}^2, e_{31}e_{12}) = W(e_{11}, e_{22}, e_{33,}, -e_{23}, e_{31}^2, e_{12}^2, -e_{31}e_{12}). \tag{1.9.1}$$

In Theorems 2 and 5 we take

$$(N_1, N_2, N_3, N_4, N_5) = (e_{11}, e_{22}, e_{33}, e_{31}^2, e_{12}^2),$$
$$(y_1, y_2) = (e_{23}, e_{31}e_{12}), \qquad (z_1, z_2) = (-e_{23}, -e_{31}e_{12}),$$

so that W must be expressible as a polynomial in the seven quantities $e_{11}, e_{22}, e_{33}, e_{12}^2, e_{23}^2, e_{31}^2, e_{12}e_{23}e_{31}$. Thus

$$W = W(e_{11}, e_{22}, e_{33}, e_{12}^2, e_{23}^2, e_{31}^2, e_{12}e_{23}e_{31}), \tag{1.9.2}$$

where W is a polynomial in the stated quantities. Since $e_{12}e_{23}e_{31}$ may be expressed as a polynomial in I_3 and the remaining quantities in (1.9.2), and is linear in I_3, we may replace (1.9.2) by the polynomial form

$$W = W(e_{11}, e_{22}, e_{33}, e_{12}^2, e_{23}^2, e_{31}^2, I_3). \tag{1.9.3}$$

1.10. The tetragonal system

For a material having tetragonal symmetry, the unit vectors $\mathbf{h}_i$ are mutually perpendicular. As in the case of the rhombic system, we take as our reference system a rectangular cartesian coordinate system x_i, the axes of which are parallel to the vectors $\mathbf{h}_i$. We shall take the x_3-axis to be the principal axis of symmetry.

The tetragonal system contains seven crystal classes which, together with the transformation groups characterizing them, are:

tetragonal-disphenoidal	$I, D_3, D_1T_3, D_2T_3,$
tetragonal-pyramidal	$I, D_3, R_1T_3, R_2T_3,$
tetragonal-dipyramidal	$I, C, R_3, D_3, R_1T_3, R_2T_3, D_1T_3, D_2T_3,$
tetragonal-scalenohedral	$I, D_1, D_2, D_3, T_3, D_1T_3, D_2T_3, D_3T_3,$
ditetragonal-pyramidal	$I, R_1, R_2, D_3, T_3, R_1T_3, R_2T_3, D_3T_3,$
tetragonal-trapezohedral	$I, D_1, D_2, D_3, CT_3, R_1T_3, R_2T_3, R_3T_3,$
ditetragonal-dipyramidal	$I, C, R_1, R_2, R_3, D_1, D_2, D_3, T_3, CT_3,$ $R_1T_3, R_2T_3, R_3T_3, D_1T_3, D_2T_3, D_3T_3.$

For each of the first three classes of the tetragonal system listed above, it is seen from § 1.6 that the limitations imposed on the form of W are

$$W(e_{11}, e_{22}, e_{33}, e_{12}, e_{23}, e_{31}) = W(e_{11}, e_{22}, e_{33}, e_{12}, -e_{23}, -e_{31})$$
$$= W(e_{22}, e_{11}, e_{33}, -e_{12}, e_{31}, -e_{23}). \quad (1.10.1)$$

It can be shown in a manner similar to that used for the monoclinic system that if W satisfies the first relation in (1.10.1) it must be expressible as a polynomial in $e_{11}, e_{22}, e_{33}, e_{23}^2, e_{31}^2, e_{12}, e_{23}e_{31}$. The second relation in (1.10.1) then becomes

$$W(e_{11}, e_{22}, e_{33}, e_{23}^2, e_{31}^2, e_{12}, e_{23}e_{31}) = W(e_{22}, e_{11}, e_{33}, e_{31}^2, e_{23}^2, -e_{12}, -e_{23}e_{31}). \quad (1.10.2)$$

We now take, in Theorems 2 and 5,

$$(y_1, y_2, \ldots, y_6) = (e_{11}, e_{22}, e_{23}^2, e_{31}^2, e_{12}, e_{23}e_{31}),$$
$$(z_1, z_2, \ldots, z_6) = (e_{22}, e_{11}, e_{31}^2, e_{23}^2, -e_{12}, -e_{23}e_{31}),$$

and $N_1 = e_{33}$, and we see that the relation (1.10.2) implies that W must be expressible as a polynomial in the twelve quantities

$$\left.\begin{array}{l} e_{11}+e_{22}, \quad e_{33}, \quad e_{23}^2+e_{31}^2, \quad e_{12}^2, \quad e_{11}e_{22} \\ e_{12}(e_{11}-e_{22}), \quad e_{23}e_{31}(e_{11}-e_{22}), \quad e_{23}e_{31}e_{12} \\ e_{12}(e_{31}^2-e_{23}^2), \quad e_{11}e_{23}^2+e_{22}e_{31}^2, \quad e_{23}e_{31}(e_{31}^2-e_{23}^2), \quad e_{23}^2e_{31}^2 \end{array}\right\}, \quad (1.10.3)$$

if we omit redundant terms. Thus, the strain energy function for a material belonging to the tetragonal-disphenoidal, tetragonal-pyramidal, or tetragonal-dipyramidal classes must be expressible as a polynomial in the quantities (1.10.3).

We now consider the last four classes of the tetragonal system listed at the beginning of this section. From §§ 1.6, 1.9, it is seen that W must first be restricted to be a polynomial of the form (1.9.2) and that, in addition,

$$W(e_{11}, e_{22}, e_{33}, e_{12}^2, e_{23}^2, e_{31}^2, e_{12}e_{23}e_{31}) = W(e_{22}, e_{11}, e_{33}, e_{12}^2, e_{31}^2, e_{23}^2, e_{12}e_{23}e_{31}).$$
(1.10.4)

We take, in Theorems 2 and 5,

$$(y_1, y_2, y_3, y_4) = (e_{11}, e_{22}, e_{23}^2, e_{31}^2),$$
$$(z_1, z_2, z_3, z_4) = (e_{22}, e_{11}, e_{31}^2, e_{23}^2),$$
$$(N_1, N_2, N_3) = (e_{33}, e_{12}^2, e_{23}e_{31}e_{12}),$$

so that W must be expressible as a polynomial in the eight quantites

$$\left.\begin{array}{l} e_{11}+e_{22}, \quad e_{33}, \quad e_{23}^2+e_{31}^2, \quad e_{12}^2, \quad e_{11}e_{22} \\ e_{23}e_{31}e_{12}, \quad e_{11}e_{23}^2+e_{22}e_{31}^2, \quad e_{23}^2e_{31}^2 \end{array}\right\},$$
(1.10.5)

if we omit redundant elements. Thus, for the tetragonal-scalenohedral, ditetragonal-pyramidal, tetragonal-trapezohedral, and ditetragonal-dipyramidal classes, the strain energy function must be expressible as a polynomial in the eight quantities (1.10.5).

1.11. The cubic system

For a material having cubic symmetry the unit vectors $\mathbf{h}_i$ are mutually perpendicular and we take as our reference system a rectangular cartesian coordinate system x_i whose axes are parallel to $\mathbf{h}_i$.

The cubic system contains five classes which, together with the symmetry transformations characterizing them, are:

tetartoidal $I, D_1, D_2, D_3, M_1, D_1M_1, D_2M_1, D_3M_1, M_2, D_1M_2, D_2M_2, D_3M_2,$

diploidal $I, C, R_1, R_2, R_3, D_1, D_2, D_3, M_1, CM_1, R_1M_1, R_2M_1, R_3M_1, D_1M_1, D_2M_1, D_3M_1, M_2, CM_2, R_1M_2, R_2M_2, R_3M_2, D_1M_2, D_2M_2, D_3M_2,$

hextetrahedral $I, D_1, D_2, D_3, T_1, D_1T_1, D_2T_1, D_3T_1, T_2, D_1T_2, D_2T_2, D_3T_2, T_3, D_1T_3, D_2T_3, D_3T_3, M_1, D_1M_1, D_2M_1, D_3M_1, M_2, D_1M_2, D_2M_2, D_3M_2,$

gyroidal $I, D_1, D_2, D_3, CT_1, R_1T_1, R_2T_1, R_3T_1, CT_2, R_1T_2, R_2T_2, R_3T_2, CT_3, R_1T_3, R_2T_3, R_3T_3, M_1, D_1M_1, D_2M_1, D_3M_1, M_2, D_1M_2, D_2M_2, D_3M_2,$

hexoctahedral I, C, R_1, R_2, R_3, D_1, D_2, D_3, T_1, CT_1, R_1T_1, R_2T_1, R_3T_1, D_1T_1, D_2T_1, D_3T_1, T_2, CT_2, R_1T_2, R_2T_2, R_3T_2, D_1T_2, D_2T_2, D_3T_2, T_3, CT_3, R_1T_3, R_2T_3, R_3T_3, D_1T_3, D_2T_3, D_3T_3, M_1, CM_1, R_1M_1, R_2M_1, R_3M_1, D_1M_1, D_2M_1, D_3M_1, M_2, CM_2, R_1M_2, R_2M_2, R_3M_2, D_1M_2, D_2M_2, D_3M_2.

From §§ 1.6, 1.9, it is seen that for each of the first two classes of the cubic system the form of W is given by (1.9.2) and in addition it is limited by the restrictions

$$W(e_{11}, e_{22}, e_{33}, e_{12}^2, e_{23}^2, e_{31}^2, e_{12}e_{23}e_{31})$$
$$= W(e_{22}, e_{33}, e_{11}, e_{23}^2, e_{31}^2, e_{12}^2, e_{12}e_{23}e_{31})$$
$$= W(e_{33}, e_{11}, e_{22}, e_{31}^2, e_{12}^2, e_{23}^2, e_{12}e_{23}e_{31}). \quad (1.11.1)$$

We take, in Theorems 4 and 5,

$$(y_1, y_2, y_3, z_1, z_2, z_3) = (e_{11}, e_{22}, e_{33}, e_{23}^2, e_{31}^2, e_{12}^2),$$
$$N_1 = e_{12}e_{23}e_{31},$$

and, omitting one redundant term, we see that for the tetartoidal and diploidal classes W must be expressible as a polynomial in the fourteen quantities

$$\left.\begin{aligned}
& e_{11}+e_{22}+e_{33}, \quad e_{22}e_{33}+e_{33}e_{11}+e_{11}e_{22} \\
& e_{11}e_{22}e_{33}, \quad e_{23}^2+e_{31}^2+e_{12}^2 \\
& e_{12}^2e_{23}^2+e_{23}^2e_{31}^2+e_{31}^2e_{12}^2, \quad e_{12}e_{23}e_{31} \\
& e_{22}e_{12}^2+e_{33}e_{23}^2+e_{11}e_{31}^2, \quad e_{31}^2e_{33}+e_{12}^2e_{11}+e_{23}^2e_{22} \\
& e_{33}e_{22}^2+e_{11}e_{33}^2+e_{22}e_{11}^2, \quad e_{12}^2e_{31}^4+e_{23}^2e_{12}^4+e_{31}^2e_{23}^4 \\
& e_{11}e_{31}^2e_{12}^2+e_{22}e_{12}^2e_{23}^2+e_{33}e_{23}^2e_{31}^2 \\
& e_{23}^2e_{22}e_{33}+e_{31}^2e_{33}e_{11}+e_{12}^2e_{11}e_{22} \\
& e_{11}e_{22}e_{31}^2+e_{22}e_{33}e_{12}^2+e_{33}e_{11}e_{23}^2 \\
& e_{23}^2e_{31}^2e_{22}+e_{31}^2e_{12}^2e_{33}+e_{12}^2e_{23}^2e_{11}
\end{aligned}\right\}. \quad (1.11.2)$$

We now consider the remaining three classes of the cubic system. For the hextetrahedral, gyroidal, and hexoctahedral classes, the form of W is given by (1.9.2) and is subject to the further limitations

$$\left.\begin{aligned}
W(e_{11}, e_{22}, e_{33}, e_{12}^2, e_{23}^2, e_{31}^2, e_{12}e_{23}e_{31}) & \\
= W(e_{11}, e_{33}, e_{22}, e_{31}^2, e_{23}^2, e_{12}^2, e_{12}e_{23}e_{31}) & \\
= W(e_{22}, e_{11}, e_{33}, e_{12}^2, e_{31}^2, e_{23}^2, e_{12}e_{23}e_{31}) & \\
= W(e_{33}, e_{22}, e_{11}, e_{23}^2, e_{12}^2, e_{31}^2, e_{12}e_{23}e_{31}) & \\
= W(e_{22}, e_{33}, e_{11}, e_{23}^2, e_{31}^2, e_{12}^2, e_{12}e_{23}e_{31}) & \\
= W(e_{33}, e_{11}, e_{22}, e_{31}^2, e_{12}^2, e_{23}^2, e_{12}e_{23}e_{31}) &
\end{aligned}\right\}. \quad (1.11.3)$$

1.12 THE STRAIN ENERGY FUNCTION

In Theorems 3 and 5 we take

$$(y_1, z_1) = (e_{11}, e_{23}^2), \quad (y_2, z_2) = (e_{22}, e_{31}^2),$$
$$(y_3, z_3) = (e_{33}, e_{12}^2), \quad N_1 = e_{12}e_{23}e_{31},$$

and, omitting one redundant term, we see that for the hextetrahedral, gyroidal, and hexoctahedral classes, W is expressible as a polynomial in the nine quantities

$$\left.\begin{array}{l} e_{11}+e_{22}+e_{33}, \quad e_{22}e_{33}+e_{33}e_{11}+e_{11}e_{22} \\ e_{11}e_{22}e_{33}, \quad e_{12}e_{23}e_{31}, \quad e_{12}^2+e_{23}^2+e_{31}^2 \\ e_{12}^2 e_{23}^2+e_{23}^2 e_{31}^2+e_{31}^2 e_{12}^2 \\ e_{22}e_{12}^2+e_{33}e_{31}^2+e_{33}e_{23}^2+e_{11}e_{12}^2+e_{11}e_{31}^2+e_{22}e_{23}^2 \\ e_{11}e_{12}^2 e_{31}^2+e_{22}e_{23}^2 e_{12}^2+e_{33}e_{31}^2 e_{23}^2 \\ e_{12}^2 e_{11}e_{22}+e_{23}^2 e_{22}e_{33}+e_{31}^2 e_{33}e_{11} \end{array}\right\} \quad (1.11.4)$$

1.12. The hexagonal system†

For a material having hexagonal symmetry, the unit vectors $\mathbf{h}_i$ are situated so that $\mathbf{h}_3$ is perpendicular to the plane defined by $\mathbf{h}_1$ and $\mathbf{h}_2$, and so that $\mathbf{h}_1$ can be made to coincide with $\mathbf{h}_2$ by a rotation of 120° about the direction of $\mathbf{h}_3$. We take as our reference system a right-handed rectangular cartesian coordinate system x_i, the x_1- and x_3-axes of which are parallel respectively to $\mathbf{h}_1$ and $\mathbf{h}_3$.

The hexagonal system contains twelve crystal classes which, together with the transformation groups characterizing them, are:

trigonal-pyramidal	$I, S_1, S_2,$
rhombohedral	$I, S_1, S_2, C, CS_1, CS_2,$
ditrigonal-pyramidal	$I, S_1, S_2, R_1, R_1S_1, R_1S_2,$
trigonal-trapezohedral	$I, S_1, S_2, D_1, D_1S_1, D_1S_2,$
hexagonal-scalenohedral	$I, S_1, S_2, C, CS_1, CS_2, R_1, R_1S_1, R_1S_2,$ $D_1, D_1S_1, D_1S_2,$
trigonal-dipyramidal	$I, S_1, S_2, R_3, R_3S_1, R_3S_2,$
hexagonal-pyramidal	$I, S_1, S_2, D_3, D_3S_1, D_3S_2,$

† In *T.E.* § 5.4 the 'Hexagonal system' should have been called 'Transversely isotropic' and has been changed to this in the reprinted and second editions. However, for classical elasticity in which W is a quadratic form the restrictions for those members of the hexagonal system which are invariant under either of the transformations R_3 or D_3 are the same as those for transverse isotropy.

hexagonal-dipyramidal	$I, S_1, S_2, C, CS_1, CS_2, R_3, R_3S_1, R_3S_2,$ $D_3, D_3S_1, D_3S_2,$
ditrigonal-dipyramidal	$I, S_1, S_2, R_1, R_1S_1, R_1S_2, R_3, R_3S_1, R_3S_2,$ $D_2, D_2S_1, D_2S_2,$
dihexagonal-pyramidal	$I, S_1, S_2, R_1, R_1S_1, R_1S_2, R_2, R_2S_1, R_2S_2,$ $D_3, D_3S_1, D_3S_2,$
hexagonal-trapezohedral	$I, S_1, S_2, D_1, D_1S_1, D_1S_2, D_2, D_2S_1, D_2S_2,$ $D_3, D_3S_1, D_3S_2,$
dihexagonal-dipyramidal	$I, S_1, S_2, C, CS_1, CS_2, R_1, R_1S_1, R_1S_2,$ $R_2, R_2S_1, R_2S_2, R_3, R_3S_1, R_3S_2, D_1,$ $D_1S_1, D_1S_2, D_2, D_2S_1, D_2S_2, D_3, D_3S_1,$ $D_3S_2.$

For each of the first two classes of the hexagonal system listed above (trigonal-pyramidal, rhombohedral) it is seen that the limitations imposed on W are

$$\left.\begin{aligned} &W(e_{11}, e_{22}, e_{33}, e_{23}, e_{31}, e_{12}) \\ &= W\Big(\tfrac{1}{4}e_{11} - \tfrac{\sqrt{3}}{2}e_{12} + \tfrac{3}{4}e_{22}, \tfrac{3}{4}e_{11} + \tfrac{\sqrt{3}}{2}e_{12} + \tfrac{1}{4}e_{22}, e_{33}, \\ &\qquad -\tfrac{\sqrt{3}}{2}e_{31} - \tfrac{1}{2}e_{23}, -\tfrac{1}{2}e_{31} + \tfrac{\sqrt{3}}{2}e_{23}, \tfrac{\sqrt{3}}{4}e_{11} - \tfrac{1}{2}e_{12} - \tfrac{\sqrt{3}}{4}e_{22}\Big) \\ &= W\Big(\tfrac{1}{4}e_{11} + \tfrac{\sqrt{3}}{2}e_{12} + \tfrac{3}{4}e_{22}, \tfrac{3}{4}e_{11} - \tfrac{\sqrt{3}}{2}e_{12} + \tfrac{1}{4}e_{22}, e_{33}, \\ &\qquad \tfrac{\sqrt{3}}{2}e_{31} - \tfrac{1}{2}e_{23}, -\tfrac{1}{2}e_{31} - \tfrac{\sqrt{3}}{2}e_{23}, -\tfrac{\sqrt{3}}{4}e_{11} - \tfrac{1}{2}e_{12} + \tfrac{\sqrt{3}}{4}e_{22}\Big) \end{aligned}\right\}. \quad (1.12.1)$$

Let

$$\left.\begin{aligned} y_1 &= e_{11} \\ y_2 &= \tfrac{1}{4}e_{11} - \tfrac{\sqrt{3}}{2}e_{12} + \tfrac{3}{4}e_{22} \\ y_3 &= \tfrac{1}{4}e_{11} + \tfrac{\sqrt{3}}{2}e_{12} + \tfrac{3}{4}e_{22} \\ z_1 &= e_{31} \\ z_2 &= -\tfrac{1}{2}e_{31} + \tfrac{\sqrt{3}}{2}e_{23} \\ z_3 &= -\tfrac{1}{2}e_{31} - \tfrac{\sqrt{3}}{2}e_{23} \\ N_1 &= e_{33} \end{aligned}\right\}, \quad (1.12.2)$$

where
$$z_1+z_2+z_3 = 0. \tag{1.12.3}$$

If W is expressible as a polynomial in the quantities e_{ij} it is also expressible as a polynomial in the quantities $y_1, y_2,..., z_3, N_1$. With (1.12.2) the limitations imposed on W as a polynomial function of these latter quantities is

$$W(y_1, y_2, y_3, z_1, z_2, z_3, N_1) = W(y_2, y_3, y_1, z_2, z_3, z_1, N_1)$$
$$= W(y_3, y_1, y_2, z_3, z_1, z_2, N_1). \tag{1.12.4}$$

Hence, using Theorems 4 and 5 and (1.12.3), and discarding numerical factors, we see that a polynomial basis for the *trigonal-pyramidal* and *rhombohedral* classes is formed by the fourteen quantities

$$\left.\begin{array}{l} e_{33}, \quad e_{11}+e_{22}, \quad e_{11}e_{22}-e_{12}^2 \\ e_{11}(e_{11}^2+6e_{11}e_{22}-12e_{12}^2+9e_{22}^2), \quad e_{31}^2+e_{23}^2 \\ e_{31}(e_{31}^2-3e_{23}^2), \quad (e_{11}-e_{22})e_{31}-2e_{12}e_{23} \\ (e_{22}-e_{11})e_{23}-2e_{12}e_{31}, \quad 3e_{12}(e_{11}-e_{22})^2-4e_{12}^3 \\ e_{23}(e_{23}^2-3e_{31}^2), \quad e_{22}e_{31}^2+e_{11}e_{23}^2-2e_{12}e_{23}e_{31} \\ e_{31}(e_{11}^2+2e_{11}e_{22}-3e_{22}^2+4e_{12}^2)-8e_{11}e_{12}e_{23} \\ e_{23}(e_{11}^2+2e_{11}e_{22}-3e_{22}^2+4e_{12}^2)+8e_{11}e_{12}e_{31} \\ (e_{11}-e_{22})e_{23}e_{31}+e_{12}(e_{23}^2-e_{31}^2) \end{array}\right\}. \tag{1.12.5}$$

We now consider the next three classes of the hexagonal system. If

$$\left.\begin{array}{l} y_1 = e_{11} \\ y_2 = \tfrac{1}{4}e_{11}-\dfrac{\sqrt{3}}{2}e_{12}+\tfrac{3}{4}e_{22} \\ y_3 = \tfrac{1}{4}e_{11}+\dfrac{\sqrt{3}}{2}e_{12}+\tfrac{3}{4}e_{22} \\ z_1 = e_{23} \\ z_2 = -\dfrac{\sqrt{3}}{2}e_{31}-\tfrac{1}{2}e_{23} \\ z_3 = \dfrac{\sqrt{3}}{2}e_{31}-\tfrac{1}{2}e_{23} \\ N_1 = e_{33} \end{array}\right\}, \tag{1.12.6}$$

we see from § 1.6 that for the ditrigonal-pyramidal, trigonal-trapezohedral, and hexagonal-scalenohedral classes W is a polynomial in the quantities (1.12.6) such that

$$\left.\begin{aligned}
W(y_1,&y_2,y_3,z_1,z_2,z_3,N_1) \\
&= W(y_1, y_3, y_2, z_1, z_3, z_2, N_1) \\
&= W(y_2, y_3, y_1, z_2, z_3, z_1, N_1) \\
&= W(y_3, y_1, y_2, z_3, z_1, z_2, N_1) \\
&= W(y_2, y_1, y_3, z_2, z_1, z_3, N_1) \\
&= W(y_3, y_2, y_1, z_3, z_2, z_1, N_1)
\end{aligned}\right\}. \qquad (1.12.7)$$

Since N_1 is invariant under the given group of transformations, and hence an element of the polynomial basis, the problem is reduced, by Theorem 5, to finding a polynomial basis for polynomials which are symmetric in the three pairs of variables (y_1, z_1), (y_2, z_2), and (y_3, z_3). A polynomial basis is then obtained immediately from Theorem 3. Using (1.12.6) we see that the polynomial basis for the *ditrigonal-pyramidal*, *trigonal-trapezohedral*, and *hexagonal-scalenohedral* classes is formed by the nine quantities

$$\left.\begin{aligned}
& e_{33}, \quad e_{11}+e_{22}, \quad e_{11}e_{22}-e_{12}^2 \\
& e_{11}(e_{11}^2+6e_{11}e_{22}-12e_{12}^2+9e_{22}^2), \quad e_{31}^2+e_{23}^2 \\
& e_{23}(e_{23}^2-3e_{31}^2), \quad (e_{22}-e_{11})e_{23}-2e_{12}e_{31} \\
& e_{11}e_{31}^2+e_{22}e_{23}^2+2e_{12}e_{23}e_{31} \\
& e_{23}(e_{11}^2+2e_{11}e_{22}-3e_{22}^2+4e_{12}^2)+8e_{11}e_{12}e_{31}
\end{aligned}\right\}. \qquad (1.12.8)$$

We now consider the trigonal-dipyramidal, hexagonal-pyramidal, and hexagonal-dipyramidal crystal classes. In a manner analogous to that adopted in discussing the monoclinic system we can show that W must be restricted to be a polynomial in e_{11}, e_{22}, e_{33}, e_{23}^2, e_{31}^2, e_{12}, $e_{23}e_{31}$. Further, if

$$\left.\begin{aligned}
y_1 &= e_{11} \\
y_2 &= \tfrac{1}{4}e_{11}-\frac{\sqrt{3}}{2}e_{12}+\tfrac{3}{4}e_{22} \\
y_3 &= \tfrac{1}{4}e_{11}+\frac{\sqrt{3}}{2}e_{12}+\tfrac{3}{4}e_{22} \\
z_1 &= e_{31}^2 \\
z_2 &= \tfrac{1}{4}e_{31}^2-\frac{\sqrt{3}}{2}e_{31}e_{23}+\tfrac{3}{4}e_{23}^2 \\
z_3 &= \tfrac{1}{4}e_{31}^2+\frac{\sqrt{3}}{2}e_{31}e_{23}+\tfrac{3}{4}e_{23}^2 \\
N_1 &= e_{33}
\end{aligned}\right\}, \qquad (1.12.9)$$

then W is expressible as a polynomial in $y_1, y_2, \ldots, N_1$ and is subject to the restrictions

$$W(y_1, y_2, y_3, z_1, z_2, z_3, N_1)$$
$$= W(y_2, y_3, y_1, z_2, z_3, z_1, N_1)$$
$$= W(y_3, y_1, y_2, z_3, z_1, z_2, N_1). \qquad (1.12.10)$$

A polynomial basis is now obtained by using Theorems 4 and 5 and (1.12.9). Hence, omitting numerical factors and making some rearrangement of terms, a polynomial basis for the *trigonal-dipyramidal hexagonal-pyramidal*, and *hexagonal-dipyramidal* classes is formed from the fourteen quantities

$$\left.\begin{aligned}
& e_{33}, \quad e_{11}+e_{22}, \quad e_{11}e_{22}-e_{12}^2 \\
& e_{11}(e_{11}^2+6e_{11}e_{22}-12e_{12}^2+9e_{22}^2), \quad e_{31}^2+e_{23}^2 \\
& e_{31}^2(e_{31}^2-3e_{23}^2)^2, \quad e_{11}e_{23}^2+e_{22}e_{31}^2-2e_{12}e_{31}e_{23} \\
& e_{12}(e_{31}^2-e_{23}^2)+(e_{22}-e_{11})e_{31}e_{23} \\
& 3e_{12}(e_{11}-e_{22})^2-4e_{12}^3, \quad e_{31}e_{23}(3e_{31}^4-10e_{31}^2e_{23}^2+3e_{23}^4) \\
& e_{11}e_{31}^4+3e_{11}e_{23}^4+2e_{22}e_{31}^4+6e_{22}e_{31}^2e_{23}^2-8e_{12}e_{31}^3e_{23} \\
& (e_{11}^2+2e_{11}e_{22}-3e_{22}^2+4e_{12}^2)e_{31}^2- \\
& \qquad\qquad -2(e_{11}^2+3e_{11}e_{22})(e_{31}^2+e_{23}^2)+8e_{11}e_{12}e_{31}e_{23} \\
& (e_{11}^2+2e_{11}e_{22}-3e_{22}^2+4e_{12}^2)e_{31}e_{23}+4e_{11}e_{12}(e_{23}^2-e_{31}^2) \\
& e_{12}(e_{31}^4+6e_{31}^2e_{23}^2-3e_{23}^4)-4(e_{11}-e_{22})e_{31}^3e_{23}
\end{aligned}\right\}. \quad (1.12.11)$$

Finally, we consider the last four classes of the hexagonal system. For the ditrigonal-dipyramidal, dihexagonal-pyramidal, hexagonal-trapezohedral, and dihexagonal-dipyramidal classes, it can be shown from § 1.6 that W can be expressed as a polynomial in the quantities (1.12.9) and that it must be restricted by the conditions

$$\left.\begin{aligned}
W(y_1, y_2, y_3, z_1, z_2, z_3, N_1) & \\
= W(y_2, y_3, y_1, z_2, z_3, z_1, N_1) & \\
= W(y_3, y_1, y_2, z_3, z_1, z_2, N_1) & \\
= W(y_1, y_3, y_2, z_1, z_3, z_2, N_1) & \\
= W(y_2, y_1, y_3, z_2, z_1, z_3, N_1) & \\
= W(y_3, y_2, y_1, z_3, z_2, z_1, N_1) &
\end{aligned}\right\}. \quad (1.12.12)$$

Since N_1 is invariant under the given group of transformations, and hence an element of the polynomial basis, the problem is reduced, by Theorem 5, to finding a polynomial basis for polynomials which are symmetric in the three pairs of variables (y_1, z_1), (y_2, z_2), and (y_3, z_3). A polynomial basis is then obtained by using Theorem 3. We see, therefore, that a polynomial basis for the *ditrigonal-dipyramidal, dihexagonal-pyramidal, hexagonal-trapezohedral,* and *dihexagonal-dipyramidal* classes is formed by the nine quantities

$$\left.\begin{aligned}
& e_{33},\quad e_{11}+e_{22},\quad e_{11}e_{22}-e_{12}^2 \\
& e_{11}(e_{11}^2+6e_{11}e_{22}-12e_{12}^2+9e_{22}^2),\quad e_{31}^2+e_{23}^2 \\
& e_{31}^2(e_{31}^2-3e_{23}^2)^2,\quad e_{11}e_{23}^2+e_{22}e_{31}^2-2e_{12}e_{31}e_{23} \\
& e_{11}e_{31}^4+3e_{11}e_{23}^4+2e_{22}e_{31}^4+6e_{22}e_{31}^2e_{23}^2-8e_{12}e_{31}^3e_{23} \\
& (e_{11}^2+2e_{11}e_{22}-3e_{22}^2+4e_{12}^2)e_{31}^2- \\
& \qquad\qquad -2(e_{11}^2+3e_{11}e_{22})(e_{31}^2+e_{23}^2)+8e_{11}e_{12}e_{31}e_{23}
\end{aligned}\right\}. \quad (1.12.13)$$

This completes the discussion of the crystal classes but we add results in the next section for a material which possesses transverse isotropy.

1.13. Transverse isotropy†

We suppose that the material is transversely isotropic with respect to a direction $\mathbf{h}_3$. We take as our reference system a right-handed rectangular cartesian coordinate system x_i, such that the x_3-axis is parallel to $\mathbf{h}_3$ and the x_1- and x_2-axes are any pair of rectangular axes in a plane perpendicular to $\mathbf{h}_3$. The transformations characterizing transverse isotropy are $\mathbf{I}$, $\mathbf{R}_1$ and the transformations

$$\left.\begin{aligned}
& x_1'+ix_2' = e^{-i\alpha}(x_1+ix_2),\quad x_3' = x_3 \\
& e_{11}'+e_{22}' = e_{11}+e_{22},\quad e_{33}' = e_{33} \\
& e_{11}'-e_{22}'+2ie_{12}' = e^{-2i\alpha}(e_{11}-e_{22}+2ie_{12}) \\
& e_{13}'+ie_{23}' = e^{-i\alpha}(e_{13}+ie_{23})
\end{aligned}\right\}, \quad (1.13.1)$$

for all values of α. The transformation (1.13.1) represents an arbitrary rigid body rotation about the x_3-axis.

It can be shown in a manner similar to that used for the monoclinic system in § 1.8 that if W is invariant under the transformation (1.13.1) with $\alpha = \pi$ then it must be expressible as a polynomial in the quantities e_{11}, e_{22}, e_{33}, e_{12}, e_{23}^2, e_{31}^2, $e_{23}e_{31}$. It follows that W can be expressed in polynomial form as

$$W = W(e_{11}+e_{22},\ e_{31}^2+e_{23}^2,\ e_{33},\ E_1,\ E_2,\ F_1,\ F_2), \quad (1.13.2)$$

where
$$\left.\begin{aligned}
E_1 &= e_{11}-e_{22},\quad E_2 = 2e_{12} \\
F_1 &= e_{31}^2-e_{23}^2,\quad F_2 = 2e_{31}e_{23}
\end{aligned}\right\}. \quad (1.13.3)$$

† A general discussion of the strain energy function W for transverse isotropy when W is not necessarily a polynomial was given by J. L. Ericksen and R. S. Rivlin, *J. rat. Mech. Analysis* **3** (1954), 281. Some of their ideas are used here. Results were stated in T.E., p. 156.

Denoting the corresponding values of E_1, E_2, F_1, F_2 in the x_i' system of axes by E_1', E_2', F_1', F_2' we see, from (1.13.1), that

$$\left. \begin{aligned} E_1' &= E_1 \cos 2\alpha + E_2 \sin 2\alpha \\ E_2' &= E_2 \cos 2\alpha - E_1 \sin 2\alpha \\ F_1' &= F_1 \cos 2\alpha + F_2 \sin 2\alpha \\ F_2' &= F_2 \cos 2\alpha - F_1 \sin 2\alpha \end{aligned} \right\}. \quad (1.13.4)$$

If W is form-invariant under the transformation (1.13.1), then

$$W(e_{11}+e_{22}, e_{31}^2+e_{32}^2, e_{33}, E_1', E_2', F_1', F_2')$$
$$= W(e_{11}+e_{22}, e_{31}^2+e_{32}^2, e_{33}, E_1, E_2, F_1, F_2), \quad (1.13.5)$$

where E_α', F_α' are related to E_α, F_α by (1.13.4), and (1.13.5) must be true for all α. We may regard (1.13.4) as relations between the components (E_1, E_2), (F_1, F_2) and (E_1', E_2'), (F_1', F_2') respectively of a pair of two-dimensional vectors under a proper orthogonal transformation. It follows from Theorems 1 and 5 that W must be expressible as a polynomial in the quantities

$$\left. \begin{matrix} e_{11}+e_{22}, & e_{33}, & e_{31}^2+e_{32}^2 \\ E_1^2+E_2^2, & F_1^2+F_2^2, & E_1F_1+E_2F_2 \end{matrix} \right\}, \quad (1.13.6)$$

$$\Delta = \begin{vmatrix} E_1 & E_2 \\ F_1 & F_2 \end{vmatrix}. \quad (1.13.7)$$

Under the transformation $\mathbf{R}_1$, the quantities (1.13.6) are unchanged while Δ is changed to $-\Delta$. With the help of Theorems 1 and 5 it therefore follows that W can be expressed as a polynomial in the functions (1.13.6) and Δ^2. Finally, since

$$\Delta^2 = \begin{vmatrix} E_1^2+E_2^2 & E_1F_1+E_2F_2 \\ E_1F_1+E_2F_2 & F_1^2+F_2^2 \end{vmatrix}, \quad (1.13.8)$$

and

$$\left. \begin{aligned} E_1^2+E_2^2 &= (e_{11}+e_{22})^2 - 4(e_{11}e_{22}-e_{12}^2) \\ F_1^2+F_2^2 &= (e_{31}^2+e_{32}^2)^2 \\ E_1F_1+E_2F_2 &= 2\,|e_{ij}| + (e_{11}+e_{22})(e_{31}^2+e_{23}^2) - 2e_{33}(e_{11}e_{22}-e_{12}^2) \end{aligned} \right\}, \quad (1.13.9)$$

W can be expressed in the form†

$$W = W(e_{11}+e_{22}, e_{11}e_{22}-e_{12}^2, e_{33}, e_{31}^2+e_{23}^2, |e_{ij}|), \quad (1.13.10)$$

where W is a polynomial.

We observe that this form of W is also invariant under the transformations $\mathbf{C}$, $\mathbf{R}_2$, $\mathbf{R}_3$, $\mathbf{D}_1$, $\mathbf{D}_2$, $\mathbf{D}_3$, $\mathbf{S}_1$, and $\mathbf{S}_2$. Moreover, since, from

† *T.E.*, p. 156.

(1.13.6)–(1.13.9), Δ can be expressed as a function of the arguments occurring in (1.13.10), the transformation (1.13.1) alone is sufficient to ensure that a reduction to the form (1.13.10) is possible; W is not then, however, necessarily a polynomial function of its arguments, although a polynomial approximation is always possible since Δ is an integral function of the remaining invariants.

It is convenient for later work to express W in an alternative form. We define K_1, K_2 by the equations

$$K_1 = e_{33}, \qquad K_2 = e_{3\alpha}e_{3\alpha}, \qquad (1.13.11)$$

where repeated Greek indices take the values 1, 2. Then, recalling (1.1.18), we see that W may be expressed as a polynomial in the form

$$W(I_1, I_2, I_3, K_1, K_2). \qquad (1.13.12)$$

From this relation we see that the transversely isotropic form of W is invariant under any transformation of the form†

$$x'_\alpha = x'_\alpha(x_1, x_2), \qquad x'_3 = \pm x_3.$$

1.14. Isotropy

If the body is initially isotropic, W must be form-invariant under all transformations of rectangular cartesian axes of the type (1.5.1). In particular, W must be a polynomial of the form (1.13.12) which is also form-invariant under the transformation $\mathbf{M}_1$. Thus

$$W(I_1, I_2, I_3, e_{33}, e_{32}^2 + e_{31}^2) = W(I_1, I_2, I_3, e_{11}, e_{13}^2 + e_{12}^2),$$

for all possible values of the strain components, so that W reduces to a polynomial of the form

$$W = W(I_1, I_2, I_3). \qquad (1.14.1)$$

It can now be verified that W is form-invariant under all transformations of the form (1.5.1).

We have said that a body is initially isotropic if W is form-invariant under all orthogonal transformations of rectangular cartesian axes. If W in (1.3.8) is a polynomial which is form-invariant only under all *proper* orthogonal transformations then it still reduces to (1.14.1), and this could be taken as the definition of isotropy. It is sometimes convenient to distinguish between isotropy in this latter sense and isotropy

---

† These relations have been employed as a definition of transverse isotropy by J. E. Adkins, *Proc. R. Soc.* A, **229** (1955), 119. The group of transformations $x'_\alpha = x'_\alpha(x_1, x_2)$, $x'_3 = x_3$ is, however, sufficient to secure the reduction to the polynomial form (1.13.10) since this includes as special cases the transformations $\mathbf{R}_1$ and (1.13.1).

with a centre of symmetry which corresponds to our first definition, but the results here are identical using either definition.

If the material is incompressible, so that $I_3 = 1$, the strain energy may be written[†]

$$W = W(I_1, I_2) \quad (I_3 = 1). \tag{1.14.2}$$

The linear form[‡]
$$W = C_1(I_1-3) + C_2(I_2-3), \tag{1.14.3}$$

where C_1 and C_2 are constants, has been found valuable in the study of large deformations of rubber, and has the advantage that a number of problems which would otherwise be difficult to handle are rendered mathematically tractable when W is restricted to have this form. It will be seen in Chapter IX that (1.14.3), although very useful, is not an entirely satisfactory form of W for rubber.

1.15. Stress–strain relations

Using (1.1.15), (1.1.33), (1.1.35), and §§ 1.7–1.14 we can obtain stress–strain relations for any of the crystal classes in a general curvilinear system of coordinates. From (1.1.15) and (1.1.33) we have, for a compressible material,

$$\tau^{ij} = \frac{1}{2\sqrt{I_3}} \left(\frac{\partial W}{\partial e_{rs}} + \frac{\partial W}{\partial e_{sr}} \right) \frac{\partial \theta^i}{\partial x^r} \frac{\partial \theta^j}{\partial x^s}. \tag{1.15.1}$$

From (1.1.35) we see that the corresponding result for an incompressible material is

$$\tau^{ij} = \frac{1}{2} \left(\frac{\partial W}{\partial e_{rs}} + \frac{\partial W}{\partial e_{sr}} \right) \frac{\partial \theta^i}{\partial x^r} \frac{\partial \theta^j}{\partial x^s} + pG^{ij}. \tag{1.15.2}$$

Equations (1.15.1) and (1.15.2) give stress–strain relations for the triclinic crystal system. We record results also for the rhombic system (called, alternatively, an *orthotropic* body), for a body which is transversely isotropic, and for an isotropic body.

---

† Restrictions on the functional form of W for an incompressible isotropic material have been discussed by M. Baker and J. L. Ericksen, *J. Wash. Acad. Sci.* **44** (1954), 33.

‡ This was suggested by M. Mooney, *J. appl. Phys.* **11** (1940), 582, on the basis of shearing experiments on rubber blocks. It is also natural to consider (1.14.3) as a particular case of W which is a polynomial in I_1-3, I_2-3.

Rhombic system (orthotropy)

From (1.15.1) and (1.9.3) we see that the stress–strain relations for the (compressible) rhombic system are

$$\tau^{ij}\sqrt{I_3} = \frac{\partial W}{\partial e_{11}}\frac{\partial \theta^i}{\partial x^1}\frac{\partial \theta^j}{\partial x^1} + \frac{\partial W}{\partial e_{22}}\frac{\partial \theta^i}{\partial x^2}\frac{\partial \theta^j}{\partial x^2} + \frac{\partial W}{\partial e_{33}}\frac{\partial \theta^i}{\partial x^3}\frac{\partial \theta^j}{\partial x^3} +$$

$$+ e_{12}\frac{\partial W}{\partial e_{12}^2}\left(\frac{\partial \theta^i}{\partial x^1}\frac{\partial \theta^j}{\partial x^2} + \frac{\partial \theta^i}{\partial x^2}\frac{\partial \theta^j}{\partial x^1}\right) +$$

$$+ e_{23}\frac{\partial W}{\partial e_{23}^2}\left(\frac{\partial \theta^i}{\partial x^2}\frac{\partial \theta^j}{\partial x^3} + \frac{\partial \theta^i}{\partial x^3}\frac{\partial \theta^j}{\partial x^2}\right) +$$

$$+ e_{31}\frac{\partial W}{\partial e_{31}^2}\left(\frac{\partial \theta^i}{\partial x^3}\frac{\partial \theta^j}{\partial x^1} + \frac{\partial \theta^i}{\partial x^1}\frac{\partial \theta^j}{\partial x^3}\right) + 2I_3\frac{\partial W}{\partial I_3}G^{ij}. \quad (1.15.3)$$

When the body is incompressible $I_3 = 1$ and (1.9.3) becomes

$$W = W(e_{11}, e_{22}, e_{33}, e_{12}^2, e_{23}^2, e_{31}^2). \quad (1.15.4)$$

The corresponding stress–strain relation is

$$\tau^{ij} = \frac{\partial W}{\partial e_{11}}\frac{\partial \theta^i}{\partial x^1}\frac{\partial \theta^j}{\partial x^1} + \frac{\partial W}{\partial e_{22}}\frac{\partial \theta^i}{\partial x^2}\frac{\partial \theta^j}{\partial x^2} + \frac{\partial W}{\partial e_{33}}\frac{\partial \theta^i}{\partial x^3}\frac{\partial \theta^j}{\partial x^3} +$$

$$+ e_{12}\frac{\partial W}{\partial e_{12}^2}\left(\frac{\partial \theta^i}{\partial x^1}\frac{\partial \theta^j}{\partial x^2} + \frac{\partial \theta^i}{\partial x^2}\frac{\partial \theta^j}{\partial x^1}\right) +$$

$$+ e_{23}\frac{\partial W}{\partial e_{23}^2}\left(\frac{\partial \theta^i}{\partial x^2}\frac{\partial \theta^j}{\partial x^3} + \frac{\partial \theta^i}{\partial x^3}\frac{\partial \theta^j}{\partial x^2}\right) +$$

$$+ e_{31}\frac{\partial W}{\partial e_{31}^2}\left(\frac{\partial \theta^i}{\partial x^3}\frac{\partial \theta^j}{\partial x^1} + \frac{\partial \theta^i}{\partial x^1}\frac{\partial \theta^j}{\partial x^3}\right) + pG^{ij}. \quad (1.15.5)$$

Transverse isotropy

When the material is compressible we have, from (1.1.18), (1.13.11), (1.13.12), and (1.15.1),

$$\tau^{ij} = \Phi g^{ij} + \Psi B^{ij} + pG^{ij} + \Theta M^{ij} + \Lambda N^{ij}, \quad (1.15.6)$$

where
$$\Phi = \frac{2}{\sqrt{I_3}}\frac{\partial W}{\partial I_1}, \quad \Psi = \frac{2}{\sqrt{I_3}}\frac{\partial W}{\partial I_2}, \quad p = 2\sqrt{I_3}\frac{\partial W}{\partial I_3}, \quad (1.15.7)$$

$$\Theta = \frac{1}{\sqrt{I_3}}\frac{\partial W}{\partial K_1}, \quad \Lambda = \frac{1}{\sqrt{I_3}}\frac{\partial W}{\partial K_2}, \quad (1.15.8)$$

$$B^{ij} = I_1 g^{ij} - g^{ir}g^{js}G_{rs} = \frac{1}{g}e^{irm}e^{jsn}g_{rs}G_{mn}, \quad (1.15.9)$$

$$M^{ij} = \frac{\partial \theta^i}{\partial x^3}\frac{\partial \theta^j}{\partial x^3}, \quad N^{ij} = \left(\frac{\partial \theta^i}{\partial x^3}\frac{\partial \theta^j}{\partial x^\alpha} + \frac{\partial \theta^i}{\partial x^\alpha}\frac{\partial \theta^j}{\partial x^3}\right)e_{\alpha 3}. \quad (1.15.10)$$

For an incompressible body (1.13.12) becomes
$$W = W(I_1, I_2, K_1, K_2) \quad (I_3 = 1). \tag{1.15.11}$$
The stress–strain relation (1.15.6) still holds but now
$$\Phi = 2\frac{\partial W}{\partial I_1}, \quad \Psi = 2\frac{\partial W}{\partial I_2}, \quad \Theta = \frac{\partial W}{\partial K_1}, \quad \Lambda = \frac{\partial W}{\partial K_2}, \tag{1.15.12}$$
and p is an unknown scalar invariant function of θ (not given by 1.15.7).

Isotropy†

For an isotropic body the strain energy reduces to $W = W(I_1, I_2, I_3)$ and (1.15.6) becomes
$$\tau^{ij} = \Phi g^{ij} + \Psi B^{ij} + p G^{ij}, \tag{1.15.13}$$
where, for a compressible material, Φ, Ψ, p are given by (1.15.7). For an incompressible material, Φ, Ψ are given by (1.15.12), and p is a scalar invariant of θ.

We also record the form of the stress tensor when referred to fixed rectangular axes in the deformed body. Using σ_{ij} for components of stress‡ at the point y_i, we have
$$\sigma_{ij} = \frac{\partial y^i}{\partial \theta^r} \frac{\partial y^j}{\partial \theta^s} \tau^{rs}. \tag{1.15.14}$$

Now
$$\left. \begin{aligned} \frac{\partial y^i}{\partial \theta^r} \frac{\partial y^j}{\partial \theta^s} g^{rs} &= \frac{\partial y^i}{\partial x^m} \frac{\partial y^j}{\partial x^m} = C_{ij} \quad \text{(say)} \\ \frac{\partial y^i}{\partial \theta^r} \frac{\partial y^j}{\partial \theta^s} g^{rm} g^{sn} G_{mn} &= C_{ik} C_{kj} \\ \frac{\partial y^i}{\partial \theta^r} \frac{\partial y^j}{\partial \theta^s} G^{rs} &= \delta_{ij} \end{aligned} \right\}, \tag{1.15.15}$$

and we observe that the invariants of C_{ij} may be taken as I_1, I_2, I_3. Hence, from (1.15.13)–(1.15.15), (1.15.6), and (1.15.9) we obtain the cartesian components of stress σ_{ij} for an isotropic body, in the form
$$\sigma_{ij} = p\delta_{ij} + (\Phi + I_1 \Psi) C_{ij} - \Psi C_{ik} C_{kj}. \tag{1.15.16}$$

1.16. Curvilinear aeolotropy

We have seen in § 1.3 that the assumption that the strain energy function W for a homogeneous aeolotropic body is expressible as a

† Results for an isotropic body were given in *T.E.*
‡ σ_{ij} are, of course, physical components of stress but are only identical with the components defined by (1.1.23) for the choice $\theta_i = y_i$.

function of the displacement gradients $\partial y_i/\partial x_j$ leads to an expression for W in terms of the strain components e_{ij}, which we usually take to be a polynomial function (1.3.8). Apart from the quantities e_{ij} this function contains only absolute constants which are dependent neither upon the deformation nor upon the position of the element of the body under consideration. Referred to any other arbitrary coordinate system θ_i, however, the expression for W contains not only the strain components γ_{ij} but also the derivatives $\partial \theta^i/\partial x^j$ which enter by virtue of the tensor transformation (1.1.15). The form of W as a function of the strain components remains constant throughout the body, only when the latter are referred to a set of rectangular cartesian axes; this ensures that the elastic properties are identical at each point.

It is possible to conceive of a material which is aeolotropic and homogeneous in the sense that its elastic properties at each point are identical provided that the axes to which these properties are referred are given a suitable orientation. In this case, if W is a given function at the point P of the strain components e_{ij}, then at any other point P' it can be expressed as the same function of strain components e'_{ij} referred to axes x'_i in the undeformed body which are obtained from the set x_i by a suitable rotation. This leads to a natural definition of a body which is elastically homogeneous and curvilinearly aeolotropic with respect to a curvilinear coordinate system θ'_i in the undeformed body as one whose strain energy function W preserves the same form at all points when expressed as a function of dimensionless strain components corresponding to the quantities e_{ij}, these components being evaluated with respect to the tangent lines at the point under consideration to the reference frame θ'_i. We may associate with this latter coordinate system base vectors $\mathbf{g}'_i$, $\mathbf{g}'^i$, $\mathbf{G}'_i$, $\mathbf{G}'^i$, metric tensors g'_{ij}, g'^{ij}, G'_{ij}, G'^{ij}, strain components γ'_{ij}, and stress tensors τ'^{ij}, s'^{ij}, π'^{ij}, t'^{ij}, which are strictly analogous to the corresponding unprimed quantities defined in § 1.1 for the system θ_i and can be obtained from them in the usual manner by tensor transformations. For simplicity we shall confine our attention to the case where the coordinate system θ'_i in the undeformed body is orthogonal, so that

$$g'^{ii} = 1/g'_{ii}, \qquad g'_{ij} = g'^{ij} = 0 \qquad (i \neq j), \tag{1.16.1}$$

where i is not summed. If ds_0, ds are corresponding elements of length in the undeformed body and the deformed body respectively, we have

$$ds^2 - ds_0^2 = 2e_{ij}\,dx^i\,dx^j = 2\gamma'_{ij}\,d\theta'^i\,d\theta'^j = 2\gamma'_{(ij)}\,ds'_i\,ds'_j,$$

where $ds'_i = \sqrt{(g'_{ii})}\, d\theta'^i$ (i not summed) is an element of length in the direction of a θ'_i curve in the undeformed body; the quantities

$$\gamma'_{(ij)} = \frac{\gamma'_{ij}}{\sqrt{(g'_{ii}g'_{jj})}} = \gamma'_{(ji)} \tag{1.16.2}$$

are therefore the required dimensionless components of strain referred to the system θ'_i in the undeformed material. These are physical quantities which correspond to the components e_{ij} for the system x_i; the suffixes in (1.16.2) are bracketed to indicate that $\gamma'_{(ij)}$ do not transform as the components of a tensor. By analogy with (1.3.8) we shall assume the strain energy function W for a material exhibiting curvilinear aeolotropy relative to the system θ'_i to be expressible as a polynomial

$$W = W(\gamma'_{(ij)}) \tag{1.16.3}$$

in the components $\gamma'_{(ij)}$ in which the coefficients are absolute constants independent of the deformation or the coordinates θ'_i.

The definition and analysis of local symmetry properties follow exactly the same lines as for the rectilinear case, the quantities $\gamma'_{(ij)}$ replacing e_{ij} in the forms of W listed in §§ 1.7 to 1.13. In particular, for materials possessing rhombic symmetry with respect to the tangent lines to the θ'_i coordinate curves at each point in the undeformed body, we have

$$W = W(\gamma'_{(11)}, \gamma'_{(22)}, \gamma'_{(33)}, \gamma'^2_{(12)}, \gamma'^2_{(23)}, \gamma'^2_{(31)}, \gamma'_{(12)}\gamma'_{(23)}\gamma'_{(31)})$$
$$= W(\gamma'^1_1, \gamma'^2_2, \gamma'^3_3, \gamma'^1_2\gamma'^2_1, \gamma'^2_3\gamma'^3_2, \gamma'^3_1\gamma'^1_3, \gamma'^1_2\gamma'^2_3\gamma'^3_1), \tag{1.16.4}$$

or $\qquad W = W(\gamma'^1_1, \gamma'^2_2, \gamma'^3_3, \gamma'^1_2\gamma'^2_1, \gamma'^2_3\gamma'^3_2, \gamma'^3_1\gamma'^1_3, I_3), \tag{1.16.5}$

corresponding respectively to (1.9.2) and (1.9.3) where

$$\gamma'^i_j = g'^{ir}\gamma'_{rj}.$$

The alternative form for W in (1.16.4) follows by using (1.16.1) and (1.16.2) if we remember that the θ'_i system is orthogonal in the undeformed body. Similarly, for materials which are transversely isotropic with respect to the θ'_3 direction at each point, W takes the form

$$W = W(I_1, I_2, I_3, K'_1, K'_2), \tag{1.16.6}$$

analogous to (1.13.12), where

$$\left.\begin{aligned} K'_1 &= \gamma'_{(33)} = \gamma'^3_3 \\ K'_2 &= \gamma'_{(3\alpha)}\gamma'_{(\alpha 3)} = \gamma'^3_\alpha \gamma'^\alpha_3 \end{aligned}\right\}. \tag{1.16.7}$$

The stress–strain relations (1.1.33), (1.1.35) may again be modified to indicate the type of material under consideration. For if, owing to the symmetry properties of the material, W can be expressed as a

function
$$W = W(\Psi_t) \tag{1.16.8}$$
of given functions†
$$\Psi_t = \Psi_t(\gamma'_{(ij)}) \qquad (t = 1, 2, ..., p) \tag{1.16.9}$$
of the strain components $\gamma'_{(ij)}$, then (1.1.33) yields, for compressible materials,
$$\tau^{ij} = \frac{1}{2\sqrt{I_3}} \sum_{t=1}^{p} \left\{ (A^{ij}_{(rs)} + A^{ij}_{(sr)}) \frac{\partial \Psi_t}{\partial \gamma'_{(rs)}} \right\} \frac{\partial W}{\partial \Psi_t}, \tag{1.16.10}$$
where
$$A^{ij}_{(rs)} = \frac{\partial \theta^i}{\partial \theta'^r} \frac{\partial \theta^j}{\partial \theta'^s} \bigg/ \sqrt{(g'_{rr} g'_{ss})} \quad (r, s \text{ not summed}), \tag{1.16.11}$$

and in (1.16.10) summation is carried out over the repeated indices r, s. The corresponding form derived from (1.1.35) for the incompressible case is
$$\tau^{ij} = \frac{1}{2} \sum_{t=1}^{p} \left\{ (A^{ij}_{(rs)} + A^{ij}_{(sr)}) \frac{\partial \Psi_t}{\partial \gamma'_{(rs)}} \right\} \frac{\partial W}{\partial \Psi_t} + pG^{ij}. \tag{1.16.12}$$

For locally triclinic materials, with a form of strain energy function corresponding to (1.7.1), the functions Ψ_t reduce to the strain components $\gamma'_{(rs)}$ and (1.16.10), (1.16.12) become
$$\tau^{ij} = \frac{1}{2\sqrt{I_3}} [A^{ij}_{(rs)} + A^{ij}_{(sr)}] \frac{\partial W}{\partial \gamma'_{(rs)}}, \tag{1.16.13}$$
$$\tau^{ij} = \tfrac{1}{2}[A^{ij}_{(rs)} + A^{ij}_{(sr)}] \frac{\partial W}{\partial \gamma'_{(rs)}} + pG^{ij}, \tag{1.16.14}$$

respectively. For locally rhombic and transversely isotropic systems we obtain, for compressible bodies,
$$\tau^{ij}\sqrt{I_3} = A^{ij}_{(11)}\frac{\partial W}{\partial \gamma'_{(11)}} + A^{ij}_{(22)}\frac{\partial W}{\partial \gamma'_{(22)}} + A^{ij}_{(33)}\frac{\partial W}{\partial \gamma'_{(33)}} +$$
$$+ (A^{ij}_{(12)}+A^{ij}_{(21)})\gamma'_{(12)}\frac{\partial W}{\partial \gamma'^2_{(12)}} + (A^{ij}_{(23)}+A^{ij}_{(32)})\gamma'_{(23)}\frac{\partial W}{\partial \gamma'^2_{(23)}} +$$
$$+ (A^{ij}_{(31)}+A^{ij}_{(13)})\gamma'_{(31)}\frac{\partial W}{\partial \gamma'^2_{(31)}} + 2I_3 \frac{\partial W}{\partial I_3} G^{ij}, \tag{1.16.15}$$
$$\tau^{ij} = \Phi g^{ij} + \Psi B^{ij} + pG^{ij} + \Theta' M'^{ij} + \Lambda' N'^{ij}, \tag{1.16.16}$$
respectively, where
$$\Theta' = \frac{1}{\sqrt{I_3}} \frac{\partial W}{\partial K'_1}, \qquad \Lambda' = \frac{1}{\sqrt{I_3}} \frac{\partial W}{\partial K'_2}, \tag{1.16.17}$$
$$M'^{ij} = A^{ij}_{(33)}, \qquad N'^{ij} = (A^{ij}_{(3a)}+A^{ij}_{(a3)})\gamma'_{(a3)}. \tag{1.16.18}$$

† For example, in the transversely isotropic case there are five independent functions Ψ, identical with the invariants $I_1, I_2, I_3, K'_1, K'_2$ occurring in (1.16.6).

Similar formulae apply for incompressible materials. In each case the stress–strain relations for curvilinearly aeolotropic bodies may be inferred from those for the corresponding rectilinearly aeolotropic case by replacing

$$e_{rs} \quad \text{by} \quad \gamma'_{(rs)} \quad \text{and} \quad \frac{\partial \theta^i}{\partial x^r} \quad \text{by} \quad \frac{\partial \theta^i}{\partial \theta'^r} \frac{1}{\sqrt{g'_{rr}}}$$

throughout, where r is not summed.

A possible generalization of the theory to the case where the curvilinear system θ'_i defining the aeolotropy is non-orthogonal has been given by Adkins.†

1.17. Geometrical constraints

Incompressible bodies provide examples of materials which cannot deform in a completely arbitrary manner; any deformation to which they are subjected must be restricted by the incompressibility conditions, a geometrical constraint which implies a functional relationship between the strain components γ_{ij}. We may imagine a constraint of a different type to occur in a body reinforced with continuous systems of thin, flexible, inextensible cords. If, for example, a cuboid of elastic material is reinforced with a system of cords which are parallel to the x_1-direction then, with suitable assumptions, we may regard the body as being inextensible in this direction, and all deformations to which the material is susceptible must satisfy the condition $e_{11} = 0$. Such problems suggest a general theory of materials whose permissible deformations are subject to geometrical constraints. For the present we shall consider such constraints to arise as an intrinsic property of the material; we consider in Chapter VII problems involving reinforcement by cords, and the relation of such problems to the present theory.

To describe a system of constraints we introduce an auxiliary reference frame θ'_i.‡ Stress, metric, and strain tensors are specified for this system which are analogous to those defined in § 1.1 for the coordinate system θ_i and are related to the latter by tensor transformations, primes being used to distinguish quantities related to the coordinates θ'_i. Systems of geometrical constraints may then be described by means of functional relationships

$$f_m(\gamma'_{ij}) = 0 \qquad (m = 1, 2, \ldots, n; n < 6) \qquad (1.17.1)$$

† J. E. Adkins, loc. cit., p. 26.
‡ This need not necessarily coincide with any reference frame used, as in § 1.16, to describe aeolotropic properties. Where several auxiliary systems are used, a numbering system may be employed to distinguish them; thus we write $^{(1)}\theta^i$, $^{(2)}\theta^i$,... .

between the strain components γ'_{ij}. The restriction upon the number of constraints arises from the fact that the deformation may be specified completely by means of the six components γ'_{ij}. Six independent constraints would imply six functionally independent relationships between these quantities, and this would require that the latter could exist only for discrete sets of values, one such set being the values $\gamma'_{ij} = 0$ appropriate to the undeformed state. Since this excludes the possibility of a continuous deformation we shall henceforth assume the number of constraints to be less than six.

If the reference frame θ'_i is chosen so that the θ'_3-curve coincides with the x_3-axis of the rectangular cartesian coordinate system x_i in the unstrained body, the relations

$$f_m(\gamma'_{\alpha\beta}) = 0 \qquad (\alpha, \beta = 1, 2) \qquad (1.17.2)$$

define a two-dimensional system of constraints which restrict the deformation in surfaces which coincide with the planes $x_3 = $ constant in the undeformed material. For a general curvilinear system θ'_i, the relations (1.17.2) restrict the deformation which can occur in the surfaces $\theta'_3 = $ constant. In either case, since only the three components $\gamma'_{11}, \gamma'_{22}, \gamma'_{12}$ enter into (1.17.2), three independent relations of this type would prohibit any continuous deformation within the surfaces $\theta'_3 = $ constant. If the body is initially unstrained, any three such conditions must be equivalent to

$$\gamma'_{11} = \gamma'_{22} = \gamma'_{12} = 0.$$

The relations (1.17.2) may evidently be employed to describe a surface constraint such as would be produced by a single layer of inextensible cords lying in a given surface $\theta'_3 = $ constant.

1.18. Stress–strain relations for materials subject to constraints

To derive stress–strain relations for materials subject to constraints we recall the rate of work equation (1.1.32), which may be written

$$W = \frac{\partial W}{\partial \gamma_{ij}}\dot{\gamma}_{ij} = \sqrt{\left(\frac{G}{g}\right)}\tau^{ij}\dot{\gamma}_{ij}. \qquad (1.18.1)$$

Time differentiation of conditions (1.17.1) yields

$$\frac{\partial f_m}{\partial \gamma'_{rs}}\dot{\gamma}'_{rs} = 0 \qquad (m = 1, 2, ..., n; n < 6).$$

By a tensor transformation, these are recognized as equivalent to the restrictions

$$\frac{\partial f_m}{\partial \gamma'_{rs}} \frac{\partial \theta^i}{\partial \theta'^r} \frac{\partial \theta^j}{\partial \theta'^s} \dot{\gamma}_{ij} = 0 \qquad (m=1,2,\ldots,n;\, n<6) \qquad (1.18.2)$$

upon $\dot{\gamma}_{ij}$. Remembering that τ^{ij} and γ_{ij}, and hence also $\dot{\gamma}_{ij}$, are symmetric tensors, we obtain from (1.18.1) and (1.18.2) the stress–strain relation

$$\tau^{ij} = \frac{1}{2\sqrt{I_3}}\left(\frac{\partial W}{\partial \gamma_{ij}} + \frac{\partial W}{\partial \gamma_{ji}}\right) + \sum_{m=1}^{n} q_m \frac{\partial f_m}{\partial \gamma'_{rs}}\left(\frac{\partial \theta^i}{\partial \theta'^r}\frac{\partial \theta^j}{\partial \theta'^s} + \frac{\partial \theta^i}{\partial \theta'^s}\frac{\partial \theta^j}{\partial \theta'^r}\right) \quad (n<6), \quad (1.18.3)$$

the Lagrangian multipliers q_m being scalar functions of the coordinates θ_i. When incompressibility is included among the constraint conditions,

$$I_3 = |\delta^r_s + 2\gamma^r_s| = |\delta^r_s + 2\gamma'^r_s| = 1,$$

and the formula for τ^{ij} becomes

$$\tau^{ij} = \left[\frac{1}{2}\left(\frac{\partial W}{\partial \gamma_{ij}} + \frac{\partial W}{\partial \gamma_{ji}}\right) + pG^{ij}\right] + \\ + \sum_{m=1}^{n} q_m \frac{\partial f_m}{\partial \gamma'_{rs}}\left(\frac{\partial \theta^i}{\partial \theta'^r}\frac{\partial \theta^j}{\partial \theta'^s} + \frac{\partial \theta^i}{\partial \theta'^s}\frac{\partial \theta^j}{\partial \theta'^r}\right) \quad (n<5). \quad (1.18.4)$$

In these formulae for τ^{ij}, the first group of terms is formally identical with those given in (1.1.33) and (1.1.35) respectively for a compressible and incompressible material not subject to other constraints. These terms can therefore be modified as indicated in §§ 1.15, 1.16 to indicate the existence of a given type of aeolotropy or the presence of symmetry properties. Where the nature of such properties is unimportant we shall denote the first group of terms by τ^{ij}_e and write equations (1.18.3) and (1.18.4) or their modified forms as

$$\tau^{ij} = \tau^{ij}_e + \sum_{m=1}^{n} q_m \frac{\partial f_m}{\partial \gamma'_{rs}}\left(\frac{\partial \theta^i}{\partial \theta'^r}\frac{\partial \theta^j}{\partial \theta'^s} + \frac{\partial \theta^i}{\partial \theta'^s}\frac{\partial \theta^j}{\partial \theta'^r}\right). \qquad (1.18.5)$$

It must be remembered, however, that the form of W which enters into the expression for τ^{ij}_e is restricted by the additional constraint conditions and may be regarded as arbitrary to that extent. Thus, if a form

$$W = W(\gamma_{ij})$$

completely describes the elastic properties of a given material, so also does

$$W' = W(\gamma_{ij}) + F(f_m), \qquad (1.18.6)$$

where F is any function of the functions $f_m(\gamma'_{ij})$ which vanishes with its arguments.

We may observe that the functional form of the relations (1.17.1) may be restricted by symmetry properties analogous to those considered in §§ 1.5–1.14. For example, in the case of isotropic constraints, the functions f_m would be functions of I_1, I_2, I_3, a particular case being the incompressibility condition $I_3 - 1 = 0$.

II

GENERAL SOLUTIONS OF PROBLEMS

In Chapter III of $T.E.$ complete solutions of a number of problems were given for isotropic bodies (mostly incompressible) in terms of a general elastic potential or strain energy function. In the present chapter we discuss further complete solutions for both isotropic and aeolotropic bodies. Results for problems considered in $T.E.$ are often included in work of the present chapter as special cases. It has, however, been found convenient to make some changes of notation from that used in $T.E.$ and this must be remembered when making comparisons with the earlier work. The first general solutions were obtained by Rivlin† but the solutions presented in this chapter are partly derived from work of other authors.‡

The mathematical aspects of the main class of problems which are examined may be characterized in the following manner. Points of the deformed body are referred to a curvilinear reference frame θ_i with which are associated metric tensors g_{ij}, g^{ij} in the undeformed material and G_{ij}, G^{ij} in the deformed body, as described in § 1.1. Points of the undeformed material are referred to a curvilinear system θ_i' with analogous associated metric tensors g_{ij}', g'^{ij}, G_{ij}', G'^{ij}. The deformation to be examined is then restricted by the following conditions:

(i) the system θ_i' is chosen so that the tensors g_{ij}', g'^{ij} are functions of a single variable θ_1';

(ii) the system θ_i is chosen similarly so that the tensors G_{ij}, G^{ij} are functions of a single variable θ_1;

(iii) the deformation is then defined by relationships between the coordinate systems which are such that θ_1 is a function of θ_1' alone, while θ_2, θ_3 are functions which are linear in the variables θ_2', θ_3'. In the

---

† See $T.E.$ for references.

‡ J. L. Ericksen and R. S. Rivlin, *J. rat. Mech. Analysis* **3** (1954), 281.

J. L. Ericksen, *Z. angew. Math. Mech.* **35** (1955), 382.

A. E. Green and E. W. Wilkes, *J. rat. Mech. Analysis* **3** (1954). 713.

A. E. Green, *Proc. R. Soc.* A, **227** (1955), 271.

J. E. Adkins, ibid. A, **229** (1955), 119; A, **231** (1955), 75; *Proc. Camb. phil. Soc. math. phys. Sci.* **50** (1954), 334; **51** (1955), 363.

See also references on p. 1.

Deformations possible in every isotropic body have been examined by J. L. Ericksen, *Z. angew. Math. Phys.* **5** (1954), 466; *J. Math. Phys.* **34** (1955), 126.

present chapter we assume also that θ_2, θ_3 are independent of θ'_1 although some progress can be made without this assumption.†

In these circumstances the derivatives $\partial \theta^i/\partial \theta'^j$ reduce to constants or functions of the single variable θ_1 (or θ'_1). Similarly, the remaining metric tensors G'_{ij}, G'^{ij}, g_{ij}, g^{ij}, and hence also the strain components γ_{ij}, γ'_{ij} referred to the systems θ_i, θ'_i, reduce also either to constants or to functions of θ'_1 (or θ_1). The incompressibility condition, where it applies, then becomes an ordinary differential equation in which either θ'_1 or θ_1 may be taken as the independent variable. If the type of aeolotropy is defined by the coordinates θ'_i the strain energy function W can be expressed entirely in terms of g'_{ij} and G'_{ij} and is independent of the coordinates θ'_2, θ'_3. From the stress–strain relations of Chapter I it follows that the stresses τ^{ij} referred to the system θ_i are also independent of θ_2, θ_3, and the equations of equilibrium reduce to ordinary differential equations, the unknowns in these equations being functions of the single variable θ'_1.

The conditions (i) and (ii) are satisfied if the curvilinear systems θ'_i, θ_i are cylindrical polar or rectangular cartesian systems of coordinates. By identifying both systems with cylindrical polar reference frames, the cylindrically symmetrical problems of §§ 2.1–2.5, 2.10, 2.12–2.14, 2.16, and 2.18 are obtained. The choice of rectangular cartesian coordinates, for the system θ'_i and polar coordinates for the system θ_i leads to the flexure problems discussed in §§ 2.7, 2.9, 2.11, 2.15, 2.17, and 2.19. If the θ_i curves form a rectangular system and the θ'_i curves are polar coordinates, the inverse type of problem is obtained in which an initially curved cuboid is bounded by plane faces in the deformed state. This is examined briefly in § 2.6, but the full investigation presents no difficulty. The homogeneous deformation of § 2.8 is obtained when the reference frames θ'_i, θ_i are both rectilinear. In the spherically symmetrical problem of § 2.20 the conditions (i) and (ii) are no longer satisfied, but the deformation is further restricted so that the incompressibility condition and the equations of equilibrium again reduce to ordinary differential equations.

The analysis of §§ 2.1–2.20 is carred out assuming general forms for the strain energy function. Incompressible bodies are studied in §§ 2.1–2.11, except for a part of § 2.8, whilst §§ 2.12–2.17 and § 2.20 are concerned with compressible bodies; both incompressible and compressible bodies are studied in §§ 2.18, 2.19. In §§ 2.21–2.24 shearing deformations of isotropic incompressible materials are examined. For these,

† See *T.E.*, § 3.7, and J. E. Adkins, loc. cit., p. 37.

solutions may be derived by restricting the strain energy function W to have the linear form suggested by Mooney† as being appropriate for rubber-like materials when the deformations are finite but not too large.

2.1. Cylindrically symmetric problems: incompressible transverse isotropy

We consider a sector of a hollow cylinder consisting of incompressible material, transversely isotropic with respect to the axis of the cylinder, which is taken to be the x_3-axis. The stress–strain relations for an incompressible body which is transversely isotropic with respect to the x_3-direction are given by (1.15.6)–(1.15.12). Results for an incompressible isotropic body are obtained immediately from these by putting $\Theta \equiv \Lambda \equiv 0$.

We choose the moving system of coordinates θ_i so that

$$(\theta_1, \theta_2, \theta_3) = (r, \theta, y_3), \tag{2.1.1}$$

where
$$y_1 = r\cos\theta, \qquad y_2 = r\sin\theta, \tag{2.1.2}$$

(r, θ, y_3) being cylindrical polar coordinates in the deformed body. If (ρ, ϑ, x_3) are cylindrical polar coordinates in the undeformed body so that
$$x_1 = \rho\cos\vartheta, \qquad x_2 = \rho\sin\vartheta, \tag{2.1.3}$$

we consider the deformation, symmetric about the x_3-axis, defined by

$$r = r(\rho), \qquad \theta = a\vartheta + bx_3, \qquad y_3 = c\vartheta + dx_3, \tag{2.1.4}$$

or

$$\rho = \rho(r), \qquad \vartheta = (d\theta - by_3)/\lambda, \qquad x_3 = (ay_3 - c\theta)/\lambda, \qquad \lambda = ad - bc, \tag{2.1.5}$$

where a, b, c, d are constants. This deformation includes as special cases simple torsion ($a = d = 1$, $c = 0$) and simple flexure ($a \neq 1$, $b = c = 0$). From (2.1.2) we have

$$\left. \begin{array}{c} G_{11} = G^{11} = G_{33} = G^{33} = 1, \qquad G_{22} = r^2, \qquad G^{22} = 1/r^2 \\ G = r^2, \qquad G_{ij} = G^{ij} = 0 \quad (i \neq j) \end{array} \right\}, \tag{2.1.6}$$

† (1.14.3)

and (2.1.3), (2.1.5) yield

$$\frac{\partial x^r}{\partial \theta^s} = \begin{bmatrix} \dfrac{\cos\vartheta}{r_\rho} & -\dfrac{d\rho\sin\vartheta}{\lambda} & \dfrac{b\rho\sin\vartheta}{\lambda} \\ \dfrac{\sin\vartheta}{r_\rho} & \dfrac{d\rho\cos\vartheta}{\lambda} & -\dfrac{b\rho\cos\vartheta}{\lambda} \\ 0 & -\dfrac{c}{\lambda} & \dfrac{a}{\lambda} \end{bmatrix}, \qquad (2.1.7)$$

$$\left|\frac{\partial x^r}{\partial \theta^s}\right| = \sqrt{g} = \frac{\rho}{\lambda r_\rho}, \qquad (2.1.8)$$

where r_ρ denotes $dr/d\rho$. From (2.1.7) and (2.1.8) we obtain

$$\frac{\partial \theta^r}{\partial x^s} = \begin{bmatrix} r_\rho\cos\vartheta & r_\rho\sin\vartheta & 0 \\ -\dfrac{a\sin\vartheta}{\rho} & \dfrac{a\cos\vartheta}{\rho} & b \\ -\dfrac{c\sin\vartheta}{\rho} & \dfrac{c\cos\vartheta}{\rho} & d \end{bmatrix}. \qquad (2.1.9)$$

Hence

$$g_{ij} = \begin{bmatrix} \dfrac{1}{r_\rho^2} & 0 & 0 \\ 0 & (c^2+d^2\rho^2)/\lambda^2 & -(ac+bd\rho^2)/\lambda^2 \\ 0 & -(ac+bd\rho^2)/\lambda^2 & (a^2+b^2\rho^2)/\lambda^2 \end{bmatrix} \qquad (2.1.10)$$

$$g^{ij} = \begin{bmatrix} r_\rho^2 & 0 & 0 \\ 0 & (a^2+b^2\rho^2)/\rho^2 & (ac+bd\rho^2)/\rho^2 \\ 0 & (ac+bd\rho^2)/\rho^2 & (c^2+d^2\rho^2)/\rho^2 \end{bmatrix}. \qquad (2.1.11)$$

If the body is incompressible,

$$\rho = \lambda r r_\rho \quad \text{or} \quad \rho^2 = \lambda r^2 + K, \qquad (2.1.12)$$

where K is a constant. From (1.1.18) and (1.13.11) with the help of (1.1.15), (2.1.6), and (2.1.9)–(2.1.12), we obtain

$$\left.\begin{aligned} I_1 &= \rho^2/(\lambda^2 r^2) + (a^2 r^2 + c^2)/\rho^2 + b^2 r^2 + d^2 \\ I_2 &= \lambda^2 r^2/\rho^2 + (c^2 + d^2\rho^2)/(\lambda^2 r^2) + (a^2 + b^2\rho^2)/\lambda^2 \end{aligned}\right\}, \qquad (2.1.13)$$

$$\left.\begin{aligned} 2K_1 &= b^2 r^2 + d^2 - 1 \\ 4K_2 &= (abr^2 + cd)^2/\rho^2 \end{aligned}\right\}, \qquad (2.1.14)$$

while (1.15.9) and (1.15.10) yield

$$\left.\begin{aligned}
B^{11} &= \{a^2r^2+c^2+\rho^2(b^2r^2+d^2)\}/(\lambda^2r^2) \\
B^{22} &= (a^2+b^2\rho^2)/(\lambda^2r^2)+\lambda^2/\rho^2 \\
B^{33} &= (c^2+d^2\rho^2)/(\lambda^2r^2)+\lambda^2r^2/\rho^2 \\
B^{23} &= (ac+bd\rho^2)/(\lambda^2r^2), \qquad B^{12} = B^{13} = 0
\end{aligned}\right\}, \quad (2.1.15)$$

$$M^{ij} = \begin{bmatrix} 0 & 0 & 0 \\ 0 & b^2 & bd \\ 0 & bd & d^2 \end{bmatrix}, \quad (2.1.16)$$

$$\left.\begin{aligned}
N^{22} &= ab(abr^2+cd)/\rho^2 \\
N^{33} &= cd(abr^2+cd)/\rho^2 \\
N^{23} &= (ad+bc)(abr^2+cd)/(2\rho^2) \\
N^{11} &= N^{12} = N^{13} = 0
\end{aligned}\right\}. \quad (2.1.17)$$

Expressions for the components of the stress tensor τ^{ij} follow by combining the results of the present section with (1.15.6). We have

$$\left.\begin{aligned}
r^2\tau^{22} &= \tau^{11}+\{r^2(a^2+b^2\rho^2)/\rho^2-\rho^2/(\lambda^2r^2)\}\Phi+ \\
&\quad +\{\lambda^2r^2/\rho^2-(c^2+d^2\rho^2)/(\lambda^2r^2)\}\Psi+ \\
&\quad +b^2r^2\Theta+\{abr^2(abr^2+cd)/\rho^2\}\Lambda \\
\tau^{33} &= \tau^{11}+\{(c^2+d^2\rho^2)/\rho^2-\rho^2/(\lambda^2r^2)\}\Phi+ \\
&\quad +\{\lambda^2r^2/\rho^2-(a^2+b^2\rho^2)/\lambda^2\}\Psi+ \\
&\quad +d^2\Theta+\{cd(abr^2+cd)/\rho^2\}\Lambda \\
\tau^{23} &= (ac+bd\rho^2)\{\Phi/\rho^2+\Psi/(\lambda^2r^2)\}+ \\
&\quad +bd\Theta+\{(ad+bc)(abr^2+cd)/(2\rho^2)\}\Lambda \\
\tau^{12} &= \tau^{13} = 0
\end{aligned}\right\}, \quad (2.1.18)$$

where r is given in terms of ρ by (2.1.12).

The equations of equilibrium (1.1.28), in the absence of body forces, become

$$\frac{\partial \tau^{11}}{\partial r}+\frac{\tau^{11}-r^2\tau^{22}}{r} = 0, \qquad \frac{\partial p}{\partial \theta} = \frac{\partial p}{\partial y_3} = 0, \quad (2.1.19)$$

showing that p is a function of r (or ρ) only. Since $\tau^{11}-r^2\tau^{22}$ is independent of p, the first equation determines τ^{11} (or p) in the form

$$\tau^{11} = \int_{r_0}^{r}(r^2\tau^{22}-\tau^{11})\frac{dr}{r}+\tau_0^{11}, \quad (2.1.20)$$

where τ_0^{11} is the value of τ^{11} at $r = r_0$ and $r^2\tau^{22} - \tau^{11}$ is given by the first of (2.1.18). The stress components then follow from the remainder of equations (2.1.18) and (2.1.19). The deformation (2.1.5) can be maintained only by surface tractions acting on the surfaces of the hollow cylinder and we consider special cases in later sections.

Since W depends only on I_1, I_2, K_1, K_2 and these invariants are functions of r (or ρ) and are independent of θ and y_3, it follows from (2.1.12)–(2.1.14) and (1.15.11) that

$$\frac{dW}{dr} = \frac{\partial W}{\partial I_1}\frac{dI_1}{dr} + \frac{\partial W}{\partial I_2}\frac{dI_2}{dr} + \frac{\partial W}{\partial K_1}\frac{dK_1}{dr} + \frac{\partial W}{\partial K_2}\frac{dK_2}{dr}$$

$$= \left\{\frac{1}{\lambda r} - \frac{\rho^2}{\lambda^2 r^3} - \frac{\lambda r(a^2r^2+c^2)}{\rho^4} + \frac{a^2 r}{\rho^2} + b^2 r\right\}\Phi +$$

$$+ \left\{\frac{\lambda^2 r}{\rho^4}(\rho^2 - \lambda r^2) + \frac{d^2}{\lambda r} - \frac{c^2 + d^2\rho^2}{\lambda^2 r^3} + \frac{b^2 r}{\lambda}\right\}\Psi +$$

$$+ b^2 r \Theta + (abr^2 + cd)\left\{\frac{abr}{\rho^2} - \frac{\lambda r}{2\rho^4}(abr^2 + cd)\right\}\Lambda.$$

When $c = 0$, $ad = \lambda$, this relation with (2.1.12) and (2.1.18) yields

$$\frac{dW}{dr} = \frac{K}{\rho^2 r}(r^2\tau^{22} - \tau^{11}) + \frac{abr^3}{\rho^2}\tau^{23},$$

or, using (2.1.19),
$$\rho^2 \frac{dW}{dr} = K\frac{d\tau^{11}}{dr} + abr^3\tau^{23}. \qquad (2.1.21)$$

Similarly, when $b = 0$, $ad = \lambda$, we obtain

$$\rho^2 \frac{dW}{dr} = K\frac{d\tau^{11}}{dr} - cdr\tau^{23}. \qquad (2.1.22)$$

Before considering special cases of the deformation (2.1.5) we mention another cylindrically symmetric deformation of an incompressible transversely isotropic body which can be maintained only by the action of surface tractions. This deformation consists of inflation, bending, extension and azimuthal shearing of an annular wedge and is defined by

$$r = \rho/\lambda^{\frac{1}{2}}, \quad \theta = a\vartheta + k\log\rho, \quad y_3 = dx_3, \quad \lambda = ad. \qquad (2.1.23)$$

The discussion of this problem is left as an exercise.

2.2. Extension and torsion of an incompressible solid circular cylinder

Solutions of the problems of torsion and flexure may be obtained from the results of the previous section by suitably choosing the parameters a, b, c, d defining the deformation. By setting

$$a = 1, \quad b = \lambda\psi, \quad c = 0, \quad d = \lambda, \qquad (2.2.1)$$

we obtain a deformation which describes the torsion about its axis of a right circular cylinder which has been subjected to an initial extension in the direction of the x_3-axis of ratio λ. The axis of the cylinder is assumed to coincide with the x_3-axis and ψ is the twist per unit length measured in the extended state. For a solid cylinder r must vanish when $\rho = 0$ so that, from (2.1.12),

$$K = 0, \qquad \rho = r\sqrt{\lambda}, \qquad (2.2.2)$$

and, if the initial and final radii of the cylinder are ρ_1 and r_1 respectively,

$$\rho_1 = r_1\sqrt{\lambda}. \qquad (2.2.3)$$

From (2.2.1) and (2.1.18) we have

$$\left.\begin{aligned} r^2\tau^{22} &= \tau^{11} + \psi^2\rho^2\{\lambda(\Phi+\Theta)+\Lambda\} \\ \tau^{33} &= \tau^{11} + \left(\lambda - \frac{1}{\lambda^2}\right)(\lambda\Phi+\Psi) + \lambda^2\Theta - \psi^2\rho^2\Psi \\ \tau^{23} &= \psi\lambda\{\lambda(\Phi+\Theta)+\Psi+\tfrac{1}{2}\Lambda\} \end{aligned}\right\}, \qquad (2.2.4)$$

and (2.1.21) reduces to

$$\frac{dW}{dr} = \psi r \tau^{23}. \qquad (2.2.5)$$

From (2.2.4) and (2.2.5) we find that

$$\frac{r^2\tau^{22} - \tau^{11}}{r} = \frac{dW}{dr} + \psi^2\lambda r(\tfrac{1}{2}\Lambda - \Psi). \qquad (2.2.6)$$

Equations (2.2.4) and (2.2.6) with (2.1.20) yield for τ^{11} the alternative expressions

$$\left.\begin{aligned} \tau^{11} &= W(r) - W(r_1) + \tfrac{1}{2}\psi^2\lambda\int_{r_1}^{r}(\Lambda - 2\Psi)r\,dr \\ \tau^{11} &= \psi^2\lambda\int_{r_1}^{r}\{\lambda(\Phi+\Theta)+\Lambda\}r\,dr \end{aligned}\right\}, \qquad (2.2.7)$$

if the surface $r = r_1$ is free from applied traction.

The resultant moment M_3 on a plane end of the cylinder, about the axis, is

$$M_3 = 2\pi \int_0^{r_1} r^3 \tau^{23}\, dr = 2\pi\psi\lambda \int_0^{r_1} \{\lambda(\Phi+\Theta)+\Psi+\tfrac{1}{2}\Lambda\}r^3\, dr, \qquad (2.2.8)$$

if we use the expression (2.2.4) for τ^{23}. Similarly, from (2.2.5),

$$M_3 = \frac{2\pi}{\psi}\int_0^{r_1} r^2 \frac{dW}{dr}\, dr = \frac{2\pi}{\psi}\left\{r_1^2 W(r_1) - 2\int_0^{r_1} rW(r)\, dr\right\}, \qquad (2.2.9)$$

if W is finite when $r=0$. Alternatively,†

$$M_3 = \frac{2\pi}{\psi\lambda}\int_0^{\rho_1} \rho^2 \frac{dW}{d\rho}\, d\rho = \frac{2\pi}{\psi\lambda}\left\{\rho_1^2 W(\rho_1) - 2\int_0^{\rho_1} \rho W(\rho)\, d\rho\right\}. \qquad (2.2.10)$$

The resultant force on a plane end of the cylinder is

$$N_3 = 2\pi \int_0^{r_1} r\tau^{33}\, dr. \qquad (2.2.11)$$

With the help of (2.1.19) and (2.2.4) this may be reduced to

$$N_3 = \frac{2\pi}{\lambda}\int_0^{\rho_1} \left\{\left(\lambda^2-\frac{1}{\lambda}\right)\left(\Phi+\frac{1}{\lambda}\Psi\right)+\lambda^2\Theta\right\}\rho\, d\rho -$$

$$-\frac{\pi\psi^2}{\lambda}\int_0^{\rho_1} \{\lambda(\Phi+\Theta)+2\Psi+\Lambda\}\rho^3\, d\rho. \qquad (2.2.12)$$

2.3. Extension, inflation, and torsion of an incompressible circular cylindrical tube

Using the values (2.2.1) for the constants a, b, c, d, we may also examine the problem of a cylindrical tube subject to extension, inflation, and torsion. The axis of the tube coincides with the x_3-axis and, as before, λ is the extension ratio in this direction and ψ the twist per unit length of deformed tube. If ρ_1, ρ_2 are the external and internal radii of the tube before deformation and r_1, r_2 the corresponding quantities after deformation then, from (2.1.12), we have

$$K = \rho^2(1-\lambda\mu^2) = \rho_1^2 - \lambda r_1^2 = \rho_2^2 - \lambda r_2^2, \qquad (2.3.1)$$

where
$$\mu = r/\rho. \qquad (2.3.2)$$

---

† Here, and subsequently, when W may be regarded either as a function of r or as a function of ρ (or x_1 in the flexure problem) the notation $W(r)$, $W(\rho)$ [or $W(x_1)$] does not necessarily imply the same functional forms.

We observe that if $\lambda < 0$ then $r_2 > r_1$ and the tube is turned inside out.

The stress components can be found by introducing the conditions (2.2.1) and (2.3.1) into (2.1.18) and making use of (2.1.20). Here, however, we restrict attention to the evaluation of the surface forces acting on the tube. The resultant couple acting on each of the plane ends is given by

$$M_3 = 2\pi \int_{r_2}^{r_1} r^3 \tau^{23} \, dr.$$

With the help of (2.1.18), (2.1.21), and (2.2.1) this may be put in the alternative forms

$$M_3 = 2\pi\psi \int_{r_2}^{r_1} \{\lambda^2(\Phi+\Theta)+\Psi/\mu^2+\lambda^2\mu^2\Lambda/2\}r^3 \, dr, \tag{2.3.3}$$

$$M_3 = \frac{2\pi}{\psi\lambda} \left\{ \int_{r_2}^{r_1} \rho^2 \frac{dW}{dr} \, dr - K \int_{r_2}^{r_1} \frac{d\tau^{11}}{dr} \, dr \right\}$$

$$= \frac{2\pi}{\psi\lambda} \left\{ \int_{\rho_2}^{\rho_1} \rho^2 \frac{dW}{d\rho} \, d\rho - K[(\tau^{11})_{r=r_1} - (\tau^{11})_{r_1=r_2}] \right\}, \tag{2.3.4}$$

with an obvious notation for the values of the normal surface stress at $r = r_1$ and $r = r_2$. For the resultant longitudinal force N_3 acting on these ends we have

$$N_3 = 2\pi \int_{r_2}^{r_1} r\tau^{33} \, dr$$

$$= 2\pi \int_{r_2}^{r_1} r\tau^{11} \, dr +$$

$$+ 2\pi \int_{r_2}^{r_1} \left\{ \left(\lambda^2 - \frac{1}{\lambda^2\mu^2}\right)(\Phi+\mu^2\Psi) + \lambda^2\Theta - \psi^2\rho^2\Psi \right\} r \, dr, \tag{2.3.5}$$

by virtue of (2.1.18) and (2.2.1). If the tube is subject to a uniform inflating pressure P and the external surface $r = r_1$ is free from applied stress, so that

$$[\tau^{11}]_{r=r_1} = 0, \quad [\tau^{11}]_{r=r_2} = -P,$$

the relations (2.1.18) and (2.1.20) with (2.2.1) yield

$$P = \int_{r_2}^{r_1} \left(\mu^2 - \frac{1}{\lambda^2\mu^2}\right)(\Phi+\lambda^2\Psi)\frac{dr}{r} + \psi^2\lambda^2 \int_{r_2}^{r_1} (\Phi+\Theta+\mu^2\Lambda)r \, dr, \tag{2.3.6}$$

and from (2.3.5)

$$N_3 = \frac{\pi}{\lambda}\int_{\rho_2}^{\rho_1}\left\{2\left[\left(\lambda^2-\frac{1}{\lambda^2\mu^2}\right)(\Phi+\mu^2\Psi)+\lambda^2\Theta\right]+\right.$$

$$\left.+\frac{1}{\lambda\mu^2}\left(\frac{\rho_2^2}{\rho^2}-1\right)\left(\mu^2-\frac{1}{\lambda^2\mu^2}\right)(\Phi+\lambda^2\Psi)\right\}\rho\,d\rho+$$

$$+\pi\psi^2\int_{\rho_2}^{\rho_1}\left\{(\rho_2^2-\rho^2)(\Phi+\Theta+\mu^2\Lambda)-\frac{2\rho^2}{\lambda}\Psi\right\}\rho\,d\rho. \quad (2.3.7)$$

If P, M_3, and N_3 have given values, (2.3.3), (2.3.6), and (2.3.7) may be regarded as three equations for the determination of λ, ψ, and K. If $\psi = 0$ then $M_3 = 0$ and (2.3.6) and (2.3.7) become equations in the unknowns λ and K. Further discussion is needed to determine whether a solution exists and is unique for a given form of strain energy function and for particular values of P, M_3, and N_3.

If $\psi \neq 0$ and if it is possible to maintain the tube in equilibrium with zero applied tractions on the curved surfaces, then

$$[\tau^{11}]_{r=r_1} = [\tau^{11}]_{r=r_2} = 0, \qquad P = 0, \quad (2.3.8)$$

and (2.3.4) yields

$$M_3 = \frac{2\pi}{\psi\lambda}\int_{\rho_2}^{\rho_1}\rho^2\frac{dW}{d\rho}\,d\rho = \frac{2\pi}{\psi\lambda}\left\{\rho_1^2W(\rho_1)-\rho_2^2W(\rho_2)-2\int_{\rho_2}^{\rho_1}\rho W(\rho)\,d\rho\right\}, \quad (2.3.9)$$

while in place of (2.3.7) we have

$$N_3 = \frac{\pi}{\lambda}\int_{\rho_2}^{\rho_1}\left\{2\left[\left(\lambda^2-\frac{1}{\lambda^2\mu^2}\right)(\Phi+\mu^2\Psi)+\lambda^2\Theta\right]-\right.$$

$$\left.-\frac{1}{\lambda\mu^2}\left(\mu^2-\frac{1}{\lambda^2\mu^2}\right)(\Phi+\lambda^2\Psi)\right\}\rho\,d\rho-$$

$$-\frac{\pi\psi^2}{\lambda}\int_{\rho_2}^{\rho_1}\{\lambda(\Phi+\Theta+\mu^2\Lambda)+2\Psi\}\rho^3\,d\rho. \quad (2.3.10)$$

Furthermore, from (2.1.20) it follows that

$$\int_{r_2}^{r_1}(r^2\tau^{22}-\tau^{11})\frac{dr}{r} = 0, \quad (2.3.11)$$

a condition which is equivalent to (2.3.6) with $P = 0$.

2.4. Incompressible circular tube turned inside out

When $\psi = 0$ (i.e., $b = 0$) and the tube is maintained in equilibrium by forces on its ends and zero tractions on its curved surfaces, equation (2.3.11) is still satisfied. Alternatively, from (2.3.9) we may replace this relation by

$$\int_{\rho_2}^{\rho} \rho^2 \frac{dW}{d\rho} d\rho = \rho_1^2 W(\rho_1) - \rho_2^2 W(\rho_2) - 2\int_{\rho_2}^{\rho_1} \rho W(\rho) d\rho = 0 \qquad (2.4.1)$$

while, from (2.3.6), with $P = 0$, $\psi = 0$, we obtain the equivalent condition

$$\int_{\rho_2}^{\rho_1} \left(1 - \frac{1}{\lambda^2 \mu^4}\right)(\Phi + \lambda^2 \Psi) \frac{d\rho}{\rho} = 0. \qquad (2.4.2)$$

Also, if $N_3 = 0$, we have from (2.3.10)

$$\int_{\rho_2}^{\rho_1} \left\{ 2\left[\left(\lambda^2 - \frac{1}{\lambda^2 \mu^2}\right)(\Phi + \mu^2 \Psi) + \lambda^2 \Theta\right] - \frac{1}{\lambda \mu^2}\left(\mu^2 - \frac{1}{\lambda^2 \mu^2}\right)(\Phi + \lambda^2 \Psi) \right\} \rho \, d\rho = 0.$$

The relations (2.4.1)–(2.4.3) are satisfied identically in the undeformed state; they are also satisfied when the tube is turned inside out and rests in a deformed state under the action of no forces over the curved surfaces and no resultant force over the ends of the tube. These equations then determine the extension ratio λ and the changes in the radii of the tube, the latter being defined by the constant K. Further discussion is needed to determine under what conditions such a state of equilibrium is possible.

2.5. Flexure of an initially curved incompressible cuboid

We suppose that the undeformed body consists of part of a cylindrical tube, bounded in the coordinate system (ρ, ϑ, x_3) by the curved surfaces $\rho = \rho_1$, $\rho = \rho_2$ and the planes $\vartheta = \pm \vartheta_0$, $x_3 = \pm L$. We consider the flexure problem obtained from (2.1.4) by putting

$$b = c = 0, \qquad ad = \lambda. \qquad (2.5.1)$$

The deformed body is then bounded by the cylindrical surfaces $r = r_1$, $r = r_2$ and the planes $\theta = \pm a\vartheta_0$, $y_3 = \pm Ld$. Also, the radii of the cylindrical surfaces in the undeformed and deformed states are related by equations (2.3.1).

Since $b = 0$ we see that (2.1.21) gives

$$\tau^{11} = (\tau^{11})_{\rho=\rho_1} + \frac{1}{K}\int_{\rho_1}^{\rho} \rho^2 \frac{dW}{d\rho} d\rho, \qquad (2.5.2)$$

and from (2.1.18)

$$\left.\begin{aligned} r^2\tau^{22} &= \tau^{11} + \left(a^2\mu^2 - \frac{1}{\lambda^2\mu^2}\right)(\Phi + d^2\Psi) \\ \tau^{33} &= \tau^{11} + \left(d^2 - \frac{1}{\lambda^2\mu^2}\right)(\Phi + a^2\mu^2\Psi) + d^2\Theta \\ \tau^{12} &= \tau^{23} = \tau^{31} = 0 \end{aligned}\right\} \qquad (2.5.3)$$

where μ is still given by (2.3.2). Furthermore, since from (2.1.19)

$$r^2\tau^{22} = \frac{d}{dr}(r\tau^{11}), \qquad (2.5.4)$$

the resultant normal force acting on the planes $\theta = \pm a\vartheta_0$ may be written

$$N_2 = 2Ld\int_{r_2}^{r_1} r^2\tau^{22}\,dr = 2Ld[r\tau^{11}]_{r_2}^{r_1}. \qquad (2.5.5)$$

The resultant couple M_2 over these planes about the y_3-axis is given by

$$M_2 = 2Ld\int_{r_2}^{r_1} r^3\tau^{22}\,dr = Ld\int_{r_2}^{r_1}\left\{\frac{d}{dr}(r^2\tau^{11}) + \frac{\rho^2 - K}{\lambda}\frac{d\tau^{11}}{dr}\right\}dr,$$

if we use (2.5.4) and (2.1.12). Hence with the help of (2.5.2) we have

$$M_2 = \frac{L}{a}\left\{[(\lambda r^2 - K)\tau^{11}]_{r_2}^{r_1} + \frac{1}{K}\int_{\rho_2}^{\rho_1}\rho^4\frac{dW}{d\rho}d\rho\right\}. \qquad (2.5.6)$$

The tractions over the plane faces $y_3 = \pm Ld$ of the deformed cylinder are equivalent to a resultant force N_3, where

$$N_3 = 2a\vartheta_0\int_{r_2}^{r_1} r\tau^{33}\,dr \qquad (2.5.7)$$

With the help of (2.5.3) and (2.5.4) the expression (2.5.7) becomes

$$N_3 = a\vartheta_0[r^2\tau^{11}]_{r_2}^{r_1} + a\vartheta_0\int_{r_2}^{r_1}\left\{\left(2d^2 - a^2\mu^2 - \frac{1}{\lambda^2\mu^2}\right)\Phi + \right.$$
$$\left. + \left(\lambda^2\mu^2 - \frac{2a^2}{\lambda^2} + \frac{d^2}{\lambda^2\mu^2}\right)\Psi + 2d^2\Theta\right\}r\,dr. \qquad (2.5.8)$$

If it is possible to maintain equilibrium with zero traction on the curved surfaces $r = r_1, r = r_2$ of the deformed cylinder then

$$(\tau^{11})_{r=r_1} = 0, \qquad (\tau^{11})_{r=r_2} = 0,$$

and from (2.5.2)
$$\int_{\rho_2}^{\rho_1} \rho^2 \frac{dW}{d\rho} d\rho = 0. \tag{2.5.9}$$

It follows from (2.5.5) that $N_2 = 0$ and the tractions on the plane faces $\theta = \pm a\vartheta_0$ reduce to the couple

$$M_2 = \frac{L}{Ka} \int_{\rho_2}^{\rho_1} \rho^4 \frac{dW}{d\rho} d\rho. \tag{2.5.10}$$

Moreover

$$N_3 = a\vartheta_0 \int_{r_2}^{r_1} \left\{ \left(2d^2 - a^2\mu^2 - \frac{1}{\lambda^2\mu^2}\right)\Phi + \left(\lambda^2\mu^2 - \frac{2a^2}{\lambda^2} + \frac{d^2}{\lambda^2\mu^2}\right)\Psi + 2d^2\Theta \right\} r \, dr. \tag{2.5.11}$$

A special case of the foregoing theory arises when the plane faces $\theta = \pm a\vartheta_0$ of the deformed cuboid meet and are joined so that the body has the form of a right circular cylindrical tube. For this, $a = \pi/\vartheta_0$.

Formulae for isotropic bodies are obtained from the results of §§ 2.1–2.5 by putting $\Theta \equiv \Delta \equiv 0$.

2.6. Straightening, stretching, and shearing of an initially curved cuboid

We consider a special case of the theory of § 2.1 when the undeformed body consists of a curved cuboid, bounded in the coordinate system (ρ, ϑ, x_3) by the curved surfaces $\rho = \rho_1$, $\rho = \rho_2$ and the planes $\vartheta = \pm\vartheta_0$, $x_3 = \pm L$. We assume that the cuboid is straightened into a rectilinear cuboid and is also stretched and sheared. We can examine this problem as a limiting case of the work of § 2.1. The results of § 2.1 are unaffected by a translation of the origin of the y_i-axes in the y_1, y_2 plane. In place of (2.1.2) we write

$$y_1 = -h + r\cos\theta, \qquad y_2 = r\sin\theta, \tag{2.6.1}$$

where h is a constant. Also, we define constants a', b', λ' by the relations

$$a' = ha, \qquad b' = hb, \qquad \lambda' = h\lambda, \qquad \lambda' = a'd - b'c, \tag{2.6.2}$$

and put
$$r = h + F(\rho), \qquad \theta' = h\theta. \tag{2.6.3}$$

Substituting from (2.6.3) into (2.6.1), we have
$$y_1 = h\{\cos(\theta'/h) - 1\} + F(\rho)\cos(\theta'/h),$$
$$y_2 = h\sin(\theta'/h) + F(\rho)\sin(\theta'/h),$$
so that, as $h \to \infty$ while θ' and $F(\rho)$ remain finite, we have
$$y_1 \to F(\rho), \qquad y_2 \to \theta'. \tag{2.6.4}$$

Remembering (2.6.2) and (2.6.3) we obtain, as a limiting case of equations (2.1.4), the relations
$$y_2 = a'\vartheta + b'x_3, \qquad y_3 = c\vartheta + dx_3. \tag{2.6.5}$$

Also, for an incompressible body, from (2.1.12), (2.6.2), and (2.6.3), we have
$$y_1 = K' + \frac{\rho^2}{2\lambda'}, \tag{2.6.6}$$
as $h \to \infty$, where
$$K + \lambda'h \to -2\lambda'K'. \tag{2.6.7}$$

The stress distribution in the deformed cylinder may be found as a limiting case of the results (2.1.18) or may be computed directly with the help of (2.6.5), (2.6.6), and the choice of moving coordinates
$$(\theta_1, \theta_2, \theta_3) = (y_1, y_2, y_3). \tag{2.6.8}$$
We leave this as an exercise.

Another limiting case of the theory of § 2.1 occurs when the initial body is a rectilinear cuboid. This is, however, discussed in § 2.11 in the more general context of a body having a certain type of aelotropy. In the next section, 2.7, we discuss the flexure and extension of a rectilinear cuboid of transversely isotropic material, *ab initio*, and not as a limiting case of the work of § 2.1.

2.7. Flexure of a cuboid of incompressible transversely isotropic material

Consider a body which in the undeformed state is a cuboid bounded by the planes
$$\left.\begin{array}{l} x_1' = A_1, \qquad x_1' = A_2, \qquad x_1' = x_1 \\ x_2' \equiv x_3 \sin\phi + x_2 \cos\phi = \pm B \\ x_3' \equiv x_3 \cos\phi - x_2 \sin\phi = \pm C \end{array}\right\}, \tag{2.7.1}$$
and let $A_1 - A_2 = 2A$. The material of the cuboid is transversely isotropic with the axis of anisotropy along the x_3-axis at an angle ϕ to

the x_3'-axis. The cuboid is now deformed symmetrically with respect to the x_1-axis so that:

(i) Each plane initially normal to the x_1-axis becomes, in the deformed state, a portion of a cylinder whose axis is the x_3'-axis.

(ii) Planes initially normal to the x_2'-axis become in the deformed state planes containing the x_3'-axis.

(iii) There is a uniform extension of ratio λ in the direction of the x_3'-axis.

We take cylindrical polar coordinates (r, θ, y_3) to define the strained body and we identify the curvilinear system θ_i with these coordinates so that
$$\left.\begin{array}{c}(\theta_1, \theta_2, \theta_3) = (r, \theta, y_3) \\ y_1 = r \cos \theta, \quad y_2 = r \sin \theta\end{array}\right\}. \tag{2.7.2}$$

If the y_i-axes and the x_i'-axes coincide, the deformation described by (i) to (iii) implies that
$$r = (2kx_1' + K)^{\frac{1}{2}}, \quad \theta = \frac{x_2'}{\lambda k}, \quad y_3 = \lambda x_3', \tag{2.7.3}$$

where k and K are constants, the forms of r, θ, y_3 being chosen so that the incompressibility condition is automatically satisfied. Since
$$x_2 = -x_3' \sin \phi + x_2' \cos \phi, \quad x_3 = x_3' \cos \phi + x_2' \sin \phi,$$
we also have
$$\left.\begin{array}{c}x_1 = \dfrac{r^2 - K}{2k} \\[6pt] x_2 = \lambda k \theta \cos \phi - \dfrac{y_3}{\lambda} \sin \phi \\[6pt] x_3 = \lambda k \theta \sin \phi + \dfrac{y_3}{\lambda} \cos \phi\end{array}\right\}. \tag{2.7.4}$$

Furthermore, from (2.7.3), if r_1 and r_2 ($r_1 > r_2$) are the radii of the curved surfaces of the body which initially are the planes $x_1 = A_1$ and $x_1 = A_2$ respectively, then
$$k = (r_1^2 - r_2^2)/(4A), \quad K = (A_1 r_2^2 - A_2 r_1^2)/(2A), \tag{2.7.5}$$
and if the planes $x_2' = \pm B$ become the planes $\theta = \pm \theta_0$,
$$\lambda k \theta_0 = B. \tag{2.7.6}$$

From (2.7.2) we have

$$G_{11} = G^{11} = G_{33} = G^{33} = 1, \quad G_{22} = r^2, \quad G^{22} = 1/r^2 \\ G_{ij} = G^{ij} = 0 \quad (i \neq j), \quad G = r^2$$
(2.7.7)

and (2.7.2) and (2.7.4) yield

$$\frac{\partial x^r}{\partial \theta^s} = \begin{bmatrix} \dfrac{r}{k} & 0 & 0 \\ 0 & \lambda k \cos\phi & -\dfrac{\sin\phi}{\lambda} \\ 0 & \lambda k \sin\phi & \dfrac{\cos\phi}{\lambda} \end{bmatrix}, \quad \left|\frac{\partial x^r}{\partial \theta^s}\right| = r.$$
(2.7.8)

Also

$$\frac{\partial \theta^r}{\partial x^s} = \begin{bmatrix} \dfrac{k}{r} & 0 & 0 \\ 0 & \dfrac{\cos\phi}{\lambda k} & \dfrac{\sin\phi}{\lambda k} \\ 0 & -\lambda \sin\phi & \lambda \cos\phi \end{bmatrix}.$$
(2.7.9)

Hence

$$g_{ij} = \begin{bmatrix} \dfrac{r^2}{k^2} & 0 & 0 \\ 0 & \lambda^2 k^2 & 0 \\ 0 & 0 & \dfrac{1}{\lambda^2} \end{bmatrix}, \quad g^{ij} = \begin{bmatrix} \dfrac{k^2}{r^2} & 0 & 0 \\ 0 & \dfrac{1}{\lambda^2 k^2} & 0 \\ 0 & 0 & \lambda^2 \end{bmatrix}.$$
(2.7.10)

From (1.1.18), (2.7.7), and (2.7.10) we have

$$I_1 = \frac{k^2}{r^2} + \frac{r^2}{\lambda^2 k^2} + \lambda^2, \quad I_2 = \frac{r^2}{k^2} + \frac{\lambda^2 k^2}{r^2} + \frac{1}{\lambda^2},$$
(2.7.11)

and (1.1.15), (2.7.7), (2.7.9), (2.7.10), with (1.13.11) yield

$$2K_1 = \frac{r^2 \sin^2\phi}{\lambda^2 k^2} + \lambda^2 \cos^2\phi - 1 \\ 4K_2 = \left(\frac{r^2}{\lambda^2 k^2} - \lambda^2\right)^2 \sin^2\phi \cos^2\phi$$
(2.7.12)

2.7 GENERAL SOLUTIONS OF PROBLEMS

Using (2.7.7)–(2.7.12) we see from (1.15.9) and (1.15.10) that

$$B^{ij} = \begin{bmatrix} \dfrac{\lambda^2 k^2}{r^2} + \dfrac{1}{\lambda^2} & 0 & 0 \\ 0 & \dfrac{1}{\lambda^2 r^2} + \dfrac{1}{k^2} & 0 \\ 0 & 0 & \dfrac{r^2}{k^2} + \dfrac{\lambda^2 k^2}{r^2} \end{bmatrix}, \quad (2.7.13)$$

$$M^{ij} = \begin{bmatrix} 0 & 0 & 0 \\ 0 & \dfrac{\sin^2\phi}{\lambda^2 k^2} & \dfrac{\sin\phi\cos\phi}{k} \\ 0 & \dfrac{\sin\phi\cos\phi}{k} & \lambda^2\cos^2\phi \end{bmatrix}, \quad (2.7.14)$$

$$\left.\begin{aligned} N^{22} &= \left(\dfrac{r^2}{\lambda^2 k^2} - \lambda^2\right)\dfrac{\cos^2\phi\sin^2\phi}{\lambda^2 k^2} \\ N^{33} &= -\left(\dfrac{r^2}{\lambda^2 k^2} - \lambda^2\right)\lambda^2\cos^2\phi\sin^2\phi \\ N^{23} &= \left(\dfrac{r^2}{\lambda^2 k^2} - \lambda^2\right)\dfrac{\sin 4\phi}{8k} \\ N^{11} &= N^{12} = N^{13} = 0 \end{aligned}\right\} \quad (2.7.15)$$

The stress tensor τ^{ij} may now be obtained from (1.15.6) and the results of the present section. Thus

$$\left.\begin{aligned} r^2\tau^{22} &= \tau^{11} + \left(\dfrac{r^2}{\lambda^2 k^2} - \dfrac{k^2}{r^2}\right)(\Phi + \lambda^2\Psi) + \\ &\quad + \dfrac{r^2\sin^2\phi}{\lambda^2 k^2}\left[\Theta + \Lambda\left(\dfrac{r^2}{\lambda^2 k^2} - \lambda^2\right)\cos^2\phi\right] \\ \tau^{33} &= \tau^{11} + \left(\lambda^2 - \dfrac{k^2}{r^2}\right)\left(\Phi + \dfrac{r^2}{\lambda^2 k^2}\Psi\right) + \\ &\quad + \lambda^2\cos^2\phi\left[\Theta - \Lambda\left(\dfrac{r^2}{\lambda^2 k^2} - \lambda^2\right)\sin^2\phi\right] \\ \tau^{23} &= \dfrac{\sin 2\phi}{4k}\left[2\Theta + \Lambda\left(\dfrac{r^2}{\lambda^2 k^2} - \lambda^2\right)\cos 2\phi\right] \\ \tau^{12} &= \tau^{13} = 0 \end{aligned}\right\} \quad (2.7.16)$$

Since W depends only upon I_1, I_2, K_1, K_2 and these invariants are independent of θ and y_3, it follows from (2.7.11), (2.7.12), and (1.15.12) that

$$r\frac{dW}{dr} = \left(\frac{r^2}{\lambda^2 k^2} - \frac{k^2}{r^2}\right)\Phi + \left(\frac{r^2}{k^2} - \frac{\lambda^2 k^2}{r^2}\right)\Psi + \frac{r^2 \sin^2\phi}{\lambda^2 k^2}\Theta +$$
$$+ \frac{r^2}{\lambda^2 k^2}\left(\frac{r^2}{\lambda^2 k^2} - \lambda^2\right)\Lambda \cos^2\phi \sin^2\phi \quad (2.7.17)$$
$$= r^2 \tau^{22} - \tau^{11}$$

When body forces are zero the equations of equilibrium again reduce to (2.1.19) so that p is independent of θ and y_3. Using (2.7.17) we then have the results

$$\tau^{11} = W + W_0, \qquad (2.7.18)$$

$$r^2 \tau^{22} = r\frac{dW}{dr} + W + W_0 = \frac{d}{dr}[r(W+W_0)], \qquad (2.7.19)$$

where W_0 is a constant.

The forces acting on the surfaces $r = r_1$ and $r = r_2$ reduce to normal stresses only and these will be zero if and only if

$$W(r_1) = W(r_2) = -W_0. \qquad (2.7.20)$$

On each of the faces $\theta = \pm\theta_0$ there acts a resultant normal force

$$N_2 = 2\lambda C \int_{r_2}^{r_1} r^2 \tau^{22}\, dr$$
$$= 2\lambda C [r(W+W_0)]_{r_2}^{r_1} \qquad (2.7.21)$$

and this is zero if the surfaces $r = r_1$, $r = r_2$ are both free from applied traction. The resultant couple M_2 acting on each of the faces $\theta = \pm\theta_0$ is

$$M_2 = 2\lambda C \int_{r_2}^{r_1} r^3 \tau^{22}\, dr,$$

and using (2.7.19) and (2.7.20) this becomes

$$M_2 = \lambda C\left\{2\int_{r_1}^{r_2} rW(r)\, dr - (r_1^2 - r_2^2) W_0\right\}. \qquad (2.7.22)$$

Alternatively, remembering (2.7.4), this may be written

$$M_2 = 2\lambda k C\left\{\int_{A_1}^{A_2} W(x_1)\, dx_1 - 2A W_0\right\}. \qquad (2.7.23)$$

In addition, shearing stresses act on the faces $\theta = \pm\theta_0$ except when $\phi = 0$ or $\phi = \frac{1}{2}\pi$, that is, if the axis of anisotropy is perpendicular to one of the faces of the undeformed body. Finally, since, in general, $\tau^{33} \neq 0$, normal forces must be applied to the planes in the deformed body which were initially at $x_3' = \pm C$.

We now restrict attention to the case when $\phi = 0$ and choose

$$k^2 = \frac{r_1 r_2}{\lambda}. \tag{2.7.24}$$

In this case it is easily verified that

$$I_1(r_1) = I_1(r_2), \qquad I_2(r_1) = I_2(r_2),$$
$$K_1(r_1) = K_1(r_2), \qquad K_2(r_1) = K_2(r_2),$$

so that the condition (2.7.20) is satisfied. Equations (2.7.5), (2.7.6), and (2.7.24) are four equations in the six unknowns k, K, λ, r_1, r_2, and θ_0, and two of these must be prescribed in order to define the deformation completely. The radius r_0 of the neutral fibre is found by putting $r_0\theta_0 = B$ giving, with (2.7.6) and (2.7.24), $r_0 = (\lambda r_1 r_2)^{\frac{1}{2}}$.

When the axis of anisotropy is along the x_1-direction we can also solve the flexure problem, but we leave this as an exercise and consider the flexure of a cuboid of a more general aeolotropic material in § 2.9.

2.8. Homogeneous strain of an aeolotropic cuboid

We next consider the problem in which a cuboid of homogeneous aeolotropic material, whose strain energy has the general polynomial form (1.3.8), is subjected to the uniform deformation

$$y_i = c_{ij} x_j, \tag{2.8.1}$$

the coefficients c_{ij} being constants. Then

$$\frac{\partial y_i}{\partial x_j} = c_{ij}, \tag{2.8.2}$$

and from (1.1.10) and (1.1.15) we have

$$\sqrt{I_3} = \left|\frac{\partial y_i}{\partial x_j}\right| = |c_{ij}|, \tag{2.8.3}$$

$$e_{rs} = \tfrac{1}{2}(c_{mr} c_{ms} - \delta_{rs}). \tag{2.8.4}$$

The stress components follow from (1.15.1), (1.15.2) and if we allow the reference frame θ_i to coincide with the rectangular cartesian system

y_i, these are then the physical components of stress σ_{ij} referred to this system. Using (2.8.2) and remembering that $G_{ij} = G^{ij} = \delta_{ij}$, we have, for compressible materials,

$$\sigma_{ij} = \frac{1}{2\sqrt{I_3}}\left(\frac{\partial W}{\partial e_{rs}} + \frac{\partial W}{\partial e_{sr}}\right) c_{ir} c_{js}, \qquad (2.8.5)$$

and for incompressible bodies,

$$\sigma_{ij} = \frac{1}{2}\left(\frac{\partial W}{\partial e_{rs}} + \frac{\partial W}{\partial e_{sr}}\right) c_{ir} c_{js} + p\, \delta_{ij}, \qquad (2.8.6)$$

the constants c_{ij} in the latter case being connected by the incompressibility condition

$$|c_{ij}| = 1. \qquad (2.8.7)$$

From (2.8.4) it follows that the strain energy function W is constant throughout the material and, in the compressible case, the stress components σ_{ij} are also constant and the equations of equilibrium in the absence of body forces are satisfied identically. In the incompressible case, the equations of equilibrium are satisfied if p is constant throughout the material, and the stresses again take constant values.

In the special case when the uniform deformation (2.8.1) reduces to a pure homogeneous strain in which the principal extension ratios are λ_i in the directions of the x_i-axes, we have

$$\left.\begin{array}{ll} c_{ii} = \lambda_i, & c_{ij} = 0 \quad (i \neq j) \\ e_{ii} = \tfrac{1}{2}(\lambda_i^2 - 1), & e_{ij} = 0 \quad (i \neq j) \end{array}\right\}, \qquad (2.8.8)$$

$$\sqrt{I_3} = \lambda_1 \lambda_2 \lambda_3, \qquad (2.8.9)$$

where i is not summed, and (2.8.5), (2.8.6) yield

$$\left.\begin{array}{l} \sigma_{ij} = \dfrac{1}{2\sqrt{I_3}}\left(\dfrac{\partial W}{\partial e_{ij}} + \dfrac{\partial W}{\partial e_{ji}}\right) \lambda_i \lambda_j \quad (i,j \text{ not summed}) \\[2mm] \sigma_{ij} = \dfrac{1}{2}\left(\dfrac{\partial W}{\partial e_{ij}} + \dfrac{\partial W}{\partial e_{ji}}\right) \lambda_i \lambda_j + p\, \delta_{ij} \quad (i,j \text{ not summed}) \end{array}\right\}, \qquad (2.8.10)$$

respectively. The stresses σ_{12}, σ_{23}, σ_{31} are zero if and only if all the derivatives $\partial W/\partial e_{ij}$ for $i \neq j$ vanish when the components e_{ij} $(i \neq j)$ are zero. This implies that W, when expressed in the polynomial form (1.3.8), contains the components e_{12}, e_{23}, e_{31} only as powers higher than the first or as products with each other. This condition is satisfied by all the crystal classes listed in Chapter I with the exception of the triclinic and monoclinic systems, and the tetragonal and hexagonal systems defined by (1.10.3), (1.12.5), (1.12.8), and (1.12.11). When the stresses σ_{ij} ($i \neq j$) are all zero, the cuboid can be maintained in its deformed state by means of surface tractions normal to its plane boundaries.

When
$$c_{11} = c_{22} = c_{33} = 1, \quad c_{12} = K, \quad c_{21} = c_{13} = c_{31} = c_{23} = c_{32} = 0, \tag{2.8.11}$$
the deformation (2.8.1) reduces to the simple shear
$$y_1 = x_1 + Kx_2, \quad y_2 = x_2, \quad y_3 = x_3, \tag{2.8.12}$$
and we have
$$\left. \begin{array}{l} e_{11} = e_{33} = e_{13} = e_{23} = 0 \\ e_{22} = \tfrac{1}{2}K^2, \quad e_{12} = e_{21} = \tfrac{1}{2}K, \quad \sqrt{I_3} = 1 \end{array} \right\}. \tag{2.8.13}$$

For incompressible materials, the stress components from (2.8.6) are given by
$$\left. \begin{array}{l} \sigma_{11} = \dfrac{\partial W}{\partial e_{11}} + K\left(\dfrac{\partial W}{\partial e_{12}} + \dfrac{\partial W}{\partial e_{21}}\right) + K^2 \dfrac{\partial W}{\partial e_{22}} + p \\[6pt] \sigma_{22} = \dfrac{\partial W}{\partial e_{22}} + p \\[6pt] \sigma_{33} = \dfrac{\partial W}{\partial e_{33}} + p \\[6pt] \sigma_{12} = K\dfrac{\partial W}{\partial e_{22}} + \dfrac{1}{2}\left(\dfrac{\partial W}{\partial e_{12}} + \dfrac{\partial W}{\partial e_{21}}\right) \\[6pt] \sigma_{13} = \dfrac{1}{2}\left(\dfrac{\partial W}{\partial e_{13}} + \dfrac{\partial W}{\partial e_{31}}\right) + \dfrac{K}{2}\left(\dfrac{\partial W}{\partial e_{23}} + \dfrac{\partial W}{\partial e_{32}}\right) \\[6pt] \sigma_{23} = \dfrac{1}{2}\left(\dfrac{\partial W}{\partial e_{23}} + \dfrac{\partial W}{\partial e_{32}}\right) \end{array} \right\}. \tag{2.8.14}$$

The stresses σ_{13}, σ_{23} again vanish only if the material has suitable symmetry properties; σ_{33} may be made to vanish by a suitable choice for p but σ_{11} and σ_{22} are not then, in general, zero. In addition to the shearing force arising from σ_{12}, normal tractions must act across the planes $x_1 = \text{constant}$, $x_2 = \text{constant}$ to maintain the deformation.

2.9. Flexure of an incompressible aeolotropic cuboid

Here we examine the flexure of a cuboid of homogeneous aeolotropic material whose strain energy function has the general polynomial form (1.3.8). The dimensions of the undeformed cuboid and the deformation to which it is subject are as described in § 2.7, except that $\phi = 0$ in

(2.7.1)–(2.7.4). Thus, allowing the coordinate systems x_i and x_i' to coincide, we see from (2.7.1) and (2.7.3) that the undeformed body is bounded by the planes

$$x_1 = A_1, \qquad x_1 = A_2, \qquad x_2 = \pm B, \qquad x_3 = \pm C, \qquad (2.9.1)$$

and the deformation is described by the relations

$$r = (2kx_1+K)^{\frac{1}{2}}, \qquad \theta = \frac{x_2}{\lambda k}, \qquad y_3 = \lambda x_3, \qquad (2.9.2)$$

where k and K are again given by (2.7.5) and (2.7.6). In view of (2.7.2) the components of the metric tensors G_{ij}, G^{ij} are then given by (2.7.7), while the derivatives $\partial x^r/\partial \theta^s$, $\partial \theta^r/\partial x^s$ are obtained from (2.7.8) and (2.7.9) respectively by putting $\phi = 0$. Hence g_{ij}, g^{ij} again take the forms (2.7.10).

From (2.7.2) and (2.9.2) it follows that

$$\frac{\partial y_r}{\partial x_s} = \begin{bmatrix} \dfrac{k}{r}\cos\theta & -\dfrac{r\sin\theta}{\lambda k} & 0 \\ \dfrac{k}{r}\sin\theta & \dfrac{r\cos\theta}{\lambda k} & 0 \\ 0 & 0 & \lambda \end{bmatrix}, \qquad (2.9.3)$$

and the strain components e_{ij} defined by (1.1.15) are therefore given by

$$\left. \begin{aligned} e_{11} &= \tfrac{1}{2}\left(\tfrac{k^2}{r^2}-1\right), \qquad e_{22} = \tfrac{1}{2}\left(\tfrac{r^2}{\lambda^2 k^2}-1\right), \qquad e_{33} = \tfrac{1}{2}(\lambda^2-1) \\ e_{12} &= e_{23} = e_{31} = 0 \end{aligned} \right\} \qquad (2.9.4)$$

Since the cuboid is incompressible the stress tensor is given by (1.15.2). Thus

$$\left. \begin{aligned} \tau^{11} &= \frac{k^2}{r^2}\frac{\partial W}{\partial e_{11}}+p, & \tau^{22} &= \frac{1}{\lambda^2 k^2}\frac{\partial W}{\partial e_{22}}+\frac{p}{r^2} \\ \tau^{33} &= \lambda^2 \frac{\partial W}{\partial e_{33}}+p, & \tau^{12} &= \frac{1}{2\lambda r}\left(\frac{\partial W}{\partial e_{12}}+\frac{\partial W}{\partial e_{21}}\right) \\ \tau^{23} &= \frac{1}{2k}\left(\frac{\partial W}{\partial e_{23}}+\frac{\partial W}{\partial e_{32}}\right), & \tau^{13} &= \frac{k\lambda}{2r}\left(\frac{\partial W}{\partial e_{13}}+\frac{\partial W}{\partial e_{31}}\right) \end{aligned} \right\} \qquad (2.9.5)$$

The strain energy function W is a polynomial of the form (1.3.8). We restrict attention to values of W in which e_{12}, e_{23}, e_{31} occur only in powers higher than the first, or as products with each other,† so that in view of (2.9.4) and (2.9.5),

$$\tau^{12} = \tau^{23} = \tau^{31} = 0. \tag{2.9.6}$$

The monoclinic, triclinic, and some of the tetragonal and hexagonal crystal classes are excluded by this condition, provided that the crystal axes coincide with the x_i-axes. The condition is satisfied by materials that are orthotropic with respect to the x_i-axes and by bodies that are transversely isotropic with the axis of anisotropy coinciding with one of the x_i-axes. The non-zero stresses in (2.9.5) are independent of θ and y_3, so that the equations of equilibrium reduce to (2.1.19). It follows that p is independent of θ and y_3 and that τ^{11}, and hence p, is given by (2.1.19).

In view of (2.9.4), the strain energy W is of the form $W(r)$, and

$$\frac{dW}{dr} = -\frac{k^2}{r^3}\frac{\partial W}{\partial e_{11}} + \frac{r}{\lambda^2 k^2}\frac{\partial W}{\partial e_{22}}. \tag{2.9.7}$$

With (2.9.5) and (2.1.19) this yields

$$\frac{dW}{dr} = \frac{r^2 \tau^{22} - \tau^{11}}{r} = \frac{d\tau^{11}}{dr} \tag{2.9.8}$$

and hence

$$\tau^{11} = W(r) + W_0, \tag{2.9.9}$$

where W_0 is a constant. We observe that this is the result (2.7.18) obtained for a transversely isotropic cuboid. The stress acting on the surfaces $r = r_1$ and $r = r_2$ reduce to normal stresses only, and these are zero if and only if the condition (2.7.20) is satisfied. The resultant normal force acting on each of the faces $\theta = \pm\theta_0$ is given by (2.7.21) and is zero if (2.7.20) can be satisfied. Further, the formulae (2.7.22) and (2.7.23) for the resultant couple on these faces also hold. Normal stresses τ^{33} must also be applied to the faces of the deformed cuboid initially at $x_3 = \pm C$.

2.10. Cylindrically symmetric problems: incompressible cylindrical aeolotropy

We examine again the cylindrically symmetrical deformation defined by (2.1.4), but this time consider more general materials possessing a suitable type of curvilinear aeolotropy. The system of moving coordinates θ_i is chosen, as in § 2.1, to coincide with the cylindrical polar

† A more general discussion is given by J. E. Adkins and A. E. Green, *Arch. ration. Mech. Analysis* 8 (1961), 9.

reference frame (r, θ, y_3) in the deformed body; the curvilinear system θ'_i which defines the type of aeolotropy is chosen similarly to coincide with the cylindrical polar coordinates (ρ, ϑ, x_3) in the undeformed body, so that
$$(\theta'_1, \theta'_2, \theta'_3) = (\rho, \vartheta, x_3). \tag{2.10.1}$$
In these circumstances, the elastic properties of the material and the deformation being examined both possess cylindrical symmetry relative to the x_3-axis.

The metric tensors G_{ij}, G^{ij}, g_{ij}, and g^{ij} are again given by (2.1.6), (2.1.10), and (2.1.11). Also, from (2.10.1) we have
$$\left. \begin{array}{l} g'_{11} = g'^{11} = g'_{33} = g'^{33} = 1, \quad g'_{22} = \rho^2, \quad g'^{22} = 1/\rho^2 \\ g'_{ij} = g'^{ij} = 0 \quad (i \neq j), \quad g' = \rho^2 \end{array} \right\}. \tag{2.10.2}$$

The metric tensors G'_{ij}, G'^{ij} may be found from the derivatives $\partial y^r/\partial \theta'^s$, $\partial \theta'^r/\partial y^s$ by a procedure analogous to that employed in calculating g_{ij}, g^{ij} in § 2.1. Only the covariant tensor G'_{ij} is required, however, and this may be derived from G_{ij} by the tensor transformation
$$G'_{ij} = \frac{\partial \theta^r}{\partial \theta'^i} \frac{\partial \theta^s}{\partial \theta'^j} G_{rs}.$$

From (2.1.4) we have
$$\frac{\partial \theta^r}{\partial \theta'^s} = \begin{bmatrix} r_\rho & 0 & 0 \\ 0 & a & b \\ 0 & c & d \end{bmatrix}, \tag{2.10.3}$$

and hence
$$G'_{ij} = \begin{bmatrix} r_\rho^2 & 0 & 0 \\ 0 & a^2 r^2 + c^2 & abr^2 + cd \\ 0 & abr^2 + cd & b^2 r^2 + d^2 \end{bmatrix}. \tag{2.10.4}$$

The definition (1.16.2) then yields for the strain components
$$\left. \begin{array}{l} \gamma'_{(11)} = \gamma'^1_1 = \tfrac{1}{2}(r_\rho^2 - 1) \\ \gamma'_{(22)} = \gamma'^2_2 = \dfrac{1}{2}\left(\dfrac{a^2 r^2 + c^2}{\rho^2} - 1\right) \\ \gamma'_{(33)} = \gamma'^3_3 = \tfrac{1}{2}(b^2 r^2 + d^2 - 1) \\ \gamma'_{(23)} = \rho \gamma'^2_3 = \gamma'^3_2/\rho = (abr^2 + cd)/(2\rho) \\ \gamma'_{(1i)} = \gamma'^i_1 = \gamma'^1_i = 0 \quad (i \neq 1) \end{array} \right\}. \tag{2.10.5}$$

We again confine attention to incompressible materials for which the conditions (2.1.12) apply and for which the strain energy W in (1.16.3) contains the components $\gamma'_{(12)}$, $\gamma'_{(13)}$ in powers higher than the first or as products with each other. This excludes the locally triclinic and tetragonal crystal classes with forms for W analogous to (1.7.1) and (1.10.3), and also the forms analogous to (1.12.5) and (1.12.11). Materials with forms for W analogous to (1.10.3) in which the suffices 1 and 3 are interchanged (i.e., in which e_{11} and e_{12} are interchanged with e_{33} and e_{23} respectively) are not, however, excluded. The stress–strain relations (1.16.14) with (2.10.2), (2.10.3), and (2.10.5) then yield

$$\left.\begin{aligned}
r^2\tau^{22} &= \tau^{11}+r^2\left[\frac{a^2}{\rho^2}\frac{\partial W}{\partial \gamma'_{(22)}}+\frac{ab}{\rho}\left(\frac{\partial W}{\partial \gamma'_{(23)}}+\frac{\partial W}{\partial \gamma'_{(32)}}\right)+b^2\frac{\partial W}{\partial \gamma'_{(33)}}\right]-\frac{\rho^2}{\lambda^2 r^2}\frac{\partial W}{\partial \gamma'_{(11)}} \\
\tau^{33} &= \tau^{11}+\frac{c^2}{\rho^2}\frac{\partial W}{\partial \gamma'_{(22)}}+\frac{cd}{\rho}\left(\frac{\partial W}{\partial \gamma'_{(23)}}+\frac{\partial W}{\partial \gamma'_{(32)}}\right)+d^2\frac{\partial W}{\partial \gamma'_{(33)}}-\frac{\rho^2}{\lambda^2 r^2}\frac{\partial W}{\partial \gamma'_{(11)}} \\
\tau^{23} &= \frac{ac}{\rho^2}\frac{\partial W}{\partial \gamma'_{(22)}}+\frac{ad+bc}{2\rho}\left(\frac{\partial W}{\partial \gamma'_{(23)}}+\frac{\partial W}{\partial \gamma'_{(32)}}\right)+bd\frac{\partial W}{\partial \gamma'_{(33)}} \\
\tau^{12} &= \tau^{13} = 0
\end{aligned}\right\}$$

(2.10.6)

Since the strain components $\gamma'_{(ij)}$ are independent of θ and y_3, and W takes the general polynomial form (1.16.3), subject to the above restriction, the strain energy function, and hence also the stresses τ^{ij}, are functions only of ρ (or r). The equations of equilibrium again take the form (2.1.19) and τ^{11} and p are determined by (2.1.20). Also, from (2.10.5) and (2.1.12) it follows that

$$\frac{dW}{dr} = \left(\frac{1}{\lambda r}-\frac{\rho^2}{\lambda^2 r^3}\right)\frac{\partial W}{\partial \gamma'_{(11)}}+\left[\frac{a^2 r}{\rho^2}-\frac{\lambda r}{\rho^4}(a^2 r^2+c^2)\right]\frac{\partial W}{\partial \gamma'_{(22)}}+$$
$$+b^2 r\frac{\partial W}{\partial \gamma'_{(33)}}+\left[\frac{abr}{\rho}-\frac{\lambda r}{2\rho^3}(abr^2+cd)\right]\left(\frac{\partial W}{\partial \gamma'_{(23)}}+\frac{\partial W}{\partial \gamma'_{(32)}}\right),$$

and with the help of (2.10.6) this becomes

$$\frac{dW}{dr} = \left(\frac{1}{r}-\frac{\lambda r}{\rho^2}\right)(r^2\tau^{22}-\tau^{11})-$$
$$-\frac{\lambda r}{\rho^2}\left[\frac{c^2}{\rho^2}\frac{\partial W}{\partial \gamma'_{(22)}}-\frac{abr^2-cd}{2\rho}\left(\frac{\partial W}{\partial \gamma'_{(23)}}+\frac{\partial W}{\partial \gamma'_{(32)}}\right)-b^2 r^2\frac{\partial W}{\partial \gamma'_{(33)}}\right].$$

When $b = 0$, $ad = \lambda$, this relation with (2.1.12), (2.1.19), and (2.10.6) yields

$$\rho^2\frac{dW}{dr} = K\frac{d\tau^{11}}{dr}-cdr\tau^{23}, \qquad (2.10.7)$$

or

$$\rho^2\frac{dW}{d\rho} = K\frac{d\tau^{11}}{d\rho}-\frac{c\rho}{a}\tau^{23}. \qquad (2.10.8)$$

Similarly, when $c = 0$, $ad = \lambda$, we obtain

$$\rho^2 \frac{dW}{dr} = K\frac{d\tau^{11}}{dr} + abr^3\tau^{23}, \tag{2.10.9}$$

or

$$\rho^2 \frac{dW}{d\rho} = K\frac{d\tau^{11}}{d\rho} + \frac{b\rho r^2}{d}\tau^{23}. \tag{2.10.10}$$

These results are independent of symmetries in the elastic material apart from the condition imposed on W concerning $\gamma'_{(12)}$ and $\gamma'_{(13)}$. The only other restrictions upon the elastic properties arise from the choice of the coordinate system θ'_i defining the type of aeolotropy.

When $b = c = 0$ we recover the flexure problem examined in § 2.5, but for a material which possesses a type of cylindrical aeolotropy. Equation (2.10.7) then yields the relation

$$\tau^{11} = [\tau^{11}]_{r=r_1} + \frac{1}{K}\int_{\rho_1}^{\rho} \rho^2 \frac{dW}{d\rho} d\rho, \tag{2.10.11}$$

identical with (2.5.2). If the undeformed body is the curved cuboid bounded by the curved surfaces $\rho = \rho_1$, $\rho = \rho_2$ and the planes $\vartheta = \pm\vartheta_0$, $x_3 = \pm L$, the subsequent formulae (2.5.4)–(2.5.10) also apply. If the deformation is to be maintained by the application of normal tractions only to the surfaces of the deformed body, so that $\tau^{12} = \tau^{23} = \tau^{31} = 0$, then the derivatives $\partial W/\partial \gamma'_{(ij)}$ for $i \neq j$ must all vanish when $\gamma'_{(12)} = \gamma'_{(23)} = \gamma'_{(31)} = 0$ and the expression for W must involve these strain components only in powers higher than the first or as products with each other.

2.11. Combined flexure and homogeneous strain of an incompressible aeolotropic cuboid

Corresponding to the cylindrically symmetrical deformation examined in §§ 2.1 and 2.10, we may consider an analogous problem in which the undeformed body is a cuboid and the cylindrical polar system (ρ, ϑ, x_3) in (2.1.4) is replaced by the cartesian reference frame (x_1, x_2, x_3). The resulting deformation, defined by

$$r = f(x_1), \qquad \theta = ax_2 + bx_3, \qquad y_3 = cx_2 + dx_3, \tag{2.11.1}$$

represents a generalization of the flexure problem of § 2.9 obtained by superposing the simple flexure upon a uniform strain.

If the undeformed body is the homogeneous incompressible (rectilinearly) aeolotropic cuboid of § 2.9, the solution of the problem represented by (2.11.1) may be obtained without difficulty by the method of that section. Alternatively, we may regard the problem defined by (2.11.1) for a rectilinearly aeolotropic material as a special case of that examined in the preceding section for cylindrically aeolotropic materials and derive its solution by a limiting procedure from the results of § 2.10.

The results of the previous section are evidently unaffected by a translation of the origin of the x_i-axes in the x_1, x_2-plane. In place of (2.1.3) we therefore write

$$x_1 = -h + \rho \cos \vartheta, \qquad x_2 = \rho \sin \vartheta, \qquad (2.11.2)$$

where h is a constant. Also, we define constants a', c', λ', k by the relations

$$\left. \begin{array}{c} a = ha', \qquad c = hc', \qquad \lambda = h\lambda' \\ \lambda' = a'd - bc' = 1/k \end{array} \right\}, \qquad (2.11.3)$$

and put

$$\rho = h + F(r), \qquad \vartheta' = h\vartheta. \qquad (2.11.4)$$

Substituting from (2.11.4) into (2.11.2), we have

$$x_1 = h\{\cos(\vartheta'/h) - 1\} + F(r)\cos(\vartheta'/h),$$
$$x_2 = h \sin(\vartheta'/h) + F(r)\sin(\vartheta'/h),$$

so that, as $h \to \infty$ while ϑ' and $F(r)$ remain finite, we have

$$\left. \begin{array}{c} x_1 \to F(r) \quad \text{or} \quad r \to f(x_1) \\ x_2 \to \vartheta' \end{array} \right\}. \qquad (2.11.5)$$

Remembering (2.11.3), (2.11.4), we thus obtain as a limiting case of equations (2.1.4) the relations

$$r = f(x_1), \qquad \theta = a'x_2 + bx_3, \qquad y_3 = c'x_2 + dx_3, \qquad (2.11.6)$$

which are identical with (2.11.1) apart from the replacement of a, c by a', c'.

When the reference frame θ_i' defining the aeolotropy is allowed to coincide with the rectangular cartesian system x_i, we have in place of (2.10.2)

$$g_{ij}' = g'^{ij} = \delta_{ij}, \qquad g' = 1, \qquad (2.11.7)$$

and the strain components $\gamma_{(ij)}'$ become identical with e_{ij}. Furthermore, as $h \to \infty$ we have, from (2.11.3) to (2.11.5), the conditions

$$\left. \begin{array}{c} r_\rho \to \dfrac{dr}{dx_1} = f', \qquad \dfrac{a}{\rho} \to a', \qquad \dfrac{c}{\rho} \to c', \qquad \dfrac{\lambda}{\rho} \to \lambda' \\ \dfrac{a}{\lambda} = \dfrac{a'}{\lambda'}, \qquad \dfrac{c}{\lambda} = \dfrac{c'}{\lambda'}, \qquad \lambda' = 1/k \end{array} \right\}. \qquad (2.11.8)$$

The incompressibility condition (2.1.12) thus becomes

$$r\frac{dr}{dx_1} = k \quad \text{or} \quad 2kx_1 + K' = r^2,\qquad(2.11.9)$$

where
$$K/h - h \to -K'/k \quad (h \to \infty).\qquad(2.11.10)$$

Introducing the limiting conditions (2.11.8) into (2.10.5) and (2.10.6) and discarding the primes from the constants in the resulting expressions we obtain

$$\left.\begin{array}{ll} e_{11} = \dfrac{1}{2}\left(\dfrac{k^2}{r^2}-1\right), & e_{22} = \tfrac{1}{2}(a^2 r^2 + c^2 - 1) \\[6pt] e_{33} = \tfrac{1}{2}(b^2 r^2 + d^2 - 1), & e_{23} = \tfrac{1}{2}(abr^2 + cd) \\[6pt] e_{12} = e_{13} = 0, & ad - bc = 1/k \end{array}\right\},\qquad(2.11.11)$$

and

$$\left.\begin{array}{l} r^2 \tau^{22} = \tau^{11} + r^2\left[a^2\dfrac{\partial W}{\partial e_{22}} + ab\left(\dfrac{\partial W}{\partial e_{23}}+\dfrac{\partial W}{\partial e_{32}}\right) + b^2\dfrac{\partial W}{\partial e_{33}}\right] - \dfrac{k^2}{r^2}\dfrac{\partial W}{\partial e_{11}} \\[10pt] \tau^{33} = \tau^{11} + c^2\dfrac{\partial W}{\partial e_{22}} + cd\left(\dfrac{\partial W}{\partial e_{23}}+\dfrac{\partial W}{\partial e_{32}}\right) + d^2\dfrac{\partial W}{\partial e_{33}} - \dfrac{k^2}{r^2}\dfrac{\partial W}{\partial e_{11}} \\[10pt] \tau^{23} = ac\dfrac{\partial W}{\partial e_{22}} + \tfrac{1}{2}(ad+bc)\left(\dfrac{\partial W}{\partial e_{23}}+\dfrac{\partial W}{\partial e_{32}}\right) + bd\dfrac{\partial W}{\partial e_{33}} \\[10pt] \tau^{12} = 0 \\[4pt] \tau^{13} = 0 \end{array}\right\}\qquad(2.11.12)$$

respectively, the strain energy function W now being a function of the components e_{ij} which contains e_{12}, e_{13} in powers higher than the first or as products with each other. This ensures that τ^{12} and τ^{13} are zero.

When $b = 0$ or $c = 0$ we see from (2.11.3), (2.11.4), and (2.11.10) that we may introduce into (2.10.7) and (2.10.9) the conditions

$$K/h^2 \to 1, \quad \rho/h \to 1, \quad cdr/\rho^2 \to 0, \quad ab/\rho^2 \to 0,\qquad(2.11.13)$$

as $h \to \infty$, giving
$$\frac{dW}{dr} = \frac{d\tau^{11}}{dr}.\qquad(2.11.14)$$

This relation, which may also be verified directly from (2.11.11) (2.11.12), and (2.1.19), yields the formula

$$\tau^{11} = W(r) + W_0,\qquad(2.11.15)$$

identical with (2.9.9).

When $b = c = 0$ the deformation reduces to the simple flexure considered in § 2.9, the constants K and λ of that section being replaced by K' and d respectively.

2.12. Cylindrical symmetry: compressible transversely isotropic material

We now examine some of the problems of the previous sections when the material is no longer incompressible. Except for the uniform deformation considered in § 2.8, it is not possible to solve these problems completely unless a special form is chosen for the strain energy function W, but certain results of general validity may be obtained without restricting the form of W.

We adopt the notation of § 2.1 and consider the deformation defined by (2.1.4), but here we restrict attention to the case

$$c = 0, \qquad ad = \lambda. \tag{2.12.1}$$

The formulae (2.1.5)–(2.1.11) still hold for the compressible material but now r as a function of ρ is *not* given by (2.1.12). The expressions (2.1.13), (2.1.14) are replaced by

$$\left. \begin{array}{l} I_1 = r_\rho^2 + \dfrac{a^2 r^2}{\rho^2} + b^2 r^2 + d^2 \\[4pt] I_2 = \dfrac{\lambda^2 r^2}{\rho^2} + r_\rho^2 \left(\dfrac{a^2 r^2}{\rho^2} + b^2 r^2 + d^2 \right) \\[4pt] I_3 = \dfrac{\lambda^2 r^2 r_\rho^2}{\rho^2} \end{array} \right\}, \tag{2.12.2}$$

$$\left. \begin{array}{l} 2K_1 = b^2 r^2 + d^2 - 1 \\[4pt] 4K_2 = \dfrac{a^2 b^2 r^4}{\rho^2} \end{array} \right\}. \tag{2.12.3}$$

Using (2.1.5)–(2.1.11), we see from (1.15.9) and (1.15.10) that the coefficients B^{ij}, M^{ij}, N^{ij} are given by

$$\left. \begin{array}{l} B^{11} = r_\rho^2 \left(\dfrac{a^2 r^2}{\rho^2} + b^2 r^2 + d^2 \right) \\[4pt] B^{22} = \dfrac{r_\rho^2}{\rho^2}(a^2 + b^2 \rho^2) + \dfrac{\lambda^2}{\rho^2} \\[4pt] B^{33} = d^2 r_\rho^2 + \dfrac{\lambda^2 r^2}{\rho^2} \\[4pt] B^{23} = bd r_\rho^2, \qquad B^{12} = B^{13} = 0 \end{array} \right\}, \tag{2.12.4}$$

$$M^{ij} = \begin{bmatrix} 0 & 0 & 0 \\ 0 & b^2 & bd \\ 0 & bd & d^2 \end{bmatrix}, \tag{2.12.5}$$

$$\left. \begin{array}{l} N^{22} = \dfrac{a^2 b^2 r^2}{\rho^2}, \qquad N^{23} = \dfrac{a^2 bd r^2}{2\rho^2} \\[4pt] N^{11} = N^{12} = N^{13} = N^{33} = 0 \end{array} \right\}. \tag{2.12.6}$$

Introducing these results with (2.1.6) and (2.1.11) into (1.15.6) we therefore obtain for the components of the stress tensor

$$\left.\begin{aligned}
\tau^{11} &= r_\rho^2\left\{\Phi+\left(\frac{a^2r^2}{\rho^2}+b^2r^2+d^2\right)\Psi\right\}+p \\
r^2\tau^{22} &= \tau^{11}+\left\{\frac{r^2}{\rho^2}(a^2+b^2\rho^2)-r_\rho^2\right\}\Phi+\left(\frac{\lambda^2 r^2}{\rho^2}-d^2 r_\rho^2\right)\Psi+ \\
&\qquad\qquad +b^2r^2\Theta+\frac{a^2b^2r^4}{\rho^2}\Lambda \\
\tau^{33} &= \tau^{11}+(d^2-r_\rho^2)\Phi+\frac{r^2}{\rho^2}\{\lambda^2-r_\rho^2(a^2+b^2\rho^2)\}\Phi+d^2\Theta \\
\tau^{23} &= bd\left\{\Phi+r_\rho^2\Psi+\Theta+\frac{a^2r^2}{2\rho^2}\Lambda\right\} \\
\tau^{12} &= \tau^{13} = 0
\end{aligned}\right\}. \quad (2.12.7)$$

All components of stress in (2.12.7) are independent of θ and y_3 so that the equations of equilibrium again reduce to the first of (2.1.19) or

$$r^2\tau^{22} = \frac{d}{dr}(r\tau^{11}) \qquad (2.12.8)$$

in the absence of body forces. If the expressions (2.12.7) for τ^{11} and τ^{22} are substituted into this equation we have a second-order non-linear differential equation for the function $r(\rho)$.

The strain energy function W is a function of I_1, I_2, I_3, K_1, K_2 and these quantities are independent of θ and y_3. Hence W is a function of r (or ρ) and

$$\frac{dW}{d\rho} = \frac{\partial W}{\partial I_1}\frac{dI_1}{d\rho}+\frac{\partial W}{\partial I_2}\frac{dI_2}{d\rho}+\frac{\partial W}{\partial I_3}\frac{dI_3}{d\rho}+\frac{\partial W}{\partial K_1}\frac{dK_1}{d\rho}+\frac{\partial W}{\partial K_2}\frac{dK_2}{d\rho}.$$

On using (1.15.7), (1.15.8), (2.12.2), and (2.12.3), this formula becomes

$$\frac{dW}{d\rho} = \frac{\lambda r r_\rho}{\rho}\left\{\left[r_\rho r_{\rho\rho}+\frac{a^2 r r_\rho}{\rho^2}-\frac{a^2 r^2}{\rho^3}+b^2 r r_\rho\right]\Phi+\right.$$
$$+\left[\frac{\lambda^2 r r_\rho}{\rho^2}-\frac{\lambda^2 r^2}{\rho^3}+r_\rho r_{\rho\rho}\left(\frac{a^2 r^2}{\rho^2}+b^2 r^2+d^2\right)+\frac{rr^3}{\rho^2}(a^2+b^2\rho^2)-\frac{a^2 r^2 r_\rho^2}{\rho^3}\right]\Psi+$$
$$\left.+b^2 r r_\rho\Theta+\frac{a^2 b^2 r^3}{2\rho^3}(2\rho r_\rho-r)\Lambda\right\}+\frac{\lambda}{\rho^2}[\rho r r_{\rho\rho}+\rho r_\rho^2-r r_\rho]p, \qquad (2.12.9)$$

and this, with the help of (2.12.7), reduces to

$$\frac{dW}{d\rho} = \frac{\lambda r r_{\rho\rho}}{\rho}\tau^{11} + \frac{\lambda r_\rho}{\rho^2}(\rho r_\rho - r)r^2\tau^{22} + \frac{\lambda b r^3 r_\rho}{d\rho^2}\tau^{23}. \qquad (2.12.10)$$

If now we make use of (2.12.8), this equation yields

$$\frac{dW}{d\rho} = \frac{\lambda}{\rho^2}\left\{\frac{d}{d\rho}[(\rho r_\rho - r)r\tau^{11}] + \frac{br^3 r_\rho}{d}\tau^{23}\right\}, \qquad (2.12.11)$$

or

$$\frac{dW}{dr} = \frac{\lambda}{\rho^2}\left\{\frac{d}{dr}[(\rho r_\rho - r)r\tau^{11}] + \frac{br^3}{d}\tau^{23}\right\}. \qquad (2.12.12)$$

2.13. Extension and torsion of a solid compressible circular cylinder

As in § 2.2 for the corresponding problem for an incompressible cylinder, we put

$$a = 1, \quad b = \lambda\psi, \quad c = 0, \quad d = \lambda, \qquad (2.13.1)$$

for a deformation describing the extension and torsion of a right circular cylinder with its axis coincident with the x_3-axis. The extension ratio parallel to the axis of the cylinder is λ and ψ is again the twist per unit length of extended cylinder.

The stress tensor τ^{ij} is given by (2.12.7) with the special values of a, b, and d given by (2.13.1), and $r(\rho)$ is determined by (2.12.8). The resultant moment M_3 about the axis on a plane end of the cylinder is

$$M_3 = 2\pi \int_0^{r_1} r^3 \tau^{23}\, dr,$$

where r_1 is the radius of the deformed cylinder, its initial radius being ρ_1. The value of τ^{23} in this expression is given by (2.12.7). Alternatively, from (2.12.12) we have

$$M_3 = \frac{2\pi}{\psi\lambda}\int_0^{r_1}\left\{\rho^2\frac{dW}{dr} - \lambda\frac{d}{dr}[(\rho r_\rho - r)r\tau^{11}]\right\}dr$$

$$= \frac{2\pi}{\psi\lambda}\left\{\int_0^{\rho_1}\rho^2\frac{dW}{d\rho}d\rho - \lambda[(\rho r_\rho - r)r\tau^{11}]_{r=0}^{r=r_1}\right\}. \qquad (2.13.2)$$

If the cylinder is extended without torsion it follows from (2.12.7) that τ^{23} and hence also M_3 is zero. Also, r is then proportional to ρ and $\rho r_\rho - r$ vanishes. Excluding this possibility it is reasonable to assume that r is an analytic function of ρ near $\rho = 0$ and that r

vanishes with ρ. If then τ^{11} is finite at $r = 0$ (in fact one would expect it to vanish on the axis) and if the surface $r = r_1$ is free from applied stress so that τ^{11} vanishes there, then (2.13.2) yields

$$M_3 = \frac{2\pi}{\psi\lambda} \int_0^{\rho_1} \rho^2 \frac{dW}{d\rho} d\rho, \qquad (2.13.3)$$

a result identical in form with (2.2.10) for an incompressible material. It should be emphasized, however, that here r is not known explicitly as a function of ρ until the equation (2.12.8) for r has been solved. In addition, a resultant force N_3 acts over each plane end of the cylinder.

2.14. Extension, inflation, and torsion of a compressible circular cylindrical tube

We consider again the problem of a cylindrical tube having its axis along the x_3-axis. The tube has external and internal radii ρ_1, ρ_2 with corresponding values r_1, r_2 after deformation. The extension ratio parallel to the axis of the tube is λ, and ψ is the twist per unit length of deformed tube. Excluding as before the case where $\rho r_\rho - r$ vanishes and taking the values (2.13.1) for a, b, c, d, we find for the couple M_3 over a plane end of the tube

$$M_3 = 2\pi \int_{r_2}^{r_1} r^3 \tau^{23} \, dr = \frac{2\pi}{\lambda\psi} \left\{ \int_{\rho_2}^{\rho_1} \rho^2 \frac{dW}{d\rho} d\rho - \lambda[(\rho r_\rho - r) r \tau^{11}]_{r=r_2}^{r=r_1} \right\}. \qquad (2.14.1)$$

If $\psi \neq 0$ and if it is possible to maintain the tube in equilibrium with zero applied tractions on the curved surfaces, then

$$M_3 = \frac{2\pi}{\lambda\psi} \int_{\rho_2}^{\rho_1} \rho^2 \frac{dW}{d\rho} d\rho, \qquad (2.14.2)$$

a result identical in form with the expression (2.3.9) for an incompressible material. The possibility of being able to maintain the deformation with zero stresses on both curved surfaces of the tube needs further investigation.

Suppose now that there is no torsion, so that $\psi = 0$, and that the tube is turned inside out and is maintained in equilibrium by tractions at its ends. If, in addition, it is possible to maintain equilibrium with the curved surfaces free from traction, then

$$\int_{\rho_2}^{\rho_1} \rho^2 \frac{dW}{d\rho} d\rho = 0, \tag{2.14.3}$$

a condition of the same form as (2.4.1) for an incompressible material.

2.15. Flexure of a cuboid of compressible aeolotropic material

As in § 2.9 we consider an undeformed cuboid bounded by the planes

$$x_1 = A_1, \qquad x_1 = A_2, \qquad x_2 = \pm B, \qquad x_3 = \pm C. \tag{2.15.1}$$

Using the coordinate system (r, θ, y_3) defined in that section, and guided by the corresponding deformation for the incompressible case, we assume that

$$r = f(x_1), \qquad \theta = \frac{x_2}{\lambda k}, \qquad y_3 = \lambda x_3, \tag{2.15.2}$$

where k and λ are constants, λ representing a constant extension ratio parallel to the x_3-axis. The equations (2.15.2) thus describe a simple flexure in which the plane faces normal to the x_1-axis are deformed into cylindrical surfaces of radii r_1, r_2 ($r_1 > r_2$) given by $r_1 = f(A_1)$, $r_2 = f(A_2)$. The components of the tensors G_{ij}, G^{ij} are given by (2.7.7) and

$$\frac{\partial x^r}{\partial \theta^s} = \begin{bmatrix} \dfrac{1}{f'} & 0 & 0 \\ 0 & \lambda k & 0 \\ 0 & 0 & \dfrac{1}{\lambda} \end{bmatrix}, \tag{2.15.3}$$

$$\frac{\partial \theta^r}{\partial x^s} = \begin{bmatrix} f' & 0 & 0 \\ 0 & \dfrac{1}{\lambda k} & 0 \\ 0 & 0 & \lambda \end{bmatrix}, \tag{2.15.4}$$

where f' denotes df/dx_1. Hence

$$g_{ij} = \begin{bmatrix} \dfrac{1}{f'^2} & 0 & 0 \\ 0 & \lambda^2 k^2 & 0 \\ 0 & 0 & \dfrac{1}{\lambda^2} \end{bmatrix}, \qquad g^{ij} = \begin{bmatrix} f'^2 & 0 & 0 \\ 0 & \dfrac{1}{\lambda^2 k^2} & 0 \\ 0 & 0 & \lambda^2 \end{bmatrix}, \tag{2.15.5}$$

and therefore

$$\left.\begin{array}{c} \gamma_{11} = \dfrac{1}{2}\left(1-\dfrac{1}{f'^2}\right), \quad \gamma_{22} = \tfrac{1}{2}(r^2-\lambda^2 k^2), \quad \gamma_{33} = \dfrac{1}{2}\left(1-\dfrac{1}{\lambda^2}\right) \\[6pt] \gamma_{12} = \gamma_{23} = \gamma_{31} = 0, \quad I_3 = \dfrac{r^2 f'^2}{k^2} \end{array}\right\} \quad (2.15.6)$$

From (1.1.15), (2.15.4), and (2.15.6) we have

$$\left.\begin{array}{c} e_{11} = \tfrac{1}{2}(f'^2-1), \quad e_{22} = \dfrac{1}{2}\left(\dfrac{r^2}{\lambda^2 k^2}-1\right), \quad e_{33} = \tfrac{1}{2}(\lambda^2-1) \\[6pt] e_{12} = e_{23} = e_{31} = 0 \end{array}\right\} \quad (2.15.7)$$

As in § 2.9 we restrict further discussion to strain energy functions W which are polynomials in e_{11}, e_{22}, e_{33}, e_{12}, e_{23}, e_{31} with e_{12}, e_{23}, e_{31} occurring only in powers higher than the first or as products of each other. The components of the stress tensor can then be written down from (1.15.1) and (2.15.4). Thus

$$\left.\begin{array}{c} \tau^{11} = \dfrac{kf'}{r}\dfrac{\partial W}{\partial e_{11}}, \quad \tau^{22} = \dfrac{1}{\lambda^2 krf'}\dfrac{\partial W}{\partial e_{22}}, \quad \tau^{33} = \dfrac{\lambda^2 k}{rf'}\dfrac{\partial W}{\partial e_{33}} \\[6pt] \tau^{12} = \tau^{23} = \tau^{31} = 0 \end{array}\right\} \quad (2.15.8)$$

The strain components e_{rs} and hence the strain energy function W are independent of θ and y_3 so that the equations of equilibrium reduce to a single relation

$$\left.\begin{array}{c} \dfrac{d\tau^{11}}{dr}+\dfrac{\tau^{11}-r^2\tau^{22}}{r} = 0 \\[6pt] r^2\tau^{22} = \dfrac{d}{dr}(r\tau^{11}) \end{array}\right\}, \quad (2.15.9)$$

or

identical in form with the first of (2.1.19).

The strain energy function W is a function of r (or x_1) only and

$$\frac{dW}{dr} = f''\frac{\partial W}{\partial e_{11}}+\frac{r}{\lambda^2 k^2}\frac{\partial W}{\partial e_{22}}.$$

Using (2.15.8) this becomes

$$\frac{dW}{dr} = \frac{rf''}{kf'}\tau^{11}+\frac{r^2 f'}{k}\tau^{22},$$

and with the help of (2.15.9) this can be written

$$\frac{dW}{dr} = \frac{1}{k}\left\{\frac{rf''}{f'}\tau^{11} + f'\frac{d}{dr}(r\tau^{11})\right\} = \frac{1}{k}\frac{d}{dr}(rf'\tau^{11}).$$

Hence
$$\tau^{11} = \frac{k}{rf'}(W+W_0), \qquad (2.15.10)$$

where W_0 is a constant. If the expressions for τ^{11} in (2.15.8) and (2.15.10) are equated we have an equation

$$\frac{\partial W}{\partial e_{11}} = \frac{1}{f'^2}(W+W_0) \qquad (2.15.11)$$

for f'.

On each of the faces $\theta = \pm\theta_0$ of the deformed cuboid there acts a resultant normal force N_2 of magnitude

$$N_2 = 2\lambda C \int_{r_2}^{r_1} r^2 \tau^{22}\,dr = 2\lambda C [r\tau^{11}]_{r_2}^{r_1}, \qquad (2.15.12)$$

and a resultant couple

$$M_2 = 2\lambda C \int_{r_2}^{r_1} r^3 \tau^{22}\,dr = 2\lambda C \int_{r_2}^{r_1} r\frac{d}{dr}(r\tau^{11})\,dr = 2\lambda C\left\{[r^2\tau^{11}]_{r_2}^{r_1} - \int_{r_2}^{r_1} r\tau^{11}\,dr\right\},$$

the formulae for N_2 and M_2 being simplified by means of (2.15.9). If we make use of (2.15.10) and (2.15.2) the latter expression yields

$$M_2 = 2\lambda C\left\{[r^2\tau^{11}]_{r_2}^{r_1} + k\int_{A_1}^{A_2} [W(x_1)+W_0]\,dx_1\right\}. \qquad (2.15.13)$$

The stresses acting on the surfaces $r = r_1$ and $r = r_2$ reduce to normal stresses only and these are zero if and only if

$$W(A_1) = W(A_2) = -W_0. \qquad (2.15.14)$$

If it is possible to satisfy this condition the resultant normal force N_2 on the faces $\theta = \pm\theta_0$ given by (2.15.12) is zero and the resultant moment on these faces, given by (2.15.13), reduces to

$$M_2 = 2\lambda kC\left\{\int_{A_1}^{A_2} W(x_1)\,dx_1 - (A_1-A_2)W_0\right\}. \qquad (2.15.15)$$

This formula is of the same form as (2.7.23) which was found for an incompressible cuboid. In the present problem, however, the deformation is not completely determined until equation (2.15.11) for $f(x_1)$ is solved. In order to maintain the deformation, normal stresses τ^{33} must also be applied over the faces of the deformed cuboid initially at $x_3 = \pm C$. The resultant of these normal stresses may be taken to be zero by a suitable choice of λ.

2.16. Cylindrically symmetric problems: compressible cylindrically aeolotropic bodies

The analysis of the deformation defined by (2.1.4) for compressible cylindrically aeolotropic bodies follows the lines indicated in § 2.10 for the incompressible case. The curvilinear system θ_i' defining the aeolotropy again coincides with the cylindrical polar coordinate system (ρ, ϑ, x_3) in the undeformed body, and the moving coordinate system θ_i with the polar coordinates (r, θ, y_3) in the deformed body.

As before, the metric tensors G_{ij}, G^{ij}, g_{ij}, and g^{ij} are given by (2.1.6), (2.1.10), and (2.1.11) and the corresponding primed quantities by (2.10.4) and (2.10.2) respectively. The formulae (2.10.3) for $\partial \theta^r / \partial \theta'^s$ and (2.10.5) for $\gamma_{(ij)}'$ also apply. The form of r as a function of ρ can, however, no longer be found from the incompressibility condition (2.1.12) and we now have

$$I_3 = \lambda^2 r^2 r_\rho^2 / \rho^2. \qquad (2.16.1)$$

From (1.16.11), (1.16.13), (2.10.2), (2.10.3), and (2.16.1) we obtain for the components τ^{ij} of the stress tensor

$$\left.\begin{aligned}
\tau^{11} &= \frac{\rho r_\rho}{\lambda r} \frac{\partial W}{\partial \gamma_{(11)}'} \\
\tau^{22} &= \frac{\rho}{\lambda r r_\rho} \left\{ \frac{a^2}{\rho^2} \frac{\partial W}{\partial \gamma_{(22)}'} + \frac{ab}{\rho}\left(\frac{\partial W}{\partial \gamma_{(23)}'} + \frac{\partial W}{\partial \gamma_{(32)}'}\right) + b^2 \frac{\partial W}{\partial \gamma_{(33)}'} \right\} \\
\tau^{33} &= \frac{\rho}{\lambda r r_\rho} \left\{ \frac{c^2}{\rho^2} \frac{\partial W}{\partial \gamma_{(22)}'} + \frac{cd}{\rho}\left(\frac{\partial W}{\partial \gamma_{(23)}'} + \frac{\partial W}{\partial \gamma_{(32)}'}\right) + d^2 \frac{\partial W}{\partial \gamma_{(33)}'} \right\} \\
\tau^{23} &= \frac{\rho}{\lambda r r_\rho} \left\{ \frac{ac}{\rho^2} \frac{\partial W}{\partial \gamma_{(22)}'} + \frac{ad+bc}{2\rho}\left(\frac{\partial W}{\partial \gamma_{(23)}'} + \frac{\partial W}{\partial \gamma_{(32)}'}\right) + bd \frac{\partial W}{\partial \gamma_{(33)}'} \right\} \\
\tau^{12} &= 0 \\
\tau^{13} &= 0
\end{aligned}\right\}, \quad (2.16.2)$$

where we assume that W does not contain terms linear in either of the components $\gamma_{(12)}'$, $\gamma_{(13)}'$ alone.

The strain components $\gamma_{(ij)}'$, the strain energy function W, and the stresses τ^{ij} are again independent of θ and y_3, so that in the absence of body forces the equations of equilibrium reduce to (2.12.8). The expressions (2.16.2) combined with this relation then yield a differential equation for the determination of r as a function of ρ.

By making use of (2.10.5) and (2.16.2) we may obtain relations analogous to (2.10.7)–(2.10.10). Thus

$$\frac{dW}{d\rho} = r_\rho r_{\rho\rho}\frac{\partial W}{\partial \gamma'_{(11)}} + \left\{\frac{a^2 rr_\rho}{\rho^2} - \frac{a^2 r^2 + c^2}{\rho^3}\right\}\frac{\partial W}{\partial \gamma'_{(22)}} + b^2 rr_\rho \frac{\partial W}{\partial \gamma'_{(33)}} +$$
$$+ \left(\frac{abrr_\rho}{\rho} - \frac{abr^2 + cd}{2\rho^2}\right)\left(\frac{\partial W}{\partial \gamma'_{(23)}} + \frac{\partial W}{\partial \gamma'_{(32)}}\right)$$
$$= \frac{\lambda}{\rho^2}\{\rho r_{\rho\rho}r\tau^{11} + r_\rho[(\rho r_\rho - r)r^2\tau^{22}]\} - \frac{c}{\rho}\left\{\frac{c}{\rho^2}\frac{\partial W}{\partial \gamma'_{(22)}} + \frac{d}{2\rho}\left(\frac{\partial W}{\partial \gamma'_{(23)}} + \frac{\partial W}{\partial \gamma'_{(32)}}\right)\right\} +$$
$$+ \frac{br^2}{\rho}\left\{\frac{a}{2\rho}\left(\frac{\partial W}{\partial \gamma'_{(23)}} + \frac{\partial W}{\partial \gamma'_{(32)}}\right) + b\frac{\partial W}{\partial \gamma'_{(33)}}\right\},$$

and when $b = 0$, $ad = \lambda$, this relation with (2.11.8) and (2.16.2) yields

$$\rho^2 \frac{dW}{d\rho} = \lambda \frac{d}{d\rho}[(\rho r_\rho - r)r\tau^{11}] - cdrr_\rho \tau^{23}, \qquad (2.16.3)$$

or
$$\rho^2 \frac{dW}{dr} = \lambda \frac{d}{dr}[(\rho r_\rho - r)r\tau^{11}] - cdr\tau^{23}. \qquad (2.16.4)$$

Similarly, when $c = 0$, we have

$$\rho^2 \frac{dW}{d\rho} = \lambda \frac{d}{d\rho}[(\rho r_\rho - r)r\tau^{11}] + abr^3 r_\rho \tau^{23}, \qquad (2.16.5)$$

or
$$\rho^2 \frac{dW}{dr} = \lambda \frac{d}{dr}[(\rho r_\rho - r)r\tau^{11}] + abr^3 \tau^{23}. \qquad (2.16.6)$$

Apart from the restriction on W involving $\gamma'_{(12)}$, $\gamma'_{(13)}$ these formulae are again independent of symmetries in the elastic material. The relations (2.10.7)–(2.10.10) for an incompressible body may be derived by introducing the conditions (2.1.12) into (2.16.3)–(2.16.6).

When $b = c = 0$ we recover the flexure problem considered in § 2.5. In the absence of the incompressibility condition it is not possible to derive the simple formulae (2.5.6), (2.5.10) for the couple M_2 on the plane faces $\theta = \pm a\vartheta_0$ of the cuboid described in that section. From (2.12.8) and (2.16.4) it again follows, however, that if equilibrium can be maintained with zero traction on the curved surfaces $r = r_1, r = r_2$

of the deformed cylinder, then the resultant normal traction N_2 on each of the faces $\theta = \pm a\vartheta_0$ is zero and

$$\int_{\rho_2}^{\rho_1} \rho^2 \frac{dW}{d\rho} d\rho = 0. \tag{2.16.7}$$

2.17. Combined flexure and homogeneous strain of a compressible aeolotropic cuboid

The combined homogeneous strain and flexure of a cuboid may be examined as a limiting case of the corresponding cylindrically symmetrical problem by the method employed in § 2.11 for incompressible materials. By means of the limiting procedure embodied in equations (2.11.2)–(2.11.5) and (2.11.8) we obtain again the deformation described by

$$r = f(x_1), \qquad \theta = ax_2 + bx_3, \qquad y_3 = cx_2 + dx_3, \tag{2.17.1}$$

where primes on a, c are omitted after the limiting process. The metric tensors g'_{ij}, g'^{ij} are given by (2.11.7), but the incompressibility condition (2.11.9) no longer applies. The conditions (2.11.8) with (2.16.1) yield

$$I_3 = \frac{r^2 f'^2}{k^2}. \tag{2.17.2}$$

The strain components e_{ij} follow by combining (2.11.8) with (2.10.5). We have

$$e_{11} = \tfrac{1}{2}(f'^2 - 1), \tag{2.17.3}$$

the remaining components taking the forms (2.11.11). Also from (2.11.8) and (2.16.2) we obtain for the components τ^{ij} of the stress tensor

$$\left.\begin{aligned}
\tau^{11} &= \frac{kf'}{r}\frac{\partial W}{\partial e_{11}} \\
\tau^{22} &= \frac{k}{rf'}\left\{a^2\frac{\partial W}{\partial e_{22}} + ab\left(\frac{\partial W}{\partial e_{23}} + \frac{\partial W}{\partial e_{32}}\right) + b^2\frac{\partial W}{\partial e_{33}}\right\} \\
\tau^{33} &= \frac{k}{rf'}\left\{c^2\frac{\partial W}{\partial e_{22}} + cd\left(\frac{\partial W}{\partial e_{23}} + \frac{\partial W}{\partial e_{32}}\right) + d^2\frac{\partial W}{\partial e_{33}}\right\} \\
\tau^{23} &= \frac{k}{rf'}\left\{ac\frac{\partial W}{\partial e_{22}} + \tfrac{1}{2}(ad+bc)\left(\frac{\partial W}{\partial e_{23}} + \frac{\partial W}{\partial e_{32}}\right) + bd\frac{\partial W}{\partial e_{33}}\right\} \\
\tau^{12} &= 0 \\
\tau^{13} &= 0
\end{aligned}\right\} \tag{2.17.4}$$

if we omit primes, and if W does not contain terms linear in either of the components e_{12}, e_{13} alone. The strain components e_{ij}, the strain energy function W, and the stresses τ^{ij} are all independent of θ and y_3 and the equations of equilibrium reduce to (2.12.8).

When $b = 0$ or $c = 0$ relations analogous to (2.11.14) and (2.11.15) follow by introducing the conditions (2.11.4) and (2.11.8) into (2.16.3) or (2.16.5). In either case we obtain

$$\frac{dW}{dx_1} = \frac{1}{k}\frac{d}{dx_1}(rf'\tau^{11}), \qquad (2.17.5)$$

and
$$\tau^{11} = \frac{k}{rf'}(W+W_0),$$

where W_0 is a constant. For incompressible bodies, these relations with (2.11.9) reduce to (2.11.14) and (2.11.15) respectively.

When $b = c = 0$ we recover the flexure problem of § 2.15 with λ there replaced by d and $ad = 1/k$.

2.18. Cylindrically symmetric problems: parametric forms

For the deformation described by (2.1.4) a number of general results, which are independent of symmetries existing in the material, may be obtained by regarding ρ and the parameters a, b, c, d defining the deformation as independent variables.† We consider a general cylindrically aeolotropic body of the type examined in §§ 2.10, 2.16 and regard the strain components $\gamma'_{(ij)}$ as functions of ρ, a, b, c, d. The strain energy function W and the stresses τ^{ij} are also treated as functions of the same variables and it is assumed that W does not contain terms linear in either of the variables $\gamma'_{(12)}$, $\gamma'_{(13)}$ alone. Furthermore, the variable r, as determined for an incompressible body from (2.1.12), is also a function of ρ, a, b, c, d. The additional arbitrary constant K may be selected independently of these quantities and need not be considered further. Similarly, for compressible bodies r is determined as a function of ρ, a, b, c, d by the equation of equilibrium (2.12.8), the additional constants introduced by integration of this

† Corresponding forms can also be derived for the deformation
$$r = r(\rho), \qquad \theta = \phi(\rho)+a\vartheta+bx_3, \qquad y_3 = w(\rho)+c\vartheta+dx_3,$$
ϕ_ρ and w_ρ being regarded as additional parameters. For details, the reader is referred to the paper by J. E. Adkins, *Proc. R. Soc.* A, **231** (1955), 75.

equation being independent of these variables. We therefore write

$$\left.\begin{array}{l} r = r(\rho, a, b, c, d) \\ \gamma'_{(ij)} = \gamma'_{(ij)}(\rho, a, b, c, d) \\ W = W(\rho, a, b, c, d) \end{array}\right\}, \qquad (2.18.1)$$

it being understood in the subsequent analysis that any relation between the parameters a, b, c, d which may arise from the boundary conditions is not employed until after the partial derivatives of r, $\gamma'_{(ij)}$, and W with respect to these quantities have been formed. It is further assumed that r is a continuous function of the variables ρ, a, b, c, d which possesses continuous first and second partial derivatives with respect to ρ, a, b, c, d. For incompressible bodies these conditions are satisfied by virtue of (2.1.12) provided $ad - bc \neq 0$.

With these assumptions we have for a compressible body, from (2.10.5),

$$\frac{\partial W}{\partial a} = r_\rho r_{\rho a} \frac{\partial W}{\partial \gamma'_{(11)}} + \left(\frac{ar^2}{\rho^2} + \frac{a^2 r r_a}{\rho^2}\right)\frac{\partial W}{\partial \gamma'_{(22)}} + b^2 r r_a \frac{\partial W}{\partial \gamma'_{(33)}} +$$
$$+ \frac{1}{2\rho}(br^2 + 2abrr_a)\left(\frac{\partial W}{\partial \gamma'_{(23)}} + \frac{\partial W}{\partial \gamma'_{(32)}}\right),$$

suffixes denoting partial differentiation, and if we make use of (2.16.2) and (2.12.8) in the form†

$$r^2 r_\rho \tau^{22} = \frac{\partial}{\partial \rho}(r\tau^{11}), \qquad (2.18.2)$$

this becomes

$$\frac{\partial W}{\partial a} = \frac{\lambda}{\rho}\frac{\partial}{\partial \rho}(r_a r \tau^{11}) + r^2\left\{\frac{a}{\rho^2}\frac{\partial W}{\partial \gamma'_{(22)}} + \frac{b}{2\rho}\left(\frac{\partial W}{\partial \gamma'_{(23)}} + \frac{\partial W}{\partial \gamma'_{(32)}}\right)\right\}. \qquad (2.18.3)$$

In a similar manner we obtain

$$\left.\begin{array}{l} \dfrac{\partial W}{\partial b} = \dfrac{\lambda}{\rho}\dfrac{\partial}{\partial \rho}(r_b r \tau^{11}) + r^2\left\{\dfrac{a}{2\rho}\left(\dfrac{\partial W}{\partial \gamma'_{(23)}} + \dfrac{\partial W}{\partial \gamma'_{(32)}}\right) + b\dfrac{\partial W}{\partial \gamma'_{(33)}}\right\} \\[4pt] \dfrac{\partial W}{\partial c} = \dfrac{\lambda}{\rho}\dfrac{\partial}{\partial \rho}(r_c r \tau^{11}) + \dfrac{c}{\rho^2}\dfrac{\partial W}{\partial \gamma'_{(22)}} + \dfrac{d}{2\rho}\left(\dfrac{\partial W}{\partial \gamma'_{(23)}} + \dfrac{\partial W}{\partial \gamma'_{(32)}}\right) \\[4pt] \dfrac{\partial W}{\partial d} = \dfrac{\lambda}{\rho}\dfrac{\partial}{\partial \rho}(r_d r \tau^{11}) + \dfrac{c}{2\rho}\left(\dfrac{\partial W}{\partial \gamma'_{(23)}} + \dfrac{\partial W}{\partial \gamma'_{(32)}}\right) + d\dfrac{\partial W}{\partial \gamma'_{(33)}} \end{array}\right\}, \qquad (2.18.4)$$

† Partial differentiation signs are used throughout this section since a, b, c, d are now regarded as independent variables. Where r and ρ have the significance of functions of position only, the change of independent variable continues to be given by

$$\partial/\partial r = (1/r_\rho)\partial/\partial \rho.$$

and

$$\frac{\partial W}{\partial \rho} = \frac{\lambda}{\rho}\frac{\partial}{\partial \rho}(r_\rho r\tau^{11}) - \frac{ar^2}{\rho}\left\{\frac{a}{\rho^2}\frac{\partial W}{\partial \gamma'_{(22)}} + \frac{b}{2\rho}\left(\frac{\partial W}{\partial \gamma'_{(23)}} + \frac{\partial W}{\partial \gamma'_{(32)}}\right)\right\} -$$

$$- \frac{c}{\rho}\left\{\frac{c}{\rho^2}\frac{\partial W}{\partial \gamma'_{(22)}} + \frac{d}{2\rho}\left(\frac{\partial W}{\partial \gamma'_{(23)}} + \frac{\partial W}{\partial \gamma'_{(32)}}\right)\right\}. \quad (2.18.5)$$

The latter formula may be reduced to the form

$$\frac{\partial W}{\partial \rho} = \frac{\lambda}{\rho^2}\frac{\partial}{\partial \rho}[(\rho r_\rho - r)r\tau^{11}] + \frac{br^2}{\rho}\left\{\frac{a}{2\rho}\left(\frac{\partial W}{\partial \gamma'_{(23)}} + \frac{\partial W}{\partial \gamma'_{(32)}}\right) + b\frac{\partial W}{\partial \gamma'_{(33)}}\right\} -$$

$$- \frac{c}{\rho}\left\{\frac{c}{\rho^2}\frac{\partial W}{\partial \gamma'_{(22)}} + \frac{d}{2\rho}\left(\frac{\partial W}{\partial \gamma'_{(23)}} + \frac{\partial W}{\partial \gamma'_{(32)}}\right)\right\}, \quad (2.18.6)$$

derived in § 2.16 by making use of (2.16.2) and (2.18.2). Combination of the results (2.18.3)–(2.18.6) and the further use of (2.16.2) now yields

$$\left.\begin{aligned}
a\frac{\partial W}{\partial a} + b\frac{\partial W}{\partial b} &= \frac{\lambda}{\rho}\left\{\frac{\partial}{\partial \rho}[(ar_a + br_b)r\tau^{11}] + r_\rho r^3\tau^{22}\right\} \\
c\frac{\partial W}{\partial c} + d\frac{\partial W}{\partial d} &= \frac{\lambda}{\rho}\left\{\frac{\partial}{\partial \rho}[(cr_c + dr_d)r\tau^{11}] + r_\rho r\tau^{33}\right\} \\
c\frac{\partial W}{\partial a} + d\frac{\partial W}{\partial b} &= \frac{\lambda}{\rho}\left\{\frac{\partial}{\partial \rho}[(cr_a + dr_b)r\tau^{11}] + r_\rho r^3\tau^{23}\right\} \\
a\frac{\partial W}{\partial c} + b\frac{\partial W}{\partial d} &= \frac{\lambda}{\rho}\left\{\frac{\partial}{\partial \rho}[(ar_c + br_d)r\tau^{11}] + r_\rho r\tau^{23}\right\}
\end{aligned}\right\}, \quad (2.18.7)$$

$$\left.\begin{aligned}
a\frac{\partial W}{\partial a} + c\frac{\partial W}{\partial c} + \rho\frac{\partial W}{\partial \rho} &= \frac{\lambda}{\rho}\left\{\frac{\partial}{\partial \rho}[(ar_a + cr_c + \rho r_\rho)r\tau^{11}] - r_\rho r\tau^{11}\right\} \\
b\frac{\partial W}{\partial b} - c\frac{\partial W}{\partial c} - \rho\frac{\partial W}{\partial \rho} &= \frac{\lambda}{\rho}\frac{\partial}{\partial \rho}[(br_b - cr_c - \rho r_\rho + r)r\tau^{11}]
\end{aligned}\right\}. \quad (2.18.8)$$

When the material is incompressible, the relations (2.1.12) apply. From the second of these, by differentiation and the use of the definition $\lambda = ad - bc$, we obtain

$$2\lambda r_a = -rd, \quad 2\lambda r_b = rc, \quad 2\lambda r_c = rb, \quad 2\lambda r_d = -ra. \quad (2.18.9)$$

With the help of these conditions and (2.1.12), the formulae (2.18.7),

(2.18.8) yield

$$a\frac{\partial W}{\partial a}+b\frac{\partial W}{\partial b}=\frac{r}{2}\frac{\partial \tau^{11}}{\partial r}$$

$$c\frac{\partial W}{\partial c}+d\frac{\partial W}{\partial d}=-\frac{r}{2}\frac{\partial \tau^{11}}{\partial r}+\tau^{33}-\tau^{11}$$

$$c\frac{\partial W}{\partial a}+d\frac{\partial W}{\partial b}=r^2\left(a\frac{\partial W}{\partial c}+b\frac{\partial W}{\partial d}\right)=r^2\tau^{23}$$

(2.18.10)

and

$$b\frac{\partial W}{\partial b}-c\frac{\partial W}{\partial c}-\rho\frac{\partial W}{\partial \rho}=-\frac{K}{\rho}\frac{\partial \tau^{11}}{\partial \rho}=-\frac{K}{\lambda r}\frac{\partial \tau^{11}}{\partial r},$$

(2.18.11)

respectively. From (2.18.10) and (2.18.11) we have the identity

$$a\frac{\partial W}{\partial a}+b\frac{\partial W}{\partial b}=\frac{\lambda r^2}{2K}\left(\rho\frac{\partial W}{\partial \rho}+c\frac{\partial W}{\partial c}-b\frac{\partial W}{\partial b}\right).$$

(2.18.12)

With (2.12.8) the first two equations of (2.18.10) give

$$\tau^{11}=2\int_{r_0}^{r}\left(a\frac{\partial W}{\partial a}+b\frac{\partial W}{\partial b}\right)\frac{dr}{r}+\tau_0^{11}$$

$$r^2\tau^{22}=\tau^{11}+2\left(a\frac{\partial W}{\partial a}+b\frac{\partial W}{\partial b}\right)$$

$$\tau^{33}=\tau^{11}+a\frac{\partial W}{\partial a}+b\frac{\partial W}{\partial b}+c\frac{\partial W}{\partial c}+d\frac{\partial W}{\partial d}$$

(2.18.13)

where τ_0^{11} is the value of τ^{11} at $r=r_0$ (or $\rho=\rho_0$). Also, from (2.18.11), we obtain the alternative formula

$$\tau^{11}=\frac{1}{K}\int_{\rho_0}^{\rho}\rho\left(\rho\frac{\partial W}{\partial \rho}+c\frac{\partial W}{\partial c}-b\frac{\partial W}{\partial b}\right)d\rho+\tau_0^{11},$$

(2.18.14)

for τ^{11}. The relations (2.18.10)–(2.18.14) for an incompressible body may also be derived directly from the formulae of § 2.10.

The relations (2.18.10), (2.18.12), and (2.12.8) may be employed to write down general formulae for the resultant forces and couples acting across radial planes $\theta=$ constant and planes normal to the y_3-axis in the deformed body. We shall suppose the undeformed body to be bounded by the cylindrical surfaces $\rho=\rho_1$, $\rho=\rho_2$ ($\rho_1>\rho_2$), these becoming $r=r_1$, $r=r_2$ respectively after deformation. The stresses across the portion of any radial plane $\theta=$ constant in the deformed body enclosed by the lines $y_3=l_1$, $y_3=l_2$ may then be regarded as giving rise to a resultant force N_2 acting normally to this plane, a shearing force S_2 acting in the direction of the y_3-axis, and a couple M_2 acting about an axis parallel to the y_3-axis. Similarly, the stresses acting on the portion of any plane $y_3 = $ constant lying between the two radial planes $\theta = \theta_1$, $\theta = \theta_2$ give rise to a resultant force N_3 parallel to the y_3-axis and a couple M_3 about this axis. From (2.12.8), writing $l = l_1 - l_2$, we then have

$$N_2 = l \int_{r_2}^{r_1} r^2 \tau^{22} \, dr = l[r\tau^{11}]_{r_2}^{r_1}, \tag{2.18.15}$$

and for incompressible bodies, from (2.18.10),

$$S_2 = l \int_{r_2}^{r_1} r\tau^{23} \, dr = \frac{l}{\lambda} \int_{\rho_2}^{\rho_1} \left(a\frac{\partial W}{\partial c} + b\frac{\partial W}{\partial d} \right) \rho \, d\rho, \tag{2.18.16}$$

$$M_2 = l \int_{r_2}^{r_1} r^3 \tau^{22} \, dr = l \left\{ \frac{1}{\lambda} \int_{\rho_2}^{\rho_1} \left(a\frac{\partial W}{\partial a} + b\frac{\partial W}{\partial b} \right) \rho \, d\rho + \tfrac{1}{2}[r^2\tau^{11}]_{r_2}^{r_1} \right\}, \tag{2.18.17}$$

$$N_3 = (\theta_1 - \theta_2) \int_{r_2}^{r_1} r\tau^{33} \, dr = (\theta_1 - \theta_2) \left\{ \frac{1}{\lambda} \int_{\rho_2}^{\rho_1} \left(c\frac{\partial W}{\partial c} + d\frac{\partial W}{\partial d} \right) \rho \, d\rho + \tfrac{1}{2}[r^2\tau^{11}]_{r_2}^{r_1} \right\}, \tag{2.18.18}$$

$$M_3 = (\theta_1 - \theta_2) \int_{r_2}^{r_1} r^3 \tau^{23} \, dr = \frac{\theta_1 - \theta_2}{\lambda} \int_{\rho_2}^{\rho_1} \left(c\frac{\partial W}{\partial a} + d\frac{\partial W}{\partial b} \right) \rho \, d\rho. \tag{2.18.19}$$

Equation (2.18.15) applies for compressible or incompressible materials. Formulae for compressible bodies corresponding to (2.18.16)–(2.18.19) may be derived from (2.18.7).

By putting $b = c = 0$, $ad = \lambda$, we may again derive results for the simple flexure of an initially curved cuboid. From (2.18.12), (2.18.13) we obtain for an incompressible body

$$\left.\begin{aligned}
\tau^{11} &= 2a \int_{r_0}^{r} \frac{1}{r} \frac{\partial W}{\partial a} \, dr + \tau_0^{11} = \frac{1}{K} \int_{\rho_0}^{\rho} \rho^2 \frac{\partial W}{\partial \rho} \, d\rho + \tau_0^{11} \\
r^2 \tau^{22} &= \tau^{11} + 2a \frac{\partial W}{\partial a} = \tau^{11} + \frac{\lambda r^2}{K} \rho \frac{\partial W}{\partial \rho} \\
\tau^{33} &= \tau^{11} + a \frac{\partial W}{\partial a} + d \frac{\partial W}{\partial d} = \tau^{11} + \frac{\lambda r^2}{2K} \rho \frac{\partial W}{\partial \rho} + d \frac{\partial W}{\partial d}
\end{aligned}\right\} \tag{2.18.20}$$

Also, if the undeformed body is bounded by the planes $\vartheta = \pm\vartheta_0$, $x_3 = \pm L$ so that
$$\theta_1 - \theta_2 = 2a\vartheta_0, \qquad l = 2Ld, \tag{2.18.21}$$
and the curved surfaces $r = r_1$, $r = r_2$ are free from applied stress, then $N_2 = 0$ and, from (2.18.17) and (2.18.18), we have

$$\left. \begin{aligned} M_2 &= 2L \int_{\rho_2}^{\rho_1} \rho \frac{\partial W}{\partial a} d\rho \\ N_3 &= 2\vartheta_0 \int_{\rho_2}^{\rho_1} \rho \frac{\partial W}{\partial d} d\rho \end{aligned} \right\}. \tag{2.18.22}$$

Again, from (2.18.10) we see that τ^{23} vanishes if $\partial W/\partial b = \partial W/\partial c = 0$ when $b = c = 0$, and the resultant force S_2 is then also zero.

By writing
$$a = 1, \qquad c = 0, \qquad d = \lambda, \qquad \vartheta_0 = \pi, \tag{2.18.23}$$
we may examine the combined extension, inflation, and torsion of a cylindrical tube. From (2.18.10), (2.18.11), (2.18.13), and (2.18.14) we obtain for the stresses in the incompressible case

$$\left. \begin{aligned} \tau^{11} &= \frac{1}{K} \int_{\rho_0}^{\rho} \rho \left(\rho \frac{\partial W}{\partial \rho} - b \frac{\partial W}{\partial b} \right) d\rho + \tau_0^{11} \\ r^2 \tau^{22} &= \tau^{11} + \frac{\lambda r^2}{K} \left(\rho \frac{\partial W}{\partial \rho} - b \frac{\partial W}{\partial b} \right) \\ \tau^{33} &= \tau^{11} + \frac{\lambda r^2}{2K} \left(\rho \frac{\partial W}{\partial \rho} - b \frac{\partial W}{\partial b} \right) + \lambda \frac{\partial W}{\partial \lambda} \\ r^2 \tau^{23} &= \lambda \frac{\partial W}{\partial b} \end{aligned} \right\}. \tag{2.18.24}$$

The torque M_3 on each of the plane ends of the tube is given by
$$M_3 = 2\pi \int_{\rho_2}^{\rho_1} \rho \frac{\partial W}{\partial b} d\rho = \frac{2\pi}{b} \left\{ \int_{\rho_2}^{\rho_1} \rho^2 \frac{\partial W}{\partial \rho} d\rho - K[\tau^{11}]_{r=r_2}^{r=r_1} \right\}, \tag{2.18.25}$$
and the resultant extending force N_3 by
$$N_3 = 2\pi \left\{ \int_{\rho_2}^{\rho_1} \rho \frac{\partial W}{\partial \lambda} d\rho + \tfrac{1}{2} [r^2 \tau^{11}]_{r=r_2}^{r=r_1} \right\}. \tag{2.18.26}$$

If the curved surfaces of the tube are free from applied traction, the resultant couple and force become

$$M_3 = 2\pi \int_{\rho_2}^{\rho_1} \rho \frac{\partial W}{\partial b}\, d\rho = \frac{2\pi}{b}\int_{\rho_2}^{\rho_1} \rho^2 \frac{\partial W}{\partial \rho}\, d\rho$$

$$N_3 = 2\pi \int_{\rho_2}^{\rho_1} \rho \frac{\partial W}{\partial \lambda}\, d\rho$$

(2.18.27)

respectively, these results being valid also for compressible materials. If $b = \lambda\psi$ so that ψ is the angle of torsion per unit length of deformed tube, we have

$$\frac{\partial W}{\partial \psi} = \frac{\partial W}{\partial b}\frac{\partial b}{\partial \psi} = \lambda \frac{\partial W}{\partial b} \qquad (2.18.28)$$

and

$$M_3 = \frac{2\pi}{\lambda}\int_{\rho_2}^{\rho_1} \rho \frac{\partial W}{\partial \psi}\, d\rho. \qquad (2.18.29)$$

Also, from (2.18.27),

$$N_3 + \psi M_3 = 2\pi \int_{\rho_2}^{\rho_1} \rho\left(\frac{\partial W}{\partial \lambda} + \psi \frac{\partial W}{\partial b}\right) d\rho = 2\pi \int_{\rho_2}^{\rho_1} \rho\left(\frac{\partial W}{\partial \lambda} + \frac{\partial W}{\partial b}\frac{\partial b}{\partial \lambda}\right) d\rho$$

$$= 2\pi \int_{\rho_2}^{\rho_1} \rho\left(\frac{\partial W}{\partial \lambda}\right)_\psi d\rho, \qquad (2.18.30)$$

the suffix ψ denoting that $\partial W/\partial \lambda$ has been evaluated keeping ψ constant. Analogous formulae for compressible bodies have been derived by Green.†

The formulae (2.18.7)–(2.18.30) have been derived for a general strain energy function W having the polynomial form (1.3.8) with e_{ij} replaced by $\gamma'_{(ij)}$ and are independent of symmetries in the material except that W does not contain terms linear in either of the variables $\gamma'_{(12)}$, $\gamma'_{(13)}$ alone. With this exception they are therefore valid for any material possessing local symmetry properties of the types described in §§ 1.7–1.14, the form of W for such a material being obtained by replacing e_{ij} by $\gamma'_{(ij)}$ in the appropriate formula of Chapter I. The effect of symmetry properties on the form of W is to group the strain components e_{ij} or $\gamma'_{(ij)}$ into permissible functional combinations which

† Loc. cit., p. 37.

we may denote by Ψ'_i ($i = 1, 2, ..., n$). For example, if a material is transversely isotropic with respect to the x_3-axis there are five independent permissible functions Ψ'_i; these are the invariants I_1, I_2, I_3, K_1, and K_2 defined by (1.1.18) and (1.13.11). Let us suppose that for a given type of material W involves the strain components $\gamma'_{(ij)}$ in permissible functional combinations

$$\Psi'_r = \Psi'_r(\gamma'_{(ij)}) \quad (r = 1, 2, ..., n)$$

so that $\qquad W = W(\Psi'_1, \Psi'_2, ..., \Psi'_n).$

The functions Ψ'_r may be expressed in terms of the parameters a, b, c, d, and ρ by making use of (2.10.5). By writing

$$\frac{\partial W}{\partial a} = \sum_{r=1}^{n} \frac{\partial W}{\partial \Psi'_r} \frac{\partial \Psi'_r}{\partial a},$$

with corresponding expressions for the derivatives of W with respect to the remaining parameters in the formulae (2.18.7)–(2.18.27), these expressions may be put in a form which exhibit the symmetry properties of the material. The formulae for isotropic bodies and for locally orthotropic or transversely isotropic materials may be recovered in this way.

2.19. Combined flexure and homogeneous strain of an aeolotropic cuboid: parametric forms

Parametric formulae for the combined uniform strain and flexure problem described in §§ 2.11, 2.17 may be derived by the procedure of the preceding section. In this case, the analysis is based upon the formulae (2.11.11) and (2.11.12) for e_{ij}, τ^{ij} when the material is incompressible; for compressible bodies the expressions (2.17.3), (2.17.4) are used. Alternatively, the parametric formulae may be deduced directly from the results of § 2.18 by using the limiting procedure of § 2.11. If the latter method is employed, we see from (2.11.3) that the operators

$$a\frac{\partial}{\partial a}, \quad ..., \quad d\frac{\partial}{\partial d}, \quad a\frac{\partial}{\partial c}, \quad ..., \quad d\frac{\partial}{\partial b},$$

are unaffected in the transition to the flexure problem, and, as before, a/ρ, c/ρ, λ/ρ, and $\partial/\partial\rho$ are replaced by a, c, $1/k$, and $\partial/\partial x_1$ respectively, primes being omitted after the limit is taken. We quote here the results for an incompressible body derived by this process.

For the deformation described by (2.11.6) and (2.11.9) the formulae

(2.18.10) and (2.18.13) for the stress components remain unchanged. In place of (2.18.11), however, we have the formula

$$\frac{\partial W}{\partial x_1} = \frac{\partial \tau^{11}}{\partial x_1}, \qquad (2.19.1)$$

or

$$\frac{\partial W}{\partial r} = \frac{\partial \tau^{11}}{\partial r}, \qquad (2.19.2)$$

corresponding to (2.11.14). The former of these relations may be inferred from (2.18.11) by introducing the limiting condition $K/h^2 \to 1$. From (2.19.2) we again derive the result

$$\tau^{11} = W(r) + W_0 \qquad (2.19.3)$$

obtained for the special cases $b = 0$ and $c = 0$ in § 2.11. Furthermore, remembering the incompressibility condition (2.11.9), we obtain from (2.19.2) and (2.18.10) the identity

$$2\left(a\frac{\partial W}{\partial a} + b\frac{\partial W}{\partial b}\right) = r\frac{\partial W}{\partial r} = \frac{r^2}{k}\frac{\partial W}{\partial x_1} = (2x_1 + K'/k)\frac{\partial W}{\partial x_1}, \qquad (2.19.4)$$

and this may be used, if desired, to modify the formulae (2.18.13) for τ^{22}, τ^{33}.

Results analogous to (2.18.15)–(2.18.19) for the resultant forces and couples across planes in the deformed body may be derived by making use of (2.18.13) and the incompressibility condition (2.11.9). The formula (2.18.15) for N_2 is unchanged and for M_2 we derive from (2.18.13), (2.19.3), and (2.19.4)

$$M_2 = l\int_{r_2}^{r_1} r^3 \tau^{22} \, dr = l\int_{r_2}^{r_1} r\left(\tau^{11} + r\frac{\partial W}{\partial r}\right) dr = \frac{l}{2}\left\{[r^2\tau^{11}]_{r=r_2}^{r=r_1} + \int_{r_2}^{r_1} r^2\frac{\partial W}{\partial r} \, dr\right\}$$

$$= \frac{l}{2}\left\{2[r_1^2 W(r_1) - r_2^2 W(r_2)] + (r_1^2 - r_2^2)W_0 - 2\int_{r_2}^{r_1} W(r) r \, dr\right\}. \qquad (2.19.5)$$

If the deformation can be maintained with the curved surfaces free from applied stress so that τ^{11} is zero when $r = r_1$ or $r = r_2$ we see from (2.18.15) that $N_2 = 0$ while from (2.19.3)

$$W(r_1) = W(r_2) = -W_0. \qquad (2.19.6)$$

The couple M_2 then becomes

$$M_2 = \frac{l}{2}\left\{2\int_{r_1}^{r_2} rW(r) \, dr - (r_1^2 - r_2^2)W_0\right\} = kl\left\{\int_A^{A_2} W(x_1) \, dx_1 - (A_1 - A_2)W_0\right\}, \qquad (2.19.7)$$

if we again make use of (2.11.9). The foregoing results are again independent of symmetries in the elastic material and apply for all of the crystal classes listed in Chapter I except for the restriction that W does not contain terms linear in either of the strains e_{12}, e_{13} alone. The formulae for the stresses and the resultant forces acting on the body may be put into forms which exhibit the symmetry properties of the material by the procedure indicated in § 2.18.

In the case of a body which is transversely isotropic with the x_3-axis as the axis of anisotropy, W is a function of the invariants I_1, I_2, K_1, and K_2, which may be derived by introducing the limiting conditions (2.11.8) into (2.1.13) and (2.1.14). Thus

$$\left.\begin{aligned} I_1 &= \frac{k^2}{r^2} + (a^2+b^2)r^2 + c^2 + d^2 \\ I_2 &= \frac{r^2}{k^2} + \frac{k^2}{r^2}\{(a^2+b^2)r^2 + c^2 + d^2\} \\ 2K_1 &= b^2 r^2 + d^2 - 1 \\ 4K_2 &= (abr^2 + cd)^2 \end{aligned}\right\} . \qquad (2.19.8)$$

The conditions (2.19.6) are then satisfied if

$$I_1(r_1) = I_1(r_2), \quad I_2(r_1) = I_2(r_2), \quad K_1(r_1) = K_1(r_2), \quad K_2(r_1) = K_2(r_2),$$
$$(2.19.9)$$

and these relations hold if

$$b = 0 \quad \text{and} \quad r_1^2 r_2^2/k^2 = (c^2+d^2)k^2 = 1/a^2. \qquad (2.19.10)$$

If the material is isotropic, so that W is a function only of I_1, and I_2, the conditions (2.19.10) may be replaced by

$$r_1^2 r_2^2/k^2 = (c^2+d^2)k^2 = 1/(a^2+b^2) \qquad (b \neq 0). \qquad (2.19.11)$$

2.20. Inversion and expansion of a spherical shell

The inflation of an isotropic incompressible sphere was considered in *T.E.* The analysis given there may also be used to discuss the following problem. A spherical shell is cut along a diametral plane. The shell is then turned inside out, after which the faces of the cut are rejoined so that the particles which were contiguous before the cut was made are so after it is rejoined. Uniform normal pressures are applied, if necessary, to the exterior and interior surfaces to maintain the spherical shape. The inversion problem has been discussed fairly fully by Ericksen† for an incompressible body. Here we indicate some general results for compressible materials.

† Loc. cit., p. 37.

2.20 GENERAL SOLUTIONS OF PROBLEMS

We assume a spherically symmetrical deformation in which a point of the shell initially at (ρ, θ, ϕ) moves to the point (r, θ, ϕ) in a system of spherical polar coordinates which has its origin at the centre of the shell; r is then a function of ρ. We assume that the external and internal radii of the shell before deformation are ρ_1, ρ_2 respectively, with corresponding values r_1, r_2 after deformation. Hence $\rho_1 > \rho_2$ but the relative magnitudes of r_1, r_2 depend on whether the problem is one of expansion or inversion.

We identify the curvilinear coordinates $(\theta_1, \theta_2, \theta_3)$ with the system of spherical polar coordinates (r, θ, ϕ) in the deformed body. The metric tensors G_{ij}, G^{ij} are given by

$$G_{ij} = \begin{bmatrix} 1 & 0 & 0 \\ 0 & r^2 & 0 \\ 0 & 0 & r^2 \sin^2\theta \end{bmatrix}, \quad G^{ij} = \begin{bmatrix} 1 & 0 & 0 \\ 0 & \dfrac{1}{r^2} & 0 \\ 0 & 0 & \dfrac{1}{r^2 \sin^2\theta} \end{bmatrix}, \quad G = r^4 \sin^2\theta, \quad (2.20.1)$$

and for g_{ij}, g^{ij} we have

$$g_{ij} = \begin{bmatrix} \dfrac{1}{r_\rho^2} & 0 & 0 \\ 0 & \rho^2 & 0 \\ 0 & 0 & \rho^2 \sin^2\theta \end{bmatrix}, \quad g^{ij} = \begin{bmatrix} r_\rho^2 & 0 & 0 \\ 0 & \dfrac{1}{\rho^2} & 0 \\ 0 & 0 & \dfrac{1}{\rho^2 \sin^2\theta} \end{bmatrix}, \quad g = (\rho^4/r_\rho^2)\sin^2\theta, \quad (2.20.2)$$

where r_ρ denotes $dr/d\rho$. The strain invariants and the tensor B^{ij} may now be found from (1.1.18) and (1.15.9). Thus

$$I_1 = r_\rho^2 + \frac{2r^2}{\rho^2}, \quad I_2 = \frac{r^4}{\rho^4} + \frac{2r^2 r_\rho^2}{\rho^2}, \quad I_3 = \frac{r^4 r_\rho^2}{\rho^4}, \quad (2.20.3)$$

$$\left. \begin{array}{c} B^{11} = \dfrac{2r^2 r_\rho^2}{\rho^2}, \qquad B^{22} = B^{33} \sin^2\theta = \dfrac{r_\rho^2}{\rho^2} + \dfrac{r^2}{\rho^4} \\[2pt] B^{12} = B^{23} = B^{31} = 0 \end{array} \right\}. \quad (2.20.4)$$

From (1.15.13), (2.20.1)–(2.20.4) we have

$$\left. \begin{array}{c} \tau^{11} = r_\rho^2 \Phi + \dfrac{2 r^2 r_\rho^2}{\rho^2} \Psi + p \\[4pt] \tau^{22} = \tau^{33} \sin^2\theta = \dfrac{1}{\rho^2} \Phi + \left(\dfrac{r^2}{\rho^4} + \dfrac{r_\rho^2}{\rho^2} \right) \Psi + \dfrac{p}{r^2} \\[4pt] \tau^{12} = \tau^{23} = \tau^{31} = 0 \end{array} \right\}, \quad (2.20.5)$$

and the equations of equilibrium reduce to

$$\frac{d\tau^{11}}{dr} + \frac{2(\tau^{11} - r^2\tau^{22})}{r} = 0,$$

or
$$\frac{d}{dr}(r^2\tau^{11}) = 2r^3\tau^{22}, \tag{2.20.6}$$

since the stresses τ^{11}, τ^{22} are independent of θ, ϕ. This is a differential equation for the unknown function $r(\rho)$.

The strain energy function W for a compressible isotropic body depends only upon the three invariants I_1, I_2, I_3 and these invariants are independent of θ, ϕ. Hence, from (1.15.7) and (2.20.3),

$$\frac{dW}{d\rho} = \left(r_\rho r_{\rho\rho} + \frac{2rr_\rho}{\rho^2} - \frac{2r^2}{\rho^3}\right)\frac{r^2 r_\rho}{\rho^2}\Phi +$$

$$+ \left(\frac{2r^3 r_\rho}{\rho^4} - \frac{2r^4}{\rho^5} + \frac{2rr_\rho^3}{\rho^2} + \frac{2r^2 r_\rho r_{\rho\rho}}{\rho^2} - \frac{2r^2 r_\rho^2}{\rho^3}\right)\frac{r^2 r_\rho}{\rho^2}\Psi +$$

$$+ \left(\frac{2r^3 r_\rho^3}{\rho^4} + \frac{r^4 r_\rho r_{\rho\rho}}{\rho^4} - \frac{2r^4 r_\rho^2}{\rho^5}\right)\frac{\rho^2}{r^2 r_\rho}p. \tag{2.20.7}$$

With the help of (2.20.5) this reduces to

$$\frac{dW}{d\rho} = \frac{r^2 r_{\rho\rho}}{\rho^2}\tau^{11} + \frac{2r^3 r_\rho}{\rho^3}(\rho r_\rho - r)\tau^{22}, \tag{2.20.8}$$

and, eliminating τ^{22} from (2.20.6) and (2.20.8), we have

$$\frac{dW}{d\rho} = \frac{r^2 r_{\rho\rho}}{\rho^2}\tau^{11} + \frac{r_\rho}{\rho^3}(\rho r_\rho - r)\frac{d}{dr}(r^2\tau^{11}) = \frac{1}{\rho^3}\frac{d}{d\rho}[(\rho r_\rho - r)r^2\tau^{11}]. \tag{2.20.9}$$

If the normal *pressures* exerted over the external and internal surfaces of the deformed shell are Π_1 and Π_2 respectively, then

$$[\tau^{11}]_{\rho=\rho_1} = -\Pi_1, \quad [\tau^{11}]_{\rho=\rho_2} = -\Pi_2, \tag{2.20.10}$$

and integration of (2.20.9) gives

$$\int_{\rho_1}^{\rho} \rho^3 \frac{dW}{d\rho}\, d\rho = (\rho r_\rho - r)r^2\tau^{11} + \Pi_1[(\rho r_\rho - r)r^2]_{\rho=\rho_1}, \tag{2.20.11}$$

and
$$\int_{\rho_1}^{\rho_2} \rho^3 \frac{dW}{d\rho}\, d\rho = \Pi_1[(\rho r_\rho - r)r^2]_{\rho=\rho_1} - \Pi_2[(\rho r_\rho - r)r^2]_{\rho=\rho_2}. \tag{2.20.12}$$

When equation (2.20.6) is solved for $r(\rho)$ then (2.20.11) gives the value of τ^{11}, the remaining stresses being given by (2.20.5). If the problem is that of the inflation of a sphere with zero pressure on the external surface, then $\Pi_1 = 0$, $\Pi_2 \neq 0$. If the spherical shell is cut, turned inside out and rejoined, and if it is possible to maintain the deformed body in the form of a spherical shell under zero internal and external pressure then, from (2.20.12),

$$\int_{\rho_2}^{\rho_1} \rho^3 \frac{dW}{d\rho} d\rho = 0. \tag{2.20.13}$$

Further discussion is required to decide whether, or not, the condition (2.20.13) can be satisfied.

When the material is incompressible we have $I_3 = 1$ or

$$r^2 r_\rho = \rho^2, \qquad \rho^3 = r^3 + K, \tag{2.20.14}$$

where K is a constant. The relation (2.20.9) then reduces to

$$\rho^3 \frac{dW}{d\rho} = K \frac{d\tau^{11}}{d\rho}, \tag{2.20.15}$$

which yields by integration

$$\tau^{11} = \frac{1}{K} \int_{\rho_0}^{\rho} \rho^3 \frac{dW}{d\rho} d\rho + \tau_0^{11}, \tag{2.20.16}$$

where τ_0^{11} is the value of τ^{11} at a given distance $\rho = \rho_0$ from the origin. A fuller analysis of this problem is given in *T.E.*

It can be shown that the results (2.20.9)–(2.20.16) continue to apply if the material is locally transversely isotropic, with the axis of anisotropy coinciding with the radial direction at each point.†

2.21. Generalizations of the shear problem‡

We shall now consider deformations of an isotropic incompressible material in which points of the unstrained body are displaced parallel to the x_3-axis through distances which are dependent only upon their positions in the x_1, x_2-plane. This problem includes as special cases the examples of simple shear and the shear of a cylindrical annulus examined in *T.E.*§ The problem may be made somewhat more general by superposing upon the shear a uniform finite extension parallel to the x_3-axis.

† J. E. Adkins, *Proc. R. Soc.* A, **229** (1955), 119.
‡ J. E. Adkins, *Proc. Camb. phil. Soc. math. phys. Sci.* **50** (1954), 334.
§ *T.E.*, chap. III.

We therefore suppose the deformation to be described by

$$y_1, y_2, y_3 = \lambda x_1, \lambda x_2, x_3/\lambda^2 + f(\lambda x_1, \lambda x_2)$$

or

$$x_1, x_2, x_3 = y_1/\lambda, y_2/\lambda, \lambda^2\{y_3 - f(y_1, y_2)\}$$

(2.21.1)

where λ is a constant and f is a function of its arguments which is as yet undetermined. Choosing the moving coordinate system θ_i to coincide with the rectangular cartesian coordinates y_i in the deformed body, so that $\theta_i = y_i$, we have

$$G_{ij} = G^{ij} = \delta_{ij}, \qquad G = 1, \tag{2.21.2}$$

$$\frac{\partial x^i}{\partial \theta^j} = \begin{bmatrix} \frac{1}{\lambda} & 0 & 0 \\ 0 & \frac{1}{\lambda} & 0 \\ -\lambda^2 \frac{\partial f}{\partial y_1} & -\lambda^2 \frac{\partial f}{\partial y_2} & \lambda^2 \end{bmatrix}, \tag{2.21.3}$$

$$g_{ij} = \begin{bmatrix} \frac{1}{\lambda^2} + \lambda^4 \left(\frac{\partial f}{\partial y_1}\right)^2 & \lambda^4 \frac{\partial f}{\partial y_1} \frac{\partial f}{\partial y_2} & -\lambda^4 \frac{\partial f}{\partial y_1} \\ \lambda^4 \frac{\partial f}{\partial y_1} \frac{\partial f}{\partial y_2} & \frac{1}{\lambda^2} + \lambda^4 \left(\frac{\partial f}{\partial y_2}\right)^2 & -\lambda^4 \frac{\partial f}{\partial y_2} \\ -\lambda^4 \frac{\partial f}{\partial y_1} & -\lambda^4 \frac{\partial f}{\partial y_2} & \lambda^4 \end{bmatrix}, \tag{2.21.4}$$

$$g^{ij} = \begin{bmatrix} \lambda^2 & 0 & \lambda^2 \frac{\partial f}{\partial y_1} \\ 0 & \lambda^2 & \lambda^2 \frac{\partial f}{\partial y_2} \\ \lambda^2 \frac{\partial f}{\partial y_1} & \lambda^2 \frac{\partial f}{\partial y_2} & \frac{1}{\lambda^4} + \lambda^2\left\{\left(\frac{\partial f}{\partial y_1}\right)^2 + \left(\frac{\partial f}{\partial y_2}\right)^2\right\} \end{bmatrix}, \tag{2.21.5}$$

and the incompressibility condition $G = g = 1$ is satisfied identically. Also, from (1.1.18),

$$\left. \begin{array}{l} I_1 = 2\lambda^2 + \dfrac{1}{\lambda^4} + \lambda^2 I \\[6pt] I_2 = \dfrac{2}{\lambda^2} + \lambda^4 + \lambda^4 I \\[6pt] I = \left(\dfrac{\partial f}{\partial y_1}\right)^2 + \left(\dfrac{\partial f}{\partial y_2}\right)^2 \end{array} \right\}. \tag{2.21.6}$$

2.21 GENERAL SOLUTIONS OF PROBLEMS

For this type of deformation the two strain invariants are not independent (since λ is regarded as a known constant) and the strain energy function may therefore be considered as a function of the single quantity I. From (2.21.6) we may then write

$$\frac{dW}{dI} = \lambda^2 \left(\frac{\partial W}{\partial I_1} + \lambda^2 \frac{\partial W}{\partial I_2} \right) = \tfrac{1}{2}\lambda^2 (\Phi + \lambda^2 \Psi). \qquad (2.21.7)$$

Combining these results with (1.15.9) we have

$$\left.\begin{aligned}
B^{11} &= \lambda^4 + \frac{1}{\lambda^2} + \lambda^4 \left(\frac{\partial f}{\partial y_2} \right)^2, & B^{22} &= \lambda^4 + \frac{1}{\lambda^2} + \lambda^4 \left(\frac{\partial f}{\partial y_1} \right)^2 \\
B^{33} &= \frac{2}{\lambda^2} + \lambda^4 I, & B^{12} &= -\lambda^4 \frac{\partial f}{\partial y_1} \frac{\partial f}{\partial y_2} \\
B^{13} &= \lambda^4 \frac{\partial f}{\partial y_1}, & B^{23} &= \lambda^4 \frac{\partial f}{\partial y_2}
\end{aligned}\right\}. \qquad (2.21.8)$$

If now we replace p by

$$p - \lambda^2 \Phi - \left(\lambda^4 + \frac{1}{\lambda^2} \right) \Psi, \qquad (2.21.9)$$

and denote by σ_{ij} the physical components of stress in the rectangular cartesian system y_i, the covariant, mixed, and contravariant components then being equal, we obtain from (1.15.13)

$$\left.\begin{aligned}
\sigma_{11} &= \lambda^4 \left(\frac{\partial f}{\partial y_2} \right)^2 \Psi + p \\
\sigma_{22} &= \lambda^4 \left(\frac{\partial f}{\partial y_1} \right)^2 \Psi + p \\
\sigma_{33} &= 2 \left\{ I - \left(1 - \frac{1}{\lambda^6} \right) \right\} \frac{dW}{dI} + p \\
\sigma_{12} &= -\lambda^4 \frac{\partial f}{\partial y_1} \frac{\partial f}{\partial y_2} \Psi \\
\sigma_{13} &= 2 \frac{\partial f}{\partial y_1} \frac{dW}{dI}, \quad \sigma_{23} = 2 \frac{\partial f}{\partial y_2} \frac{dW}{dI}
\end{aligned}\right\}. \qquad (2.21.10)$$

We observe that if W is regarded as a function of $\partial f/\partial y_1$, $\partial f/\partial y_2$, the expressions for σ_{13} and σ_{23} may be written as

$$\sigma_{13} = \partial W / \partial (\partial f/\partial y_1), \qquad \sigma_{23} = \partial W / \partial (\partial f/\partial y_2),$$

a result which is also valid for some aeolotropic materials.[†]

[†] J. E. Adkins, *Proc. R. Soc.* A, **231** (1955), 75.

Remembering that I_1, I_2, I, and W are functions only of y_1, y_2, the equations of equilibrium with (2.21.10) yield

$$\left.\begin{aligned}\lambda^4\left\{\frac{\partial}{\partial y_1}\left[\left(\frac{\partial f}{\partial y_2}\right)^2\Psi\right]-\frac{\partial}{\partial y_2}\left[\frac{\partial f}{\partial y_1}\frac{\partial f}{\partial y_2}\Psi\right]\right\}+\frac{\partial p}{\partial y_1}&=0\\ \lambda^4\left\{\frac{\partial}{\partial y_2}\left[\left(\frac{\partial f}{\partial y_1}\right)^2\Psi\right]-\frac{\partial}{\partial y_1}\left[\frac{\partial f}{\partial y_1}\frac{\partial f}{\partial y_2}\Psi\right]\right\}+\frac{\partial p}{\partial y_2}&=0\\ 2\left\{\frac{\partial}{\partial y_1}\left[\frac{\partial f}{\partial y_1}\frac{dW}{dI}\right]+\frac{\partial}{\partial y_2}\left[\frac{\partial f}{\partial y_2}\frac{dW}{dI}\right]\right\}+\frac{\partial p}{\partial y_3}&=0\end{aligned}\right\}. \quad (2.21.11)$$

Since all the terms which occur in (2.21.11) are independent of y_3, it follows that p must take the form

$$p = \kappa y_3 + F(y_1, y_2), \quad (2.21.12)$$

where κ is a constant. Moreover, by eliminating p from the first and second of equations (2.21.11) we find that these are consistent if

$$\frac{\partial^2}{\partial y_1 \partial y_2}\left\{\left[\left(\frac{\partial f}{\partial y_1}\right)^2-\left(\frac{\partial f}{\partial y_2}\right)^2\right]\Psi\right\}-\left(\frac{\partial^2}{\partial y_1^2}-\frac{\partial^2}{\partial y_2^2}\right)\left[\frac{\partial f}{\partial y_1}\frac{\partial f}{\partial y_2}\Psi\right]=0. \quad (2.21.13)$$

The third of equations (2.21.11) and equation (2.21.13) are thus two differential equations for $f(y_1, y_2)$ which must be satisfied simultaneously. The solution of these, if it exists, subject to specified boundary conditions, when introduced into the first and second of (2.21.11) yields a pair of equations for the determination of p. Before proceeding further it is, in general, necessary to know the form of W as a function of I_1, I_2. It may, however, be observed that (2.21.13) is satisfied identically if

$$f = \phi(k_1 y_1 + k_2 y_2) \quad \text{or} \quad f = \psi(y_1^2 + y_2^2), \quad (2.21.14)$$

where k_1 and k_2 are constants and ϕ and ψ are arbitrary functions of the arguments $k_1 y_1 + k_2 y_2$ and $y_1^2 + y_2^2$ respectively. The third of equations (2.21.11) then becomes an ordinary differential equation for the determination of ϕ or ψ.

A considerable simplification of the theory results if the strain energy function is linear in the invariants I_1 and I_2 and we shall henceforth assume that W has the Mooney form (1.14.3); Φ and Ψ then take the constant values $2C_1$ $2C_2$ respectively. Putting

$$\kappa = -8A\lambda^2(C_1 + \lambda^2 C_2) \quad (2.21.15)$$

in (2.21.12), where A is a constant, and remembering (2.21.7), we find that the last of equations (2.21.11) reduces to the Poisson equation

$$\frac{\partial^2 f}{\partial y_1^2}+\frac{\partial^2 f}{\partial y_2^2} = 4A, \qquad (2.21.16)$$

and this is sufficient to ensure that (2.21.13) is satisfied identically. With (2.21.15), the first and second of equations (2.21.11) yield

$$p = \lambda^4 C_2\left[8Af-\left(\frac{\partial f}{\partial y_1}\right)^2-\left(\frac{\partial f}{\partial y_2}\right)^2\right]-8A\lambda^2(C_1+\lambda^2 C_2)y_3+B, \quad (2.21.17)$$

where B is a constant of integration. Introducing this value of p into (2.21.10) and making use of (2.21.6), we obtain for the stress components

$$\left.\begin{aligned}\sigma_{11} &= -\lambda^4 C_2\left[\left(\frac{\partial f}{\partial y_1}\right)^2-\left(\frac{\partial f}{\partial y_2}\right)^2\right]+q \\ \sigma_{22} &= \lambda^4 C_2\left[\left(\frac{\partial f}{\partial y_1}\right)^2-\left(\frac{\partial f}{\partial y_2}\right)^2\right]+q \\ \sigma_{33} &= \lambda^2(2C_1+\lambda^2 C_2)\left[\left(\frac{\partial f}{\partial y_1}\right)^2+\left(\frac{\partial f}{\partial y_2}\right)^2\right]+q-D \\ \sigma_{12} &= -2\lambda^4 C_2\frac{\partial f}{\partial y_1}\frac{\partial f}{\partial y_2} \\ \sigma_{13} &= 2\lambda^2(C_1+\lambda^2 C_2)\frac{\partial f}{\partial y_1} \\ \sigma_{23} &= 2\lambda^2(C_1+\lambda^2 C_2)\frac{\partial f}{\partial y_2}\end{aligned}\right\}, \quad (2.21.18)$$

where

$$\left.\begin{aligned}q &= 8A\lambda^2[\lambda^2 C_2 f-(C_1+\lambda^2 C_2)y_3]+B \\ D &= 2\lambda^2\left(1-\frac{1}{\lambda^6}\right)(C_1+\lambda^2 C_2)\end{aligned}\right\}. \quad (2.21.19)$$

When (2.21.16) has been solved for f, subject to specified boundary conditions, the stresses are obtained by substitution of the result into (2.20.18).

2.22. Generalized shear: complex variable

To obtain solutions of (2.21.16) it is often convenient to introduce the complex variables

$$z = y_1+iy_2, \qquad \bar{z} = y_1-iy_2. \qquad (2.22.1)$$

With the resulting coordinate system $(\theta_1, \theta_2, \theta_3) = (z, \bar{z}, y_3)$ we may associate the system of complex stresses T^{ij} defined by†

$$\left.\begin{aligned} T^{11} &= \bar{T}^{22} = \sigma_{11} - \sigma_{22} + 2i\sigma_{12} \\ T^{12} &= \sigma_{11} + \sigma_{22} \\ T^{13} &= \bar{T}^{23} = \sigma_{13} + i\sigma_{23} \\ T^{33} &= \sigma_{33} \end{aligned}\right\}. \qquad (2.22.2)$$

Since
$$\frac{\partial}{\partial y_1} = \frac{\partial}{\partial z} + \frac{\partial}{\partial \bar{z}}, \qquad \frac{\partial}{\partial y_2} = i\left(\frac{\partial}{\partial z} - \frac{\partial}{\partial \bar{z}}\right), \qquad (2.22.3)$$

we then have, from (2.21.18),

$$\left.\begin{aligned} T^{11} &= \bar{T}^{22} = -8\lambda^4 C_2 \left(\frac{\partial f}{\partial \bar{z}}\right)^2 \\ T^{12} &= 2q \\ T^{13} &= \bar{T}^{23} = 4\lambda^2 (C_1 + \lambda^2 C_2) \frac{\partial f}{\partial \bar{z}} \\ T^{33} &= 4\lambda^2 (2C_1 + \lambda^2 C_2) \frac{\partial f}{\partial z} \frac{\partial f}{\partial \bar{z}} + q - D \end{aligned}\right\}, \qquad (2.22.4)$$

while (2.21.16) becomes
$$\frac{\partial^2 f}{\partial z \, \partial \bar{z}} = A. \qquad (2.22.5)$$

The solution of this equation may be written

$$f(z, \bar{z}) = Az\bar{z} + \Omega(z) + \bar{\Omega}(\bar{z}), \qquad (2.22.6)$$

since f is real, where $\Omega(z)$ is an arbitrary function of z.

Suppose now that the body is a cylinder with its generators parallel to the y_3-axis, its cross-section in the deformed state being defined by a region R in the y_1, y_2-plane. For single-valued stresses and displacements it follows from (2.22.6), (2.21.1), and (2.22.4) that

$$[\Omega(z) + \bar{\Omega}(\bar{z})]_C = 0, \qquad [\Omega'(z)]_C = 0, \qquad (2.22.7)$$

where $[\ ]_C$ denotes the change in value of the quantity inside the brackets during a complete circuit of a closed contour C lying entirely inside the region R. If the region R is finite and simply connected, we assume that $\Omega(z)$ is holomorphic in the open region R and that $\Omega(z)$ and $\Omega'(z)$ are continuous on to the boundary. The displacements and stresses are then single-valued. When R is multiply-connected,

† See § 3.5 or T.E., § 6.4.

we suppose it to be bounded by one or more smooth non-intersecting contours $L_0, L_1, \ldots, L_n$ of which L_0 surrounds all the others (Fig. 2.1). The contour L_0 may be absent (i.e., it may be reducible to the point at infinity) in which case the region R will be infinite with the contours $L_1, L_2, \ldots, L_n$ as internal boundaries. If we require the stresses and displacements to be single-valued, we assume, in view of (2.22.7), that

$$\Omega(z) = \sum_{k=1}^{n} a_k \log(z-z_k) + \Omega_0(z), \qquad (2.22.8)$$

where a_k are real constants, the point z_k is inside the contour L_k (and therefore outside R), and where $\Omega_0(z)$ is holomorphic in R except possibly at infinity. The displacement and stresses are then single-valued.

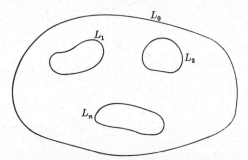

Fig. 2.1. Multiply-connected region.

2.23. Generalized shear: conformal transformations

The usefulness of the complex variable solution (2.22.6) may be extended with the aid of a conformal transformation. If

$$\zeta = \xi + i\eta, \qquad \bar{\zeta} = \xi - i\eta, \qquad (2.23.1)$$

where (ξ, η) is a system of orthogonal curvilinear coordinates in the y_1, y_2-plane, the regular function

$$z = z(\zeta) \qquad (2.23.2)$$

defines a conformal mapping between the systems z, ζ. For this transformation

$$\frac{dz}{d\zeta} = Je^{i\alpha}, \qquad (2.23.3)$$

where J and α are real quantities given by

$$J^2 = \frac{dz}{d\zeta}\frac{d\bar{z}}{d\bar{\zeta}}, \qquad e^{-2i\alpha} = \frac{d\zeta}{dz}\frac{d\bar{z}}{d\bar{\zeta}}. \qquad (2.23.4)$$

The solution (2.22.6) for f may then be written
$$f(z,\bar{z}) = f_0(\zeta,\bar{\zeta}) = A z(\zeta)\bar{z}(\bar{\zeta}) + \omega(\zeta) + \bar{\omega}(\bar{\zeta}), \qquad (2.23.5)$$
where
$$\omega(\zeta) = \Omega[z(\zeta)], \qquad (2.23.6)$$
and the conditions (2.22.7) for single-valued stresses and displacements become
$$[\omega(\zeta)+\bar{\omega}(\bar{\zeta})]_C = 0, \qquad \left[\frac{d\omega(\zeta)}{d\zeta}\frac{d\zeta}{dz}\right]_C = 0. \qquad (2.23.7)$$

If we use the suffixes ξ, η, 3 to distinguish the physical components of stress referred to the orthogonal system (ξ, η, y_3) we have the relations†

$$\left.\begin{aligned}\sigma_{\xi\xi}-\sigma_{\eta\eta}+2i\sigma_{\xi\eta} &= e^{-2i\alpha}T^{11}\\ \sigma_{\xi\xi}+\sigma_{\eta\eta} &= T^{12}\\ \sigma_{\xi 3}+i\sigma_{\eta 3} &= e^{-i\alpha}T^{13}\\ \sigma_{33} &= T^{33}\end{aligned}\right\}, \qquad (2.23.8)$$

and these, with (2.22.4) and (2.23.3), yield

$$\left.\begin{aligned}\sigma_{\xi\xi}-\sigma_{\eta\eta}+2i\sigma_{\xi\eta} &= -8\lambda^4 C_2 M^2 \left(\frac{\partial f_0}{\partial\bar{\zeta}}\right)^2\\ \sigma_{\xi\xi}+\sigma_{\eta\eta} &= 2q\\ \sigma_{\xi 3}+i\sigma_{\eta 3} &= 4M\lambda^2(C_1+\lambda^2 C_2)\frac{\partial f_0}{\partial\bar{\zeta}}\\ \sigma_{33} &= 4M^2\lambda^2(2C_1+\lambda^2 C_2)\frac{\partial f_0}{\partial\zeta}\frac{\partial f_0}{\partial\bar{\zeta}}+q-D\end{aligned}\right\}, \qquad (2.23.9)$$

where
$$M = \frac{1}{J} = \left(\frac{d\zeta}{dz}\frac{d\bar{\zeta}}{d\bar{z}}\right)^{\frac{1}{2}}. \qquad (2.23.10)$$

These stresses may be expressed in terms of the complex potential function $\omega(\zeta)$ by means of (2.23.5).

In some instances it is desirable to know the resultant longitudinal force F acting in the direction of the y_3-axis on a cylindrical boundary surface defined by a curve C in the plane $y_3 = $ constant and the normals to this plane through C. We choose the coordinate system θ_i so that $\theta_3 = y_3$ and the θ_1, θ_2 curves form a curvilinear system in the y_1, y_2-plane, this choice being identical with that of § 3.1. If ds is an element

† See, for example, T.E., § 6.5, for a fuller treatment.

of the curve C, we obtain with the help of (1.1.24)

$$F = \int \mathbf{t}\cdot\mathbf{G}_3 \, ds = \int n_\alpha \tau^{\alpha 3} \, ds = \int \sqrt{G}(\tau^{13} \, d\theta^2 - \tau^{23} \, d\theta^1), \qquad (2.23.11)$$

where
$$\sqrt{G} = \left|\frac{\partial y^i}{\partial \theta^j}\right| = \left|\frac{\partial y^\alpha}{\partial \theta^\beta}\right|.$$

Greek indices are restricted to take the values 1, 2, and we have used the formulae

$$n_1 = \sqrt{G}\frac{d\theta^2}{ds}, \qquad n_2 = -\sqrt{G}\frac{d\theta^1}{ds},$$

which are derived by a tensor transformation from the components dy_2/ds, $-dy_1/ds$ of the normal $\mathbf{n}$ to the curve C in the cartesian system y_α.†

If now we allow the curvilinear reference frame to coincide with the complex system z, $\bar{z}$ so that $\sqrt{G} = \tfrac{1}{2}i$, equations (2.22.4), (2.23.5), and (2.23.11) yield

$$\left.\begin{aligned}
F &= \tfrac{1}{2}i\int (T^{13}\, d\bar{z} - T^{23}\, dz) \\
&= 2i\lambda^2(C_1+\lambda^2 C_2)\int \left(\frac{\partial f}{\partial \bar{z}}\, d\bar{z} - \frac{\partial f}{\partial z}\, dz\right) \\
&= 2i\lambda^2(C_1+\lambda^2 C_2)\int \left(\frac{\partial f_0}{\partial \bar{\zeta}}\, d\bar{\zeta} - \frac{\partial f_0}{\partial \zeta}\, d\zeta\right) \\
&= 2i\lambda^2(C_1+\lambda^2 C_2)\int \{A[z(\zeta)\bar{z}'(\bar{\zeta})\, d\bar{\zeta} - z'(\zeta)\bar{z}(\bar{\zeta})\, d\zeta] + \\
&\qquad\qquad + \bar{\omega}'(\bar{\zeta})\, d\bar{\zeta} - \omega'(\zeta)\, d\zeta\}
\end{aligned}\right\} \quad (2.23.12)$$

If the curve C coincides with the arc of the curve $\xi = \xi_1$ lying between $\eta = \eta_1$ and $\eta = \eta_2$, we obtain by putting $d\zeta = -d\bar{\zeta} = i\, d\eta$ in (2.23.12)

$$F = F_\xi = 2\lambda^2(C_1+\lambda^2 C_2)\int_{\eta_2}^{\eta_1}[A\{z(\zeta)\bar{z}'(\bar{\zeta}) + z'(\zeta)\bar{z}(\bar{\zeta})\} + \\
\qquad\qquad + \omega'(\zeta) + \bar{\omega}'(\bar{\zeta})]_{\xi=\xi_1} d\eta. \qquad (2.23.13)$$

Similarly, if the boundary surface is defined by the arc of the curve $\eta = \eta_1$ lying between $\xi = \xi_1$ and $\xi = \xi_2$, we have $d\zeta = d\bar{\zeta} = d\xi$ and (2.23.12) yields

$$F = F_\eta = 2i\lambda^2(C_1+\lambda^2 C_2)\int_{\xi_2}^{\xi_1}[A\{z(\zeta)\bar{z}'(\bar{\zeta}) - z'(\zeta)\bar{z}(\bar{\zeta})\} + \\
\qquad\qquad + \bar{\omega}'(\bar{\zeta}) - \omega'(\zeta)]_{\eta=\eta_1} d\xi. \qquad (2.23.14)$$

† T.E., p. 186.

2.24. Cylindrical annuli

The class of generalized shear problem in which the displacement function f takes constant values over closed cylindrical surfaces with generators parallel to the x_3-axis, one of which completely encloses the other, is of some practical importance. When $\lambda = 1$ this problem represents an approximation to the situation which arises during the longitudinal shear of cylindrical rubber bush mountings. In these mountings the curved cylindrical surfaces are bonded to metal cylinders, which are effectively rigid, and the deformation is produced by the displacement of one of these metal cylinders in the longitudinal direction relative to the other.

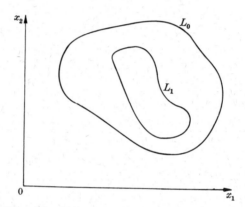

FIG. 2.2. Cross section of a cylindrical annulus.

Fig. 2.2 represents a section of a body of this type by a plane normal to the x_3-axis, the elastic material being contained between the closed surfaces L_0, L_1. Since the deformation arises from the displacement of all points of L_0 relative to those of L_1, the solution of the problem merely depends upon that of the Poisson equation (2.22.5) subject to the boundary conditions

$$f = K \text{ on } L_0, \qquad f = 0 \text{ on } L_1, \qquad (2.24.1)$$

K being a constant. When the boundaries L_0, L_1 are irregular in shape, this problem may be solved numerically by relaxation methods.† We consider here some examples involving regular boundaries for which the solution can be derived analytically, but allow λ to be different from unity for greater generality.

† R. V. Southwell, *Relaxation Methods in Theoretical Physics*. (Clarendon Press, Oxford, 1946.)

(i) *Concentric circular cylinders*

We introduce polar coordinates (r, θ) in the z-plane by means of the relations
$$z = y_1 + iy_2 = re^{i\theta}, \tag{2.24.2}$$
and suppose the elastic material to be bounded initially by concentric circular cylinders of radii r_1/λ, r_2/λ so that the boundaries of the deformed body become the surfaces $r = r_1$, $r = r_2$. If the outer surface $r = r_1$ is displaced through a distance K parallel to the x_3-axis relative to the inner surface, the boundary conditions for f become
$$\left.\begin{array}{ll} f = K & \text{when} \quad r = r_1 \\ f = 0 & \text{when} \quad r = r_2 \end{array}\right\}. \tag{2.24.3}$$
These conditions together with (2.22.7) can be satisfied if in (2.22.6) we put
$$\Omega(z) = b \log z + c, \tag{2.24.4}$$
where
$$\left.\begin{array}{l} b = \dfrac{K - A(r_1^2 - r_2^2)}{2 \log(r_1/r_2)} \\[2mm] c = \dfrac{A(r_1^2 \log r_2 - r_2^2 \log r_1) - K \log r_2}{2 \log(r_1/r_2)} \end{array}\right\}. \tag{2.24.5}$$

Equation (2.22.6) then yields
$$f = Ar^2 + 2b \log r + 2c. \tag{2.24.6}$$

The stress components σ_{ij} can be determined by combining (2.24.6) and (2.24.2) with (2.22.4) and (2.22.2); the physical components of stress referred to the reference frame (r, θ, y_3) can then be obtained by a rotation of the axes of reference at the point (r, θ, y_3) through an angle θ about the y_3-axis so that they coincide with the directions of the (r, θ, y_3) coordinate lines at that point. Alternatively, these components may be evaluated by the method of § 2.23 by observing that under the conformal transformation
$$\zeta = \log z = \log(re^{i\theta}), \tag{2.24.7}$$
we have
$$\xi = \log r, \quad \eta = \theta, \tag{2.24.8}$$
so that the physical components of stress in the coordinate system (ξ, η, y_3) are also the required physical components in the polar coordinate system (r, θ, y_3). In the notation of § 2.23 we have
$$J = r, \quad \alpha = 0, \tag{2.24.9}$$
and, since f is now a function of r ($= e^{\frac{1}{2}(\zeta+\bar\zeta)}$) alone,
$$\frac{\partial f}{\partial \zeta} = \frac{\partial f}{\partial \bar\zeta} = \frac{1}{2} r \frac{df}{dr}.$$

From (2.23.9) we thus obtain

$$\left.\begin{aligned}\sigma_{rr} &= -\lambda^4 C_2 \left(\frac{df}{dr}\right)^2 + q \\ \sigma_{\theta\theta} &= \lambda^4 C_2 \left(\frac{df}{dr}\right)^2 + q \\ \sigma_{33} &= \lambda^2(2C_1 + \lambda^2 C_2)\left(\frac{df}{dr}\right)^2 + q - D \\ \sigma_{r3} &= 2\lambda^2(C_1 + \lambda^2 C_2)\frac{df}{dr} \\ \sigma_{r\theta} &= \sigma_{\theta 3} = 0 \end{aligned}\right\}, \qquad (2.24.10)$$

the suffixes ξ, η being replaced by r, θ respectively. The complete expressions for the stress components may now be obtained by combining these formulae with (2.24.5) and (2.24.6).

When $A = 0$ we have from (2.21.19), (2.24.5), and (2.24.6)

$$\left.\begin{aligned} q &= B, \quad b = K/\{2\log(r_1/r_2)\}, \quad c = -b\log r_2 \\ f &= \frac{K\log(r/r_2)}{\log(r_1/r_2)} = 2b\log(r/r_2) \end{aligned}\right\}, \qquad (2.24.11)$$

and (2.24.10) yield for the non-zero stresses

$$\left.\begin{aligned} \sigma_{rr} &= -4b^2\lambda^4 C_2/r^2 + B \\ \sigma_{\theta\theta} &= 4b^2\lambda^4 C_2/r^2 + B \\ \sigma_{33} &= 4b^2\lambda^2(2C_1 + \lambda^2 C_2)/r^2 + B - D \\ \sigma_{r3} &= 4b\lambda^2(C_1 + \lambda^2 C_2)/r \end{aligned}\right\}. \qquad (2.24.12)$$

This is a particular case of the problem examined in § 3.7 of *T.E.* in which the stresses are independent of the longitudinal coordinate y_3.

The resultant longitudinal force F per unit length of deformed surface acting parallel to the x_3-axis on the cylindrical surface $r = r_1$ is

$$F = \int_0^{2\pi} [\sigma_{r3} r]_{r=r_1}\, d\theta = 2\pi r_1 [\sigma_{r3}]_{r=r_1}. \qquad (2.24.13)$$

We observe that this is equal to the resultant force F on the inner cylinder $r = r_2$ only when $A = 0$; otherwise the difference in these forces is balanced by the difference in the normal surface tractions over the ends $y_3 = $ constant of the deformed annulus.

(ii) *Conformal transformations*

When the conformal transformation (2.23.2) is such that the lines $\xi = $ constant represent simple closed contours in the planes $x_3 = $ constant (or $y_3 = $ constant) the inequalities $\xi_1 \geqslant \xi \geqslant \xi_2$ define an annulus of shape dependent upon (2.23.2). The conditions

$$\left. \begin{array}{ll} f = K & \text{when} \quad \xi = \xi_1 \\ f = 0 & \text{when} \quad \xi = \xi_2 \end{array} \right\} \quad (2.24.14)$$

then define a boundary value problem of the type described by (2.24.1). We confine attention to the case where $A = 0$, so that the stresses (2.23.9) are independent of y_3 and the resultant forces parallel to the y_3-axis on the surfaces $\xi = \xi_1$ and $\xi = \xi_2$ are equal and opposite. The boundary conditions (2.24.14) with (2.23.7) can be satisfied if, in (2.23.5), we put

$$\omega(\zeta) = b\zeta + c, \quad (2.24.15)$$

where
$$b = \frac{K}{2(\xi_1 - \xi_2)}, \quad c = -\frac{K\xi_2}{2(\xi_1 - \xi_2)}. \quad (2.24.16)$$

Hence
$$f_0 = b(\zeta + \bar{\zeta}) + 2c = 2(b\xi + c), \quad (2.24.17)$$

and the stresses (2.23.9) reduce to

$$\left. \begin{array}{l} \sigma_{\xi\xi} = -4\lambda^4 C_2 M^2 b^2 + B \\ \sigma_{\eta\eta} = 4\lambda^4 C_2 M^2 b^2 + B \\ \sigma_{33} = 4\lambda^2(2C_1 + \lambda^2 C_2)M^2 b^2 + B - D \\ \sigma_{\xi 3} = 4\lambda^2(C_1 + \lambda^2 C_2)Mb \\ \sigma_{\xi\eta} = \sigma_{\eta 3} = 0 \end{array} \right\}. \quad (2.24.18)$$

Also, from (2.23.13), the resultant longitudinal force on a portion of the surface $\xi = \xi_1$ (or $\xi = \xi_2$) lying between $\eta = \eta_1$ and $\eta = \eta_2$ becomes
$$F = 4\lambda^2(C_1 + \lambda^2 C_2)b(\eta_1 - \eta_2). \quad (2.24.19)$$

If, in (2.23.2), $z(\zeta)$ is a periodic function of η with period 2π so that $z(\zeta) = z(\zeta + 2\pi i)$, any closed contour $\xi = $ constant is described once as η increases from 0 to 2π. The resultant longitudinal force F on either of the boundary surfaces $\xi = \xi_1$ or $\xi = \xi_2$ is then given in magnitude by
$$F = 8\pi\lambda^2 b(C_1 + \lambda^2 C_2). \quad (2.24.20)$$

The solution of many boundary value problems can thus be reduced to the determination of a suitable conformal transformation to describe the cylindrical boundaries of the elastic material and the evaluation of

M once this transformation has been found. Under the transformation

$$z = a \cosh \zeta,$$

for example, the lines $\xi = $ constant represent ellipses in the planes $x_3 = $ constant in the undeformed body with foci at $(\pm a/\lambda, 0)$ and major and minor axes $(2a/\lambda)\cosh \xi$, $(2a/\lambda)\sinh \xi$ respectively. Equations (2.24.14) to (2.24.20) therefore yield the solution to the problem in which an elliptic annulus bounded by the surfaces $\xi = \xi_1$, $\xi = \xi_2$ is subjected to a shearing deformation in which the surface $\xi = \xi_1$ is displaced through a distance K in a direction parallel to the x_3-axis relative to the inner surface $\xi = \xi_2$, this distance being measured relative to the deformed state, that is, after the uniform extension λ has been applied. The value of M required is now

$$M = \frac{1}{a}\left[\frac{2}{\cosh 2\xi - \cos 2\eta}\right]^{\frac{1}{2}}.$$

Similarly, the transformation

$$z = a \coth \tfrac{1}{2}\zeta$$

describes the corresponding problem for an annulus bounded by eccentrically situated circular cylinders of radii in the undeformed body r_1, r_2 and axes at distance d apart, where

$$r_1 = (a/\lambda)\operatorname{cosech} \xi_1, \qquad r_2 = (a/\lambda)\operatorname{cosech} \xi_2,$$

$$d = (a/\lambda)(\coth \xi_2 - \coth \xi_1).$$

In this case the value of M to be inserted in equations (2.24.18) becomes

$$M = \frac{1}{a}(\cosh \xi - \cos \eta).$$

III
FINITE PLANE STRAIN

In classical elasticity a marked reduction in complexity occurs both in the formulation of the theory and in the solution of particular problems when attention is confined to two-dimensional deformations. Corresponding simplifying features also appear in the present theory under the assumption of plane strain.† Provided that the normal to the planes of movement can be regarded as a principal direction for the elastic material, the theory can be formulated so that the strain components, the invariants, the strain energy function, and the stress components all become functions of two only of the coordinate variables. Under these circumstances the equations of equilibrium can be satisfied by the introduction of an Airy stress function which is analogous to that of classical elasticity but is related to points in the deformed body.

The restriction to two-dimensional deformations makes it possible, as in the infinitesimal case, to formulate the theory in complex variable notation. For large strains, however, it is possible to relate the complex coordinates either to points in the undeformed body or to points in the deformed material. Both methods are of value, but the advantages of the complex variable approach have so far been realized mainly in the approximation procedure of Chapter VI where the equations are linearized and can be integrated along the lines followed in the classical infinitesimal theory.

For isotropic and transversely isotropic bodies a further simplification results from a reduction of the number of independent invariants in plane strain. When the material is incompressible there then follows a reciprocal theorem from which it is possible to deduce a solution of the elastic equations in which coordinates in the deformed body appear as independent variables, from a corresponding solution in which coordinates in the undeformed body play this role.

When the elastic material is isotropic and incompressible and has the linear Mooney form of strain energy function the equations governing

---

† The theory of finite plane strain has been developed in the following papers:
J. E. Adkins, A. E. Green, and R. T. Shield, *Phil. Trans. R. Soc.* A, **246** (1953), 181.
A. E. Green and E. W. Wilkes, *J. rat. Mech. Analysis* **3** (1954), 713.
J. E. Adkins, *Proc. Camb. phil. Soc. math. phys. Sci.* **51** (1955), 363; *J. Mech. Phys. Solids* **6** (1958), 267.
See also W. W. Klingbeil and R. T. Shield, *Z. angew. Math. Phys.* **17** (1966), 489.

the deformation can be expressed in symmetric forms which do not involve the elastic constants. Some solutions of these equations are given.

3.1. Plane strain superposed on uniform finite extension

We use the notation of § 1.1 and suppose that an elastic body which is deformed by a uniform finite extension parallel to the x_3-axis with constant extension ratio λ, subsequently receives a finite plane strain parallel to the x_1, x_2-plane. The curvilinear coordinate system θ_i is chosen so that $\theta_3 = y_3$ and then

$$\left. \begin{array}{c} x_\alpha = x_\alpha(\theta_1, \theta_2), \quad y_\alpha = y_\alpha(\theta_1, \theta_2, t) \\ x_3 = y_3/\lambda = \theta_3/\lambda \end{array} \right\}, \quad (3.1.1)$$

where, here and subsequently, Greek indices take the values 1, 2. The position vectors $\mathbf{r}$, $\mathbf{R}$ of points in the undeformed body and in the deformed body respectively may now be written in the special forms

$$\left. \begin{array}{c} \mathbf{r} = \mathbf{b}(x_1, x_2) + x_3 \mathbf{a}_3 = \mathbf{b}(\theta_1, \theta_2) + \dfrac{\theta_3 \mathbf{a}_3}{\lambda} \\ \mathbf{R} = \mathbf{B}(y_1, y_2) + y_3 \mathbf{A}_3 = \mathbf{B}(\theta_1, \theta_2, t) + \theta_3 \mathbf{A}_3 \end{array} \right\}, \quad (3.1.2)$$

where $\mathbf{a}_3$, $\mathbf{A}_3$ are constant unit vectors perpendicular to the plane surface $\theta_3 = $ constant, and $\mathbf{b}$, $\mathbf{B}$ are vectors in the plane $\theta_3 = 0$. From these results it follows that

$$\left. \begin{array}{c} \mathbf{g}_\alpha = \dfrac{\partial \mathbf{r}}{\partial \theta^\alpha} = \dfrac{\partial \mathbf{b}}{\partial \theta^\alpha} = \mathbf{a}_\alpha, \quad \mathbf{g}_3 = \dfrac{\partial \mathbf{r}}{\partial \theta^3} = \dfrac{\mathbf{a}_3}{\lambda} \\ \mathbf{G}_\alpha = \dfrac{\partial \mathbf{R}}{\partial \theta^\alpha} = \dfrac{\partial \mathbf{B}}{\partial \theta^\alpha} = \mathbf{A}_\alpha, \quad \mathbf{G}_3 = \dfrac{\partial \mathbf{R}}{\partial \theta^3} = \mathbf{A}_3 \end{array} \right\}, \quad (3.1.3)$$

where $\mathbf{a}_\alpha$, $\mathbf{A}_\alpha$ are the covariant base vectors associated with the θ_α curves in the plane $\theta_3 = 0$ of the undeformed body and the deformed body respectively. Hence

$$\left. \begin{array}{c} g_{\alpha\beta} = \mathbf{a}_\alpha \cdot \mathbf{a}_\beta = a_{\alpha\beta}, \quad g_{\alpha 3} = \dfrac{\mathbf{a}_\alpha \cdot \mathbf{a}_3}{\lambda} = 0, \quad g_{33} = \dfrac{\mathbf{a}_3 \cdot \mathbf{a}_3}{\lambda^2} = \dfrac{1}{\lambda^2} \\ g = |g_{ik}| = |a_{\alpha\beta}|/\lambda^2 = a/\lambda^2 \\ g^{\alpha\beta} = \mathbf{a}^\alpha \cdot \mathbf{a}^\beta = a^{\alpha\beta}, \quad g^{\alpha 3} = \mathbf{a}^\alpha \cdot \mathbf{a}^3 \lambda = 0, \quad g^{33} = \mathbf{a}^3 \cdot \mathbf{a}^3 \lambda^2 = \lambda^2 \end{array} \right\}, \quad (3.1.4)$$

and similarly

$$\left. \begin{array}{c} G_{\alpha\beta} = A_{\alpha\beta}, \quad G_{\alpha 3} = 0, \quad G_{33} = 1 \\ G = |G_{ik}| = |A_{\alpha\beta}| = A \\ G^{\alpha\beta} = A^{\alpha\beta}, \quad G^{\alpha 3} = 0, \quad G^{33} = 1 \end{array} \right\}, \quad (3.1.5)$$

$a_{\alpha\beta}$, $a^{\alpha\beta}$ being the covariant and contravariant metric tensors associated with curvilinear coordinates θ_α in the plane $x_3 = 0$ of the undeformed body, and $A_{\alpha\beta}$, $A^{\alpha\beta}$ being the corresponding metric tensors in the plane $y_3 = 0$ of the deformed body. These metric tensors, like the base vectors $\mathbf{a}_\alpha$, $\mathbf{a}^\alpha$, $\mathbf{A}_\alpha$, and $\mathbf{A}^\alpha$, are therefore dependent only upon the two coordinates θ_α.

The strain components γ_{ij} are given by

$$\gamma_{\alpha\beta} = \tfrac{1}{2}(A_{\alpha\beta}-a_{\alpha\beta}), \qquad \gamma_{\alpha 3} = 0, \qquad \gamma_{33} = \tfrac{1}{2}\left(1-\frac{1}{\lambda^2}\right), \qquad (3.1.6)$$

or, referred to the cartesian system x_i,

$$e_{\alpha\beta} = \tfrac{1}{2}\left(\frac{\partial y_\lambda}{\partial x_\alpha}\frac{\partial y_\lambda}{\partial x_\beta}-\delta_{\alpha\beta}\right), \qquad e_{\alpha 3} = 0, \qquad e_{33} = \tfrac{1}{2}(\lambda^2-1). \qquad (3.1.7)$$

For the quantities I_r, K_α defined by (1.1.18), (1.13.11) we have

$$\left.\begin{aligned} I_1 &= \lambda^2 + a^{\alpha\beta}A_{\alpha\beta} \\ I_2 &= \lambda^2(A/a)a_{\alpha\beta}A^{\alpha\beta} + A/a \\ I_3 &= \lambda^2 A/a \end{aligned}\right\}, \qquad (3.1.8)$$

$$K_1 = \tfrac{1}{2}(\lambda^2-1), \qquad K_2 = 0. \qquad (3.1.9)$$

The invariants I_r are not, however, independent for

$$\begin{aligned} a_{\alpha\beta}A^{\alpha\beta}A/a &= (_0\epsilon_{\alpha\rho})(_0\epsilon_{\beta\nu})(\epsilon^{\alpha\lambda}\epsilon^{\beta\mu})a^{\rho\nu}A_{\lambda\mu}A/a \\ &= \epsilon_{\alpha\rho}\epsilon_{\beta\nu}\epsilon^{\alpha\lambda}\epsilon^{\beta\mu}a^{\rho\nu}A_{\lambda\mu} \\ &= a^{\alpha\beta}A_{\alpha\beta}, \end{aligned}$$

where

$$\left.\begin{aligned} _0\epsilon^{\alpha\beta}\sqrt{a} = \frac{_0\epsilon_{\alpha\beta}}{\sqrt{a}} = \epsilon^{\alpha\beta}\sqrt{A} = \frac{\epsilon_{\alpha\beta}}{\sqrt{A}} &= 1 \quad \text{if } \alpha=1, \beta=2 \\ &= -1 \quad \text{if } \alpha=2, \beta=1 \\ &= 0 \quad \text{otherwise} \end{aligned}\right\}. \qquad (3.1.10)$$

and $\qquad \epsilon_{\alpha\rho}\epsilon^{\alpha\lambda} = (_0\epsilon_{\alpha\rho})(_0\epsilon^{\alpha\lambda}) = \delta^\lambda_\rho$

We therefore have $\qquad I_3 - \lambda^2 I_2 + \lambda^4 I_1 - \lambda^6 = 0. \qquad (3.1.11)$

The alternative set of invariants J_r may now be obtained from (1.1.20) and between these there exists, by virtue of (3.1.11), the relation

$$J_3 + (1-\lambda^2)J_2 + (1-\lambda^2)^2 J_1 + (1-\lambda^2)^3 = 0. \qquad (3.1.12)$$

In subsequent work we require to make changes of the independent variables. If θ'_α is a second system of curvilinear coordinates in the x_1, x_2 plane we have the relations

$$\frac{\partial \theta^\alpha}{\partial \theta'^\mu} \frac{\partial \theta'^\mu}{\partial \theta^\beta} = \delta^\alpha_\beta, \qquad \frac{\partial \theta'^\alpha}{\partial \theta^\mu} \frac{\partial \theta^\mu}{\partial \theta'^\beta} = \delta^\alpha_\beta,$$

which may be solved to yield

$$\left(\frac{\partial \theta^1}{\partial \theta'^1}, \frac{\partial \theta^1}{\partial \theta'^2}, \frac{\partial \theta^2}{\partial \theta'^1}, \frac{\partial \theta^2}{\partial \theta'^2}\right) = \left(\frac{\partial \theta'^2}{\partial \theta^2}, -\frac{\partial \theta'^1}{\partial \theta^2}, -\frac{\partial \theta'^2}{\partial \theta^1}, \frac{\partial \theta'^1}{\partial \theta^1}\right)\Delta, \qquad (3.1.13)$$

where
$$\Delta = \frac{\partial(\theta^1, \theta^2)}{\partial(\theta'^1, \theta'^2)}. \qquad (3.1.14)$$

We observe that (3.1.13) may be written more concisely as

$$\left.\begin{aligned} \epsilon_{\alpha\gamma} \frac{\partial \theta^\gamma}{\partial \theta'^\beta} &= -\epsilon'_{\beta\rho} \frac{\partial \theta'^\rho}{\partial \theta^\alpha} \\ {}_0\epsilon_{\alpha\gamma} \frac{\partial \theta^\gamma}{\partial \theta'^\beta} &= -{}_0\epsilon'_{\beta\rho} \frac{\partial \theta'^\rho}{\partial \theta^\alpha} \end{aligned}\right\} \qquad (3.1.15)$$

or

where the tensors ${}_0\epsilon'_{\alpha\beta}$, $\epsilon'_{\alpha\beta}$ are defined for the curvilinear system θ'_α by relations analogous to (3.1.10).

3.2. Airy's stress function

Consider the state of stress in a body which is such that

$$\tau^{\alpha 3} = 0, \qquad (3.2.1)$$

while all other components $\tau^{\alpha\beta}$, τ^{33} of the stress tensor are independent of θ_3. We shall see in § 3.3 that this situation arises under conditions of plane strain for isotropic bodies and for materials with suitable symmetry properties. The stresses may then be expressed in terms of an Airy stress function by a procedure analogous to that employed in *T.E.* for the classical infinitesimal theory.† From (1.1.26) and (3.1.5) we now have

$$\left.\begin{aligned} \mathbf{T}_\alpha &= \sqrt{(AA^{\alpha\alpha})}\mathbf{t}_\alpha = \sqrt{(A)}\tau^{\alpha\beta}\mathbf{A}_\beta \\ \mathbf{T}_3 &= \sqrt{(A)}\mathbf{t}_3 = \sqrt{(A)}\tau^{33}\mathbf{A}_3 \end{aligned}\right\}, \qquad (3.2.2)$$

and since the vectors $\mathbf{T}_i$ are independent of θ_3, the equations of motion (1.1.27) with $\mathbf{F} = \mathbf{0}$, $\mathbf{f} = \mathbf{0}$, yield

$$\mathbf{T}_{\alpha,\alpha} = \mathbf{0}. \qquad (3.2.3)$$

† Chapter VI. Body forces are omitted from the present discussion.

The equation of equilibrium (3.2.3) may then be satisfied by putting
$$\mathbf{T}_\alpha = \sqrt{(A)}\epsilon^{\gamma\alpha}\boldsymbol{\chi}_{,\gamma}, \tag{3.2.4}$$
where $\boldsymbol{\chi}$ is a vector in the plane $\theta_3 = 0$. If
$$\boldsymbol{\chi} = \chi^\beta \mathbf{A}_\beta, \tag{3.2.5}$$
then
$$\boldsymbol{\chi}_{,\gamma} = \chi^\beta\|_\gamma \mathbf{A}_\beta, \tag{3.2.6}$$
and hence
$$\mathbf{T}_\alpha = \sqrt{(A)}\epsilon^{\gamma\alpha}\chi^\beta\|_\gamma \mathbf{A}_\beta, \tag{3.2.7}$$
or, using (3.2.2),
$$\tau^{\alpha\beta}\mathbf{A}_\beta = \epsilon^{\gamma\alpha}\chi^\beta\|_\gamma \mathbf{A}_\beta. \tag{3.2.8}$$
Taking the scalar product of both sides of this equation with the base vector $\mathbf{A}^\lambda$, and replacing λ by β, we obtain
$$\tau^{\alpha\beta} = \epsilon^{\gamma\alpha}\chi^\beta\|_\gamma. \tag{3.2.9}$$
Since the tensor $\tau^{\alpha\beta}$ is symmetric, $\epsilon^{\gamma\alpha}\chi^\beta\|_\gamma = \epsilon^{\gamma\beta}\chi^\alpha\|_\gamma$ and we may put
$$\chi^\beta = \epsilon^{\rho\beta}\phi_{,\rho}, \tag{3.2.10}$$
and hence
$$\tau^{\alpha\beta} = \epsilon^{\gamma\alpha}\epsilon^{\rho\beta}\phi\|_{\rho\gamma} = \epsilon^{\alpha\gamma}\epsilon^{\beta\rho}\phi\|_{\gamma\rho}, \tag{3.2.11}$$
where ϕ is a scalar invariant function of θ_1, θ_2 and is Airy's stress function for the deformed body. Equations (3.2.11) may also be written
$$\tau^\alpha_\beta = \delta^{\alpha\gamma}_{\beta\rho}\phi\|^\rho_\gamma, \tag{3.2.12}$$
or, if we solve for ϕ,
$$\left.\begin{aligned}\phi\|_{\alpha\beta} &= \epsilon_{\alpha\gamma}\epsilon_{\beta\rho}\tau^{\gamma\rho}\\ &= (_0\epsilon_{\alpha\gamma})(_0\epsilon_{\beta\rho})(A/a)\tau^{\gamma\rho}\\ \phi\|^\alpha_\beta &= \delta^{\alpha\gamma}_{\beta\rho}\tau^\rho_\gamma\end{aligned}\right\}. \tag{3.2.13}$$

Expressions for the force and couple resultants across a plane curve lying in the plane $\theta_3 = 0$ have been derived in *T.E.* for the classical infinitesimal theory of elasticity and by Adkins, Green, and Shield† for finite plane strain. The analysis for a large deformation is exactly analogous to that for the infinitesimal theory given in *T.E.* (§ 6.2) provided that all quantities are referred to the *deformed* body. Thus, consider any curve AB which lies in the plane $y_3 = 0$ of the deformed body and does not intersect itself, and suppose that the positive direction along the curve is from A to B (Fig. 3.1). The curve AB separates the plane into two regions 1 and 2 which lie immediately adjacent to it, and the force exerted by the region 1 on the region 2 across an element of AB is $-\mathbf{t}\,ds$, measured per unit length of the y_3-axis. The unit

† Loc. cit., p. 101.

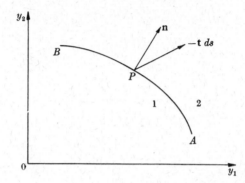

Fig. 3.1. Force vector on curve separating regions 1 and 2.

normal in the plane $\theta_3 = 0$ at a point P of the curve AB, directed from the region 1 to the region 2, is

$$\mathbf{n} = n_\alpha \mathbf{A}^\alpha = n^\alpha \mathbf{A}_\alpha, \qquad (3.2.14)$$

so that, from (1.1.24), the stress vector $\mathbf{t}$, which lies in the plane $y_3 = 0$, is given by

$$\mathbf{t} = \frac{n_\alpha \mathbf{T}^\alpha}{\sqrt{A}}. \qquad (3.2.15)$$

The total force $\mathbf{P}$ exerted across a part AP of the curve AB, measured per unit length of the y_3-axis, is therefore

$$\mathbf{P} = -\int_A^P \frac{n_\alpha \mathbf{T}^\alpha}{\sqrt{A}}\, ds$$

$$= \boldsymbol{\chi} = \epsilon^{\rho\beta} \phi_{,\rho} \mathbf{A}_\beta, \qquad (3.2.16)$$

apart from an arbitrary constant vector which may be absorbed into $\boldsymbol{\chi}$ since this does not affect the stresses.

The moment about the x_3-axis of the forces exerted by region 1 on the region 2 across AP, measured per unit length of the y_3-axis in the deformed body, is

$$\mathbf{M} = \int_A^P [\mathbf{B} \times \boldsymbol{\chi}_{,\beta}] \frac{d\theta^\beta}{ds}\, ds$$

$$= (B^\alpha \phi_{,\alpha} - \phi) \mathbf{A}_3, \qquad (3.2.17)$$

where
$$\mathbf{B} = B^\alpha \mathbf{A}_\alpha = B_\alpha \mathbf{A}^\alpha \qquad (3.2.18)$$

is the vector defined by (3.1.2), and to derive the second form for $\mathbf{M}$, transformations of the type given in $T.E.$, § 6.2, have been employed.

The couple is therefore of magnitude
$$M = B^\alpha \phi_{,\alpha} - \phi. \tag{3.2.19}$$

If the curve AB is a single bounding curve of a body and this curve is entirely free from applied stresses, then, provided body forces are zero,
$$\mathbf{\chi} = \mathbf{0}, \tag{3.2.20}$$
or
$$\phi_{,1} = 0, \quad \phi_{,2} = 0 \tag{3.2.21}$$
at all points of AB. This implies that ϕ is constant on AB, and this constant may be taken to be zero if we remember (3.2.19).

3.3. Stress–strain relations: equations of equilibrium

For the deformation defined by (3.1.1), the stress–strain relations (1.15.1) and (1.15.2) yield†

$$\left.\begin{aligned}\tau^{\alpha\beta} &= \frac{1}{2\sqrt{I_3}}\left(\frac{\partial W}{\partial e_{\mu\rho}}+\frac{\partial W}{\partial e_{\rho\mu}}\right)\frac{\partial \theta^\alpha}{\partial x^\mu}\frac{\partial \theta^\beta}{\partial x^\rho}\\ \tau^{\alpha 3} &= 0 \\ \tau^{33} &= \frac{\lambda^2}{\sqrt{I_3}}\frac{\partial W}{\partial e_{33}}\end{aligned}\right\}, \tag{3.3.1}$$

and
$$\left.\begin{aligned}\tau^{\alpha\beta} &= \frac{1}{2}\left(\frac{\partial W}{\partial e_{\mu\rho}}+\frac{\partial W}{\partial e_{\rho\mu}}\right)\frac{\partial \theta^\alpha}{\partial x^\mu}\frac{\partial \theta^\beta}{\partial x^\rho}+pA^{\alpha\beta}\\ \tau^{\alpha 3} &= 0 \\ \tau^{33} &= \lambda^2\frac{\partial W}{\partial e_{33}}+p\end{aligned}\right\}, \tag{3.3.2}$$

for compressible and incompressible bodies respectively, provided the derivatives $\partial W/\partial e_{\alpha 3}$, $\partial W/\partial e_{3\alpha}$ vanish when $e_{13} = e_{23} = 0$. This occurs if the expression for W in terms of e_{ij} does not contain terms which are linear either in e_{13} or e_{23} alone. This condition is satisfied by materials possessing rhombic, tetragonal, or cubic symmetry, by the hexagonal crystal classes specified by (1.12.11) and (1.12.13), by materials which are transversely isotropic with respect to the x_3-direction, by isotropic bodies, and by materials which are monoclinic with the x_3-axis as the preferred direction, that is, with the vectors $\mathbf{h}_1$, $\mathbf{h}_3$ interchanged in § 1.8 and the strain components e_{12}, e_{23} interchanged in equation (1.8.2).

† A similar theory of plane strain can be developed for curvilinearly aeolotropic bodies using the formulae of § 1.16. In this case the auxiliary system θ_i' defining the aeolotropy is chosen so that $\theta_3' = x_3$ and the coordinate curves θ_1', θ_2' form a curvilinear system in the x_1, x_2-plane.

By combining (3.3.1) with (3.2.13) we have

$$\phi\|_{\alpha\beta} = \frac{\sqrt{I_3}}{\lambda^2}\left[\frac{\partial W}{\partial e_{11}}\frac{\partial x^2}{\partial \theta^\alpha}\frac{\partial x^2}{\partial \theta^\beta} + \frac{\partial W}{\partial e_{22}}\frac{\partial x^1}{\partial \theta^\alpha}\frac{\partial x^1}{\partial \theta^\beta} - \frac{1}{2}\left(\frac{\partial W}{\partial e_{12}} + \frac{\partial W}{\partial e_{21}}\right)\left(\frac{\partial x^1}{\partial \theta^\alpha}\frac{\partial x^2}{\partial \theta^\beta} + \frac{\partial x^2}{\partial \theta^\alpha}\frac{\partial x^1}{\partial \theta^\beta}\right)\right],$$

(3.3.3)

where, in deriving this result, we have used the expression (3.1.8) for I_3 and the relations derived by putting $\theta'_\alpha = x_\alpha$ in (3.1.13). For an incompressible body, we have similarly

$$\phi\|_{\alpha\beta} = \frac{1}{\lambda^2}\left[\frac{\partial W}{\partial e_{11}}\frac{\partial x^2}{\partial \theta^\alpha}\frac{\partial x^2}{\partial \theta^\beta} + \frac{\partial W}{\partial e_{22}}\frac{\partial x^1}{\partial \theta^\alpha}\frac{\partial x^1}{\partial \theta^\beta} - \frac{1}{2}\left(\frac{\partial W}{\partial e_{12}} + \frac{\partial W}{\partial e_{21}}\right)\left(\frac{\partial x^1}{\partial \theta^\alpha}\frac{\partial x^2}{\partial \theta^\beta} + \frac{\partial x^2}{\partial \theta^\alpha}\frac{\partial x^1}{\partial \theta^\beta}\right)\right] + pA_{\alpha\beta}.$$

(3.3.4)

Rhombic system

From (1.9.2) the strain energy has the polynomial form

$$W = W(e_{11}, e_{22}, e_{33}, e_{12}^2, e_{23}^2, e_{31}^2, e_{12}e_{23}e_{31}).$$

Hence, in view of (3.1.7), the stress–strain relations for compressible bodies become

$$\left.\begin{array}{l}\tau^{\alpha\beta}\sqrt{I_3} = \dfrac{\partial W}{\partial e_{11}}\dfrac{\partial \theta^\alpha}{\partial x^1}\dfrac{\partial \theta^\beta}{\partial x^1} + \dfrac{\partial W}{\partial e_{22}}\dfrac{\partial \theta^\alpha}{\partial x^2}\dfrac{\partial \theta^\beta}{\partial x^2} + e_{12}\dfrac{\partial W}{\partial e_{12}^2}\left(\dfrac{\partial \theta^\alpha}{\partial x^1}\dfrac{\partial \theta^\beta}{\partial x^2} + \dfrac{\partial \theta^\alpha}{\partial x^2}\dfrac{\partial \theta^\beta}{\partial x^1}\right) \\[2mm] \tau^{\alpha 3} = 0 \\[2mm] \tau^{33} = \dfrac{\lambda^2}{\sqrt{I_3}}\dfrac{\partial W}{\partial e_{33}}\end{array}\right\},$$ (3.3.5)

and for incompressible bodies

$$\left.\begin{array}{l}\tau^{\alpha\beta} = \dfrac{\partial W}{\partial e_{11}}\dfrac{\partial \theta^\alpha}{\partial x^1}\dfrac{\partial \theta^\beta}{\partial x^1} + \dfrac{\partial W}{\partial e_{22}}\dfrac{\partial \theta^\alpha}{\partial x^2}\dfrac{\partial \theta^\beta}{\partial x^2} + e_{12}\dfrac{\partial W}{\partial e_{12}^2}\left(\dfrac{\partial \theta^\alpha}{\partial x^1}\dfrac{\partial \theta^\beta}{\partial x^2} + \dfrac{\partial \theta^\alpha}{\partial x^2}\dfrac{\partial \theta^\beta}{\partial x^1}\right) + pA^{\alpha\beta} \\[2mm] \tau^{\alpha 3} = 0 \\[2mm] \tau^{33} = \lambda^2\dfrac{\partial W}{\partial e_{33}} + p\end{array}\right\}$$ (3.3.6)

In both (3.3.5) and (3.3.6) W reduces to†

$$W = W(e_{11}, e_{22}, e_{33}, e_{12}^2).$$ (3.3.7)

† For an incompressible body $4e_{12}^2 = (1+2e_{11})(1+2e_{22}) - 1/\lambda^2$ so that
$$W = W_1(e_{11}, e_{22}, \lambda^2)$$
and then $\tau^{\alpha\beta}$ reduces to
$$\tau^{\alpha\beta} = \frac{\partial W_1}{\partial e_{11}}\frac{\partial \theta^\alpha}{\partial x^1}\frac{\partial \theta^\beta}{\partial x^1} + \frac{\partial W_1}{\partial e_{22}}\frac{\partial \theta^\alpha}{\partial x^2}\frac{\partial \theta^\beta}{\partial x^2} + pA^{\alpha\beta}.$$
However, τ^{33} must still be found from (3.3.6).

The formulae for $\tau^{\alpha\beta}$ may again be combined with (3.2.13) to give

$$\phi\|_{\alpha\beta} = \frac{\sqrt{I_3}}{\lambda^2}\left[\frac{\partial W}{\partial e_{11}}\frac{\partial x^2}{\partial \theta^\alpha}\frac{\partial x^2}{\partial \theta^\beta}+\frac{\partial W}{\partial e_{22}}\frac{\partial x^1}{\partial \theta^\alpha}\frac{\partial x^1}{\partial \theta^\beta}-e_{12}\frac{\partial W}{\partial e_{12}^2}\left(\frac{\partial x^1}{\partial \theta^\alpha}\frac{\partial x^2}{\partial \theta^\beta}+\frac{\partial x^2}{\partial \theta^\alpha}\frac{\partial x^1}{\partial \theta^\beta}\right)\right], \quad (3.3.8)$$

and

$$\phi\|_{\alpha\beta} = \frac{1}{\lambda^2}\left[\frac{\partial W}{\partial e_{11}}\frac{\partial x^2}{\partial \theta^\alpha}\frac{\partial x^2}{\partial \theta^\beta}+\frac{\partial W}{\partial e_{22}}\frac{\partial x^1}{\partial \theta^\alpha}\frac{\partial x^1}{\partial \theta^\beta}\right.$$
$$\left.-e_{12}\frac{\partial W}{\partial e_{12}^2}\left(\frac{\partial x^1}{\partial \theta^\alpha}\frac{\partial x^2}{\partial \theta^\beta}+\frac{\partial x^2}{\partial \theta^\alpha}\frac{\partial x^1}{\partial \theta^\beta}\right)\right]+pA_{\alpha\beta}, \quad (3.3.9)$$

for compressible and incompressible materials respectively.

Transverse isotropy

The stress–strain relations for transversely isotropic bodies follow by introducing the values of the metric tensors and invariants derived in § 3.1 into the relations of § 1.15. The tensor B^{ij} then becomes

$$B^{\alpha\beta} = (_0\epsilon^{\alpha\rho})(_0\epsilon^{\beta\nu})a_{\rho\nu}\lambda^2+\epsilon^{\alpha\rho}\epsilon^{\beta\nu}A_{\rho\nu}A/a$$
$$= \lambda^2 a^{\alpha\beta}+AA^{\alpha\beta}/a,$$
$$B^{\alpha 3} = 0,$$
$$B^{33} = \lambda^2(I_1-\lambda^2),$$

while the only non-zero component of the tensors M^{ij}, N^{ij} is

$$M^{33} = \lambda^2.$$

The stress–strain relation (1.15.6) therefore yields

$$\left.\begin{array}{l}\tau^{\alpha\beta} = \mathscr{H}a^{\alpha\beta}+\mathscr{L}A^{\alpha\beta}\\ \tau^{\alpha 3} = 0\\ \tau^{33} = \lambda^2\Phi+\lambda^2(I_1-\lambda^2)\Psi+\mathscr{F}+p\end{array}\right\}, \quad (3.3.10)$$

where

$$\left.\begin{array}{l}\mathscr{H} = \Phi+\lambda^2\Psi = \dfrac{2}{\sqrt{I_3}}\left(\dfrac{\partial W}{\partial I_1}+\lambda^2\dfrac{\partial W}{\partial I_2}\right)\\[6pt] \mathscr{L} = p+\dfrac{I_3\Psi}{\lambda^2} = 2\sqrt{(I_3)}\left(\dfrac{\partial W}{\partial I_3}+\dfrac{1}{\lambda^2}\dfrac{\partial W}{\partial I_2}\right)\\[6pt] \mathscr{F} = \lambda^2\Theta = \dfrac{\lambda^2}{\sqrt{I_3}}\dfrac{\partial W}{\partial K_1}\end{array}\right\}. \quad (3.3.11)$$

In these relations W is regarded as a function of the five quantities I_r, K_α. By virtue of (3.1.9) and (3.1.11) W may equally well be regarded as a function of three invariants I, J, K defined by

$$\left.\begin{aligned} I &= I_1 = \lambda^2 + a^{\alpha\beta} A_{\alpha\beta} \\ J &= I_3 = \lambda^2 A/a \\ K &= K_1 = \tfrac{1}{2}(\lambda^2 - 1) \end{aligned}\right\}, \qquad (3.3.12)$$

since
$$I_2 = \lambda^2 (I - \lambda^2) + J/\lambda^2, \qquad (3.3.13)$$

and
$$K_2 = 0.$$

Putting
$$W(I_1, I_2, I_3, K_1, K_2) = W_1(I, J, K), \qquad (3.3.14)$$

we have

$$\left.\begin{aligned} \frac{\partial W_1}{\partial I} &= \frac{\partial W}{\partial I_1} + \frac{\partial W}{\partial I_2}\frac{\partial I_2}{\partial I} = \frac{\partial W}{\partial I_1} + \lambda^2 \frac{\partial W}{\partial I_2} \\ \frac{\partial W_1}{\partial J} &= \frac{\partial W}{\partial I_3} + \frac{\partial W}{\partial I_2}\frac{\partial I_2}{\partial J} = \frac{\partial W}{\partial I_3} + \frac{1}{\lambda^2} \frac{\partial W}{\partial I_2} \\ \frac{\partial W_1}{\partial K} &= \frac{\partial W}{\partial K_1} \end{aligned}\right\}, \qquad (3.3.15)$$

so that

$$\mathscr{H} = \frac{2}{\sqrt{J}} \frac{\partial W_1}{\partial I}, \qquad \mathscr{L} = 2\sqrt{(J)} \frac{\partial W_1}{\partial J}, \qquad \mathscr{F} = \frac{\lambda^2}{\sqrt{J}} \frac{\partial W_1}{\partial K}. \qquad (3.3.16)$$

We may observe that τ^{33} can then be written

$$\tau^{33} = (2\lambda^2 - I + J/\lambda^4)\Phi + (I - \lambda^2 - J/\lambda^4)\mathscr{H} + \mathscr{L} + \mathscr{F}, \qquad (3.3.17)$$

where Φ is regarded as a function of the five invariants I_r, K_α, while $\mathscr{H}$, $\mathscr{L}$, and $\mathscr{F}$ are regarded as functions only of I, J, K.

Alternatively, the invariants J_1, J_2, J_3 defined by (1.1.20) may be employed in place of I_1, I_2, I_3. Writing

$$W(I_1, I_2, I_3, K_1, K_2) = W_2(J_1, J_2, J_3, K), \qquad (3.3.18)$$

we then have

$$\left.\begin{aligned} \frac{\partial W}{\partial I_1} &= \frac{\partial W_2}{\partial J_1} - 2\frac{\partial W_2}{\partial J_2} + \frac{\partial W_2}{\partial J_3} \\ \frac{\partial W}{\partial I_2} &= \frac{\partial W_2}{\partial J_2} - \frac{\partial W_2}{\partial J_3} \\ \frac{\partial W}{\partial I_3} &= \frac{\partial W_2}{\partial J_3} \end{aligned}\right\}, \qquad (3.3.19)$$

and $\mathscr{H}$, $\mathscr{L}$, and τ^{33} become

$$\left.\begin{aligned}\mathscr{H} &= \frac{2}{\sqrt{I_3}}\left[\frac{\partial W_2}{\partial J_1}+(\lambda^2-2)\frac{\partial W_2}{\partial J_2}-(\lambda^2-1)\frac{\partial W_2}{\partial J_3}\right] \\ \mathscr{L} &= \frac{2\sqrt{I_3}}{\lambda^2}\left[\frac{\partial W_2}{\partial J_2}+(\lambda^2-1)\frac{\partial W_2}{\partial J_3}\right]\end{aligned}\right\}, \quad (3.3.20)$$

$$\tau^{33} = \frac{2}{\sqrt{I_3}}\left\{\lambda^2\frac{\partial W_2}{\partial J_1}+\lambda^2(J_1+1-\lambda^2)\frac{\partial W_2}{\partial J_2}+\right.$$
$$\left.+[(1-\lambda^2)(J_1+1-\lambda^2)+J_2+J_3]\frac{\partial W_2}{\partial J_3}\right\}+\mathscr{F}, \quad (3.3.21)$$

respectively.

If the material is incompressible, $I_3 = J = 1$ and W becomes a function of I_1, I_2, K_1, and K_2 or alternatively of I and K. In this case

$$\left.\begin{aligned}\mathscr{H} &= 2\left(\frac{\partial W}{\partial I_1}+\lambda^2\frac{\partial W}{\partial I_2}\right) = 2\frac{\partial W_1}{\partial I} \\ \mathscr{F} &= \lambda^2\frac{\partial W}{\partial K_1} = \lambda^2\frac{\partial W_1}{\partial K}\end{aligned}\right\}, \quad (3.3.22)$$

and p (and hence $\mathscr{L}$) is a scalar function which is determined from the equations of equilibrium and boundary conditions. Also, (3.3.17) then becomes

$$\tau^{33} = (2\lambda^2-I+1/\lambda^4)\Phi+(I-\lambda^2-1/\lambda^4)\mathscr{H}+\mathscr{L}+\mathscr{F}. \quad (3.3.23)$$

In terms of the Airy stress function we have, from (3.2.13),

$$\begin{aligned}\phi\|_{\alpha\beta} &= (A/a)({}_0\epsilon_{\alpha\gamma})({}_0\epsilon_{\beta\rho})a^{\gamma\rho}\mathscr{H}+\epsilon_{\alpha\gamma}\epsilon_{\beta\rho}A^{\gamma\rho}\mathscr{L} \\ &= (A/a)a_{\alpha\beta}\mathscr{H}+A_{\alpha\beta}\mathscr{L},\end{aligned} \quad (3.3.24)$$

or, for an incompressible body, when $I_3 = 1$,

$$\phi\|_{\alpha\beta} = a_{\alpha\beta}\mathscr{H}/\lambda^2+A_{\alpha\beta}\mathscr{L}. \quad (3.3.25)$$

In the latter case we may eliminate $\mathscr{L}$ from the equations of equilibrium. Thus

$$2\mathscr{L}+(I-\lambda^2)\mathscr{H} = A^{\alpha\beta}\phi\|_{\alpha\beta} = \phi\|_\alpha^\alpha$$
$$= \frac{1}{\sqrt{A}}\{\sqrt{(A)}A^{\rho\gamma}\phi_{,\rho}\}_{,\gamma}, \quad (3.3.26)$$

and then (3.3.25) yields

$$2\phi\|_{\alpha\beta} = A_{\alpha\beta}\phi\|_\rho^\rho+[2a_{\alpha\beta}/\lambda^2-(I-\lambda^2)A_{\alpha\beta}]\mathscr{H}, \quad (3.3.27)$$

or
$$2\phi\|_\beta^\alpha = \delta_\beta^\alpha\phi\|_\rho^\rho+[2A^{\alpha\gamma}a_{\beta\gamma}/\lambda^2-(I-\lambda^2)\delta_\beta^\alpha]\mathscr{H}. \quad (3.3.28)$$

Also, from (3.3.23) and (3.3.26) we have

$$2\tau^{33} = 2(2\lambda^2 - I + 1/\lambda^4)\Phi + (I - \lambda^2 - 2/\lambda^4)\mathscr{H} + 2\mathscr{F} + \phi\|_\alpha^\alpha. \quad (3.3.29)$$

Isotropic bodies

Differences in the forms of the stress–strain relations for transversely isotropic bodies and isotropic materials are due entirely to the absence of the terms in K_1, K_2 in the strain energy function in the latter case. Thus, for isotropic bodies we may write

$$W = W(I_1, I_2, I_3) = W_1(I, J) = W_2(J_1, J_2, J_3). \quad (3.3.30)$$

With this form for W, the formulae

$$\left.\begin{array}{l} \tau^{\alpha\beta} = \mathscr{H}a^{\alpha\beta} + \mathscr{L}A^{\alpha\beta} \\ \phi\|_{\alpha\beta} = (A/a)a_{\alpha\beta}\mathscr{H} + A_{\alpha\beta}\mathscr{L} \end{array}\right\} \quad (3.3.31)$$

given in (3.3.10) and (3.3.24) still hold, with

$$\left.\begin{array}{l} \mathscr{H} = \Phi + \lambda^2 \Psi = \dfrac{2}{\sqrt{J}}\dfrac{\partial W_1}{\partial I} = \dfrac{2}{\sqrt{I_3}}\left[\dfrac{\partial W_2}{\partial J_1} + (\lambda^2 - 2)\dfrac{\partial W_2}{\partial J_2} - (\lambda^2 - 1)\dfrac{\partial W_2}{\partial J_3}\right] \\[6pt] \mathscr{L} = p + I_3\Psi/\lambda^2 = 2\sqrt{(J)}\dfrac{\partial W_1}{\partial J} = \dfrac{2\sqrt{I_3}}{\lambda^2}\left[\dfrac{\partial W_2}{\partial J_2} + (\lambda^2 - 1)\dfrac{\partial W_2}{\partial J_3}\right] \end{array}\right\}, \quad (3.3.32)$$

but now the formulae (3.3.10), (3.3.17), and (3.3.21) for τ^{33} yield

$$\begin{aligned} \tau^{33} &= \lambda^2 \Phi + \lambda^2(I_1 - \lambda^2)\Psi + p \\ &= (2\lambda^2 - I + J/\lambda^4)\Phi + (I - \lambda^2 - J/\lambda^4)\mathscr{H} + \mathscr{L} \\ &= \dfrac{2}{\sqrt{I_3}}\left\{\lambda^2 \dfrac{\partial W_2}{\partial J_1} + \lambda^2(J_1 + 1 - \lambda^2)\dfrac{\partial W_2}{\partial J_2} + [(1 - \lambda^2)(J_1 + 1 - \lambda^2) + J_2 + J_3]\dfrac{\partial W_2}{\partial J_3}\right\}. \end{aligned}$$

$$(3.3.33)$$

As before, in the incompressible case, p and therefore $\mathscr{L}$ become arbitrary scalar functions of the coordinates but W is a function of the single invariant I and

$$\mathscr{H} = 2\dfrac{dW_1(I)}{dI}. \quad (3.3.34)$$

Equations (3.3.25) to (3.3.28) again apply, but in the formulae (3.3.23) and (3.3.29) for τ^{33} the function $\mathscr{F}$ is omitted.

3.4. Complex variable formulation: geometrical relations

The solution of two-dimensional problems in the classical linear theory of elasticity is often greatly simplified by the introduction of complex variables.† A similar formulation is possible for finite

---

† See, for example, *T.E.*, chaps. VI–IX.

deformations, although the advantages of this approach have so far mainly been realized in its application to the approximation methods of Chapter VI. Also, for finite deformations, the complex coordinate reference frame may be related to points either in the undeformed body or in the deformed body, and the relevant equations for either coordinate system may be derived by an appropriate choice for the curvilinear coordinate system θ_α in the relations of the preceding sections. Since the resulting expressions are simpler for complex coordinates in the deformed body, we consider this case first; the corresponding formulae for complex coordinates in the undeformed body may then be derived by a simple change of independent variable.

Accordingly, we introduce complex coordinates $(\zeta, \bar{\zeta})$, $(z, \bar{z})$ in the undeformed body and in the deformed body respectively by means of the relations

$$\left.\begin{array}{ll} \zeta = x_1 + ix_2, & \bar{\zeta} = x_1 - ix_2 \\ z = y_1 + iy_2, & \bar{z} = y_1 - iy_2 \end{array}\right\}, \tag{3.4.1}$$

so that

$$\left.\begin{array}{ll} x_1 = \tfrac{1}{2}(\zeta + \bar{\zeta}), & x_2 = -\tfrac{1}{2}i(\zeta - \bar{\zeta}) \\ y_1 = \tfrac{1}{2}(z + \bar{z}), & y_2 = -\tfrac{1}{2}i(z - \bar{z}) \end{array}\right\}. \tag{3.4.2}$$

These yield the differential formulae

$$\left.\begin{array}{ll} \dfrac{\partial}{\partial \zeta} = \dfrac{1}{2}\left(\dfrac{\partial}{\partial x_1} - i\dfrac{\partial}{\partial x_2}\right), & \dfrac{\partial}{\partial \bar{\zeta}} = \dfrac{1}{2}\left(\dfrac{\partial}{\partial x_1} + i\dfrac{\partial}{\partial x_2}\right) \\[2mm] \dfrac{\partial}{\partial x_1} = \dfrac{\partial}{\partial \zeta} + \dfrac{\partial}{\partial \bar{\zeta}}, & \dfrac{\partial}{\partial x_2} = i\left(\dfrac{\partial}{\partial \zeta} - \dfrac{\partial}{\partial \bar{\zeta}}\right) \end{array}\right\}, \tag{3.4.3}$$

with corresponding results for the coordinates in the deformed body. Furthermore, if we interpret (3.1.13) in terms of the complex coordinate systems $(\zeta, \bar{\zeta})$ and $(z, \bar{z})$ and introduce the resulting relations into the formulae of the type

$$\frac{\partial}{\partial z} = \frac{\partial \zeta}{\partial z}\frac{\partial}{\partial \zeta} + \frac{\partial \bar{\zeta}}{\partial z}\frac{\partial}{\partial \bar{\zeta}},$$

we obtain

$$\frac{\partial}{\partial z} = \Delta\left(\frac{\partial \bar{z}}{\partial \bar{\zeta}}\frac{\partial}{\partial \zeta} - \frac{\partial \bar{z}}{\partial \zeta}\frac{\partial}{\partial \bar{\zeta}}\right), \qquad \frac{\partial}{\partial \bar{z}} = \Delta\left(-\frac{\partial z}{\partial \bar{\zeta}}\frac{\partial}{\partial \zeta} + \frac{\partial z}{\partial \zeta}\frac{\partial}{\partial \bar{\zeta}}\right), \tag{3.4.4}$$

and similarly

$$\frac{\partial}{\partial \zeta} = \left(\frac{\partial \bar{\zeta}}{\partial \bar{z}}\frac{\partial}{\partial z} - \frac{\partial \bar{\zeta}}{\partial z}\frac{\partial}{\partial \bar{z}}\right)\!\Big/\Delta, \qquad \frac{\partial}{\partial \bar{\zeta}} = \left(-\frac{\partial \zeta}{\partial \bar{z}}\frac{\partial}{\partial z} + \frac{\partial \zeta}{\partial z}\frac{\partial}{\partial \bar{z}}\right)\!\Big/\Delta, \tag{3.4.5}$$

where

$$\Delta = \frac{\partial(\zeta, \bar{\zeta})}{\partial(z, \bar{z})} = 1\Big/\!\left[\frac{\partial(z, \bar{z})}{\partial(\zeta, \bar{\zeta})}\right] = \frac{\lambda}{\sqrt{I_3}}. \tag{3.4.6}$$

If the components of displacement referred to the x_α-axes are (u, v), the complex displacement function D is defined by

$$D = u+iv, \qquad \bar{D} = u-iv, \tag{3.4.7}$$

and then
$$z = \zeta+D, \qquad \bar{z} = \bar{\zeta}+\bar{D}. \tag{3.4.8}$$

Denoting the covariant and contravariant base vectors in the system of complex coordinates z, $\bar{z}$ by $\mathbf{A}_\alpha$, $\mathbf{A}^\alpha$ respectively, the position vector $\mathbf{B}$ of a point in the plane $y_3 = 0$ may be written

$$\mathbf{B} = z^\alpha \mathbf{A}_\alpha = z_\alpha \mathbf{A}^\alpha. \tag{3.4.9}$$

By tensor transformations

$$z^1 = \frac{\partial z}{\partial y_1} y_1 + \frac{\partial z}{\partial y_2} y_2 = y_1 + iy_2 = z,$$

$$z^2 = \frac{\partial \bar{z}}{\partial y_1} y_1 + \frac{\partial \bar{z}}{\partial y_2} y_2 = y_1 - iy_2 = \bar{z},$$

so that the complex coordinates z, $\bar{z}$ may also be denoted by z^α. Similarly, the complex coordinates ζ, $\bar{\zeta}$ may be denoted by ζ^α and to preserve uniformity of notation we may write $x_3 = \zeta^3$, $y_3 = z^3$.

We now allow the curvilinear reference frame θ_α to coincide with the system of complex coordinates z, $\bar{z}$ so that

$$\theta_1 = z^1 = z, \qquad \theta_2 = z^2 = \bar{z} \tag{3.4.10}$$

and the base vectors $\mathbf{A}_\alpha$ then have the significance indicated in (3.1.3). The covariant and contravariant metric tensors $A_{\alpha\beta}$, $A^{\alpha\beta}$ become

$$\left.\begin{array}{ll} A_{12} = \tfrac{1}{2}, & A_{11} = A_{22} = 0 \\ A^{12} = 2, & A^{11} = A^{22} = 0 \\ A = |A_{\alpha\beta}| = -\tfrac{1}{4}, & \sqrt{A} = \tfrac{1}{2}i \end{array}\right\}, \tag{3.4.11}$$

while, remembering (3.1.8), we have

$$\sqrt{a} = \frac{i\lambda}{2\sqrt{I_3}} = \frac{\partial(x_1, x_2)}{\partial(\theta_1, \theta_2)} = \frac{i}{2}\frac{\partial(\zeta, \bar{\zeta})}{\partial(z, \bar{z})}$$

$$= \frac{i}{2}\left(1 - \frac{\partial D}{\partial z}\frac{\partial \bar{D}}{\partial \bar{z}} + \frac{\partial D}{\partial z}\frac{\partial \bar{D}}{\partial \bar{z}} - \frac{\partial D}{\partial \bar{z}}\frac{\partial \bar{D}}{\partial z}\right), \tag{3.4.12}$$

$$\left.\begin{array}{l} a_{11} = \bar{a}_{22} = \left(\dfrac{\partial x_1}{\partial z}\right)^2 + \left(\dfrac{\partial x_2}{\partial z}\right)^2 = \dfrac{\partial \zeta}{\partial z}\dfrac{\partial \bar{\zeta}}{\partial z} = \dfrac{\partial \bar{D}}{\partial z}\left(\dfrac{\partial D}{\partial z} - 1\right) \\[2mm] a_{12} = \dfrac{1}{2}\left(\dfrac{\partial \zeta}{\partial z}\dfrac{\partial \bar{\zeta}}{\partial \bar{z}} + \dfrac{\partial \zeta}{\partial \bar{z}}\dfrac{\partial \bar{\zeta}}{\partial z}\right) = \dfrac{1}{2}\left(1 - \dfrac{\partial D}{\partial z}\dfrac{\partial \bar{D}}{\partial \bar{z}} + \dfrac{\partial D}{\partial z}\dfrac{\partial \bar{D}}{\partial \bar{z}} + \dfrac{\partial D}{\partial \bar{z}}\dfrac{\partial \bar{D}}{\partial z}\right) \\[2mm] \phantom{a_{12}} = \dfrac{\lambda}{2\sqrt{I_3}} + \dfrac{\partial D}{\partial \bar{z}}\dfrac{\partial \bar{D}}{\partial z} \\[2mm] a^{11} = \bar{a}^{22} = a_{22}/a, \qquad a^{12} = -a_{12}/a \end{array}\right\}, \tag{3.4.13}$$

3.4 FINITE PLANE STRAIN

and the strain components $\gamma_{\alpha\beta}$ are given by

$$\left.\begin{aligned}\gamma_{11} &= \bar{\gamma}_{22} = \frac{1}{2}\frac{\partial \bar{D}}{\partial z}\left(1-\frac{\partial D}{\partial z}\right) \\ \gamma_{12} &= \frac{1}{4}\left(\frac{\partial D}{\partial z}+\frac{\partial \bar{D}}{\partial \bar{z}}-\frac{\partial D}{\partial z}\frac{\partial \bar{D}}{\partial \bar{z}}-\frac{\partial \bar{D}}{\partial z}\frac{\partial D}{\partial \bar{z}}\right)\end{aligned}\right\}. \qquad (3.4.14)$$

In addition, we require the strain components Γ_{ij} referred to the complex coordinates ζ^i in the undeformed body. These may be determined from the transformation

$$\Gamma_{ij} = \frac{\partial x^r}{\partial \zeta^i}\frac{\partial x^s}{\partial \zeta^j}e_{rs}, \qquad (3.4.15)$$

which, with (3.4.2), yields

$$\left.\begin{aligned}\Gamma_{11} &= \bar{\Gamma}_{22} = \tfrac{1}{4}(e_{11}-e_{22}-2ie_{12}) \\ \Gamma_{12} &= \tfrac{1}{4}(e_{11}+e_{22}) \\ \Gamma_{13} &= \bar{\Gamma}_{23} = \tfrac{1}{2}(e_{13}-ie_{23}) = 0 \\ \Gamma_{33} &= e_{33} = \tfrac{1}{2}(\lambda^2-1)\end{aligned}\right\}. \qquad (3.4.16)$$

For these we also have the reciprocal relations

$$\left.\begin{aligned}e_{11} &= \Gamma_{11}+\bar{\Gamma}_{22}+2\Gamma_{12} \\ e_{22} &= -\Gamma_{11}-\bar{\Gamma}_{22}+2\Gamma_{12} \\ e_{12} &= i(\Gamma_{11}-\bar{\Gamma}_{22})\end{aligned}\right\}. \qquad (3.4.17)$$

From (3.4.15) and (1.1.15) we have

$$\Gamma_{ij} = \frac{1}{2}\left(\frac{\partial y^r}{\partial \zeta^i}\frac{\partial y^r}{\partial \zeta^j}-\frac{\partial x^r}{\partial \zeta^i}\frac{\partial x^r}{\partial \zeta^j}\right),$$

or, for the components $\Gamma_{\alpha\beta}$,

$$\left.\begin{aligned}\Gamma_{11} &= \bar{\Gamma}_{22} = \frac{1}{2}\frac{\partial z}{\partial \zeta}\frac{\partial \bar{z}}{\partial \zeta} \\ \Gamma_{12} &= \frac{1}{4}\left(\frac{\partial z}{\partial \zeta}\frac{\partial \bar{z}}{\partial \bar{\zeta}}+\frac{\partial z}{\partial \bar{\zeta}}\frac{\partial \bar{z}}{\partial \zeta}-1\right)\end{aligned}\right\}. \qquad (3.4.18)$$

From (3.1.8) and (3.4.11) to (3.4.13) we obtain

$$\left.\begin{aligned}I_1 &= \lambda^2+a^{12} = \lambda^2+\frac{2\sqrt{I_3}}{\lambda}+\frac{4I_3}{\lambda^2}\frac{\partial D}{\partial \bar{z}}\frac{\partial \bar{D}}{\partial z} \\ I_2 &= -\frac{\lambda^2 a_{12}}{a}-\frac{1}{4a} = 2\lambda\sqrt{I_3}+4I_3\frac{\partial D}{\partial \bar{z}}\frac{\partial \bar{D}}{\partial z}+\frac{I_3}{\lambda^2} \\ I_3 &= -\frac{\lambda^2}{4a} = \lambda^2\bigg/\left[\frac{\partial(\zeta,\bar{\zeta})}{\partial(z,\bar{z})}\right]^2\end{aligned}\right\}, \qquad (3.4.19)$$

and then equations (1.1.20) yield

$$\begin{aligned}
J_1 &= \lambda^2-3-\frac{a_{12}}{a} = \lambda^2+\frac{2\sqrt{I_3}}{\lambda}+\frac{4I_3}{\lambda^2}\frac{\partial D}{\partial \bar{z}}\frac{\partial \bar{D}}{\partial z}-3 \\
J_2 &= 3-2\lambda^2-(\lambda^2-2)\frac{a_{12}}{a}-\frac{1}{4a} \\
&= 3-2\lambda^2+\frac{I_3}{\lambda^2}+2(\lambda^2-2)\left[\frac{\sqrt{I_3}}{\lambda}+\frac{2I_3}{\lambda^2}\frac{\partial D}{\partial \bar{z}}\frac{\partial \bar{D}}{\partial z}\right] \\
J_3 &= (\lambda^2-1)\left(1+\frac{a_{12}}{a}-\frac{1}{4a}\right) \\
&= (\lambda^2-1)\left[\left(1-\frac{\sqrt{I_3}}{\lambda}\right)^2-4\frac{I_3}{\lambda^2}\frac{\partial D}{\partial \bar{z}}\frac{\partial \bar{D}}{\partial z}\right]
\end{aligned} \right\}. \qquad (3.4.20)$$

For incompressible bodies, $I_3 = 1$, and equations (3.4.19) give

$$\left. \begin{aligned}
I &= I_1 = \lambda^2+\frac{2}{\lambda}+\frac{4}{\lambda^2}\frac{\partial D}{\partial \bar{z}}\frac{\partial \bar{D}}{\partial z} \\
I_2 &= \frac{1}{\lambda^2}+2\lambda+4\frac{\partial D}{\partial \bar{z}}\frac{\partial \bar{D}}{\partial z} \\
\frac{\partial(\zeta, \bar{\zeta})}{\partial(z, \bar{z})} &= \lambda
\end{aligned} \right\}. \qquad (3.4.21)$$

In the preceding formulae the independent variables may be changed from $z, \bar{z}$ to $\zeta, \bar{\zeta}$ by making use of (3.4.4), or, where appropriate, from $\zeta, \bar{\zeta}$ to $z, \bar{z}$ by means of (3.4.5).

3.5. Complex variable: stresses; equations of equilibrium

The analysis of § 3.2 may be interpreted in terms of complex variables to yield expressions for the complex stresses and force and couple resultants. Denoting by T^{ij} the contravariant components of the stress tensor referred to the complex coordinates z^i in the deformed body, and by σ^{ij} ($= \sigma_{ij}$) the corresponding (physical) components referred to the real system y_i, we have by tensor transformations

$$T^{ij} = \frac{\partial z^i}{\partial y^r}\frac{\partial z^j}{\partial y^s}\sigma^{rs} \qquad (\sigma^{rs} = \sigma_{rs}), \qquad (3.5.1)$$

or, remembering (3.4.1),

$$\left.\begin{aligned} T^{11} &= \bar{T}^{22} = \sigma_{11}-\sigma_{22}+2i\sigma_{12} \\ T^{12} &= \sigma_{11}+\sigma_{22} \\ T^{13} &= \bar{T}^{23} = \sigma_{13}+i\sigma_{23} \\ T^{33} &= \sigma_{33} \end{aligned}\right\}, \qquad (3.5.2)$$

and in the present case $T^{13} = 0$. Corresponding systems† T'^{ij}, σ'^{ij} may be defined for the complex and real systems ζ^i, x^i respectively which form curvilinear systems of coordinates in the undeformed body. These may be expressed in terms of each other by relations analogous to (3.5.2) and may be related to the systems T^{ij}, σ^{ij} by means of tensor transformations. For example,

$$\sigma'^{ij} = \frac{\partial x^i}{\partial y^r}\frac{\partial x^j}{\partial y^s}\sigma^{rs}, \qquad T'^{ij} = \frac{\partial \zeta^i}{\partial z^r}\frac{\partial \zeta^j}{\partial z^s}T^{rs}. \qquad (3.5.3)$$

It is readily seen that again $\sigma'^{\alpha 3} = 0$, $T'^{\alpha 3} = 0$, but it should be noted that the quantities σ'^{rs} defined by (3.5.3) are not physical components.

With the choice $\theta^\alpha = z^\alpha$ for the curvilinear coordinate system, the components (3.4.11) of the metric tensor in the deformed body are constants, the corresponding Christoffel symbols are zero, and covariant differentiation in the deformed body therefore reduces to partial differentiation. In the absence of body forces (3.2.11) thus yields

$$T^{11} = \bar{T}^{22} = -4\frac{\partial^2 \phi}{\partial \bar{z}^2}, \qquad T^{12} = 4\frac{\partial^2 \phi}{\partial z\,\partial \bar{z}}. \qquad (3.5.4)$$

If the resultant force **P** across any arc AP of a curve in a y_1, y_2-plane of the deformed body has components (X, Y) along the y_1-, y_2-axes respectively, a simple tensor transformation gives

$$\mathbf{P} = (X+iY)\mathbf{A}_1 + (X-iY)\mathbf{A}_2 = P\mathbf{A}_1 + \bar{P}\mathbf{A}_2, \qquad (3.5.5)$$

and remembering (3.4.11) we may interpret (3.2.16) in complex coordinates to get

$$P = 2i\frac{\partial \phi}{\partial \bar{z}}. \qquad (3.5.6)$$

Also, from (3.2.19), (3.4.9), and (3.4.10), the couple about the origin is

$$M = z\frac{\partial \phi}{\partial z} + \bar{z}\frac{\partial \phi}{\partial \bar{z}} - \phi. \qquad (3.5.7)$$

From (3.5.6), or directly from (3.2.21), we have at all points of a

† T'^{ij} denote components of the stress tensor across curves in the deformed body which were originally defined by complex coordinates $(\zeta, \bar{\zeta})$ in the undeformed body.

boundary curve which is entirely free from applied stress

$$\frac{\partial \phi}{\partial z} = 0, \qquad (3.5.8)$$

together with the complex conjugate of this equation.

The stress–strain relations for any given material may be put into complex variable form by combining the relations of § 3.4 with those of § 3.3. In the general (triclinic) case W may, by virtue of (3.4.17), be regarded as a function of the strain components $\Gamma_{\alpha\beta}$ and Γ_{33} (or e_{33}). From (3.3.1) and (3.4.16) we then have for compressible materials

$$T^{\alpha\beta} = \frac{1}{2\sqrt{I_3}}\left(\frac{\partial W}{\partial \Gamma_{\lambda\rho}} + \frac{\partial W}{\partial \Gamma_{\rho\lambda}}\right)\frac{\partial z^{\alpha}}{\partial \zeta^{\lambda}}\frac{\partial z^{\beta}}{\partial \zeta^{\rho}}, \qquad (3.5.9)$$

while (3.3.2) yields for the incompressible case

$$T^{\alpha\beta} = \frac{1}{2}\left(\frac{\partial W}{\partial \Gamma_{\lambda\rho}} + \frac{\partial W}{\partial \Gamma_{\rho\lambda}}\right)\frac{\partial z^{\alpha}}{\partial \zeta^{\lambda}}\frac{\partial z^{\beta}}{\partial \zeta^{\rho}} + pA^{\alpha\beta}. \qquad (3.5.10)$$

In these relations ζ^{α}, z^{α} represent the complex variables $(\zeta, \bar{\zeta})$, $(z, \bar{z})$; the formulae for τ^{33} $(= T^{33})$ are unchanged. Also, the equations of equilibrium (3.3.3) and (3.3.4) may now be written

$$\frac{\partial^2 \phi}{\partial z^{\alpha} \partial z^{\beta}} = -\frac{\sqrt{I_3}}{4\lambda^2}\left[\frac{\partial W}{\partial \Gamma_{11}}\frac{\partial \zeta^2}{\partial z^{\alpha}}\frac{\partial \zeta^2}{\partial z^{\beta}} + \frac{\partial W}{\partial \Gamma_{22}}\frac{\partial \zeta^1}{\partial z^{\alpha}}\frac{\partial \zeta^1}{\partial z^{\beta}} - \right.$$
$$\left. -\frac{1}{2}\left(\frac{\partial W}{\partial \Gamma_{12}} + \frac{\partial W}{\partial \Gamma_{21}}\right)\left(\frac{\partial \zeta^1}{\partial z^{\alpha}}\frac{\partial \zeta^2}{\partial z^{\beta}} + \frac{\partial \zeta^2}{\partial z^{\alpha}}\frac{\partial \zeta^1}{\partial z^{\beta}}\right)\right], \qquad (3.5.11)$$

and

$$\frac{\partial^2 \phi}{\partial z^{\alpha} \partial z^{\beta}} = -\frac{1}{4\lambda^2}\left[\frac{\partial W}{\partial \Gamma_{11}}\frac{\partial \zeta^2}{\partial z^{\alpha}}\frac{\partial \zeta^2}{\partial z^{\beta}} + \frac{\partial W}{\partial \Gamma_{22}}\frac{\partial \zeta^1}{\partial z^{\alpha}}\frac{\partial \zeta^1}{\partial z^{\beta}} - \right.$$
$$\left. -\frac{1}{2}\left(\frac{\partial W}{\partial \Gamma_{12}} + \frac{\partial W}{\partial \Gamma_{21}}\right)\left(\frac{\partial \zeta^1}{\partial z^{\alpha}}\frac{\partial \zeta^2}{\partial z^{\beta}} + \frac{\partial \zeta^2}{\partial z^{\alpha}}\frac{\partial \zeta^1}{\partial z^{\beta}}\right)\right] + pA_{\alpha\beta}, \qquad (3.5.12)$$

respectively. We may observe that alternative expressions for the stresses which exhibit z, $\bar{z}$ as independent variables may be derived from these relations by utilizing (3.4.5).

Rhombic system

For orthotropic materials in plane strain W may conveniently be expressed in terms of the functions

$$\left.\begin{aligned}\Gamma_1 &= \Gamma_{11} + \Gamma_{22} = \tfrac{1}{2}(e_{11} - e_{22}) \\ \Gamma_2 &= \Gamma_{11}\Gamma_{22} = \tfrac{1}{16}[(e_{11} - e_{22})^2 + 4e_{12}^2] \\ \Gamma_{12} &= \tfrac{1}{4}(e_{11} + e_{22})\end{aligned}\right\}, \qquad (3.5.13)$$

and Γ_{33} ($= e_{33}$), for then

$$e_{11} = 2\Gamma_{12}+\Gamma_1, \qquad e_{22} = 2\Gamma_{12}-\Gamma_1 \atop e_{12}^2 = 4\Gamma_2-\Gamma_1^2 \Bigg\}, \qquad (3.5.14)$$

and (3.5.9) and (3.5.11) yield

$$T^{\alpha\beta} = \frac{1}{\sqrt{I_3}}\Bigg[\left(\frac{\partial z^\alpha}{\partial \zeta^1}\frac{\partial z^\beta}{\partial \zeta^1}+\frac{\partial z^\alpha}{\partial \zeta^2}\frac{\partial z^\beta}{\partial \zeta^2}\right)\frac{\partial W}{\partial \Gamma_1}+\left(\frac{\partial z^\alpha}{\partial \zeta^1}\frac{\partial z^\beta}{\partial \zeta^1}\Gamma_{22}+\frac{\partial z^\alpha}{\partial \zeta^2}\frac{\partial z^\beta}{\partial \zeta^2}\Gamma_{11}\right)\frac{\partial W}{\partial \Gamma_2}+$$
$$+\frac{1}{2}\left(\frac{\partial z^\alpha}{\partial \zeta^1}\frac{\partial z^\beta}{\partial \zeta^2}+\frac{\partial z^\alpha}{\partial \zeta^2}\frac{\partial z^\beta}{\partial \zeta^1}\right)\frac{\partial W}{\partial \Gamma_{12}}\Bigg], \qquad (3.5.15)$$

and

$$\frac{\partial^2 \phi}{\partial z^\alpha \partial z^\beta} = -\frac{\sqrt{I_3}}{4\lambda^2}\Bigg[\left(\frac{\partial \zeta^1}{\partial z^\alpha}\frac{\partial \zeta^1}{\partial z^\beta}+\frac{\partial \zeta^2}{\partial z^\alpha}\frac{\partial \zeta^2}{\partial z^\beta}\right)\frac{\partial W}{\partial \Gamma_1}+\left(\frac{\partial \zeta^1}{\partial z^\alpha}\frac{\partial \zeta^1}{\partial z^\beta}\Gamma_{11}+\frac{\partial \zeta^2}{\partial z^\alpha}\frac{\partial \zeta^2}{\partial z^\beta}\Gamma_{22}\right)\frac{\partial W}{\partial \Gamma_2}-$$
$$-\frac{1}{2}\left(\frac{\partial \zeta^1}{\partial z^\alpha}\frac{\partial \zeta^2}{\partial z^\beta}+\frac{\partial \zeta^2}{\partial z^\alpha}\frac{\partial \zeta^1}{\partial z^\beta}\right)\frac{\partial W}{\partial \Gamma_{12}}\Bigg], \qquad (3.5.16)$$

respectively. Corresponding formulae for incompressible materials may be derived from (3.5.10) and (3.5.12).

Transversely isotropic and isotropic bodies

For isotropic and transversely isotropic bodies the equations of equilibrium become formally identical but the forms of W which occur in these relations are different. From (3.3.24) or (3.3.31) with (3.4.11) to (3.4.13) we have, for compressible bodies,

$$\left.\begin{aligned}\frac{\partial^2 \phi}{\partial z^2} &= \frac{I_3}{\lambda^2}\frac{\partial \bar{D}}{\partial z}\left(\frac{\partial D}{\partial z}-1\right)\mathcal{H} \\ \frac{\partial^2 \phi}{\partial z\, \partial \bar{z}} &= \left(\frac{\sqrt{I_3}}{2\lambda}+\frac{I_3}{\lambda^2}\frac{\partial D}{\partial \bar{z}}\frac{\partial \bar{D}}{\partial z}\right)\mathcal{H}+\tfrac{1}{2}\mathcal{L}\end{aligned}\right\}. \qquad (3.5.17)$$

For incompressible materials, if we put $\alpha = 1$, $\beta = 1$ in (3.3.27), we have

$$\lambda^2 \frac{\partial^2 \phi}{\partial z^2} = \frac{\partial \bar{D}}{\partial z}\left(\frac{\partial D}{\partial z}-1\right)\mathcal{H}. \qquad (3.5.18)$$

The values $\alpha = 2$, $\beta = 2$ yield the complex conjugate of this equation; when $\alpha = 1$, $\beta = 2$ (3.3.27) is identically satisfied. Expressions for $T^{\alpha\beta}$ follow from (3.5.4); the formulae of § 3.3 for τ^{33} (or T^{33}) are again unchanged.

The results of the present section may be put in a form in which ζ, $\bar{\zeta}$ appear as independent variables by making use of (3.4.4).

3.6. Incompressible materials: reciprocal equations

For incompressible isotropic or transversely isotropic bodies we have from (3.5.18) and (3.4.21) in the absence of body forces

$$\lambda^2 \frac{\partial^2 \phi}{\partial z^2} = 2 \frac{\partial \zeta}{\partial z} \frac{\partial \bar{\zeta}}{\partial z} \frac{\partial W}{\partial I} \quad [W = W(I, K)], \tag{3.6.1}$$

$$\frac{\partial(\zeta, \bar{\zeta})}{\partial(z, \bar{z})} = \lambda, \tag{3.6.2}$$

$$I = \lambda^2 + \frac{2}{\lambda} + \frac{4}{\lambda^2} \frac{\partial \zeta}{\partial \bar{z}} \frac{\partial \bar{\zeta}}{\partial z}, \quad K = \tfrac{1}{2}(\lambda^2 - 1). \tag{3.6.3}$$

From the first of these equations ϕ may be eliminated by differentiating twice with respect to $\bar{z}$ and equating the resulting expression to its complex conjugate, giving

$$\mathscr{I}\left[\frac{\partial W}{\partial I} \frac{\partial^2}{\partial \bar{z}^2}\left(\frac{\partial \zeta}{\partial z} \frac{\partial \bar{\zeta}}{\partial z}\right) + \frac{4}{\lambda^2} \frac{\partial^2 W}{\partial I^2}\left\{\frac{\partial \zeta}{\partial z} \frac{\partial \bar{\zeta}}{\partial z} \frac{\partial^2}{\partial \bar{z}^2}\left(\frac{\partial \zeta}{\partial \bar{z}} \frac{\partial \bar{\zeta}}{\partial z}\right) + 2\frac{\partial}{\partial \bar{z}}\left(\frac{\partial \zeta}{\partial z} \frac{\partial \bar{\zeta}}{\partial z}\right) \frac{\partial}{\partial \bar{z}}\left(\frac{\partial \zeta}{\partial \bar{z}} \frac{\partial \bar{\zeta}}{\partial z}\right)\right\} + \right.$$
$$\left. + \frac{16}{\lambda^4} \frac{\partial^3 W}{\partial I^3} \frac{\partial \zeta}{\partial z} \frac{\partial \bar{\zeta}}{\partial z}\left\{\frac{\partial}{\partial \bar{z}}\left(\frac{\partial \zeta}{\partial \bar{z}} \frac{\partial \bar{\zeta}}{\partial z}\right)\right\}^2\right] = 0, \tag{3.6.4}$$

where $\mathscr{I}$ denotes the imaginary part of the expression enclosed in square brackets. By using (3.1.13) and (3.4.4) to change the independent variables we obtain from (3.6.2) to (3.6.4)

$$\frac{\partial(z, \bar{z})}{\partial(\zeta, \bar{\zeta})} = \frac{1}{\lambda}, \tag{3.6.5}$$

$$I = \lambda^2 + \frac{2}{\lambda} + 4 \frac{\partial z}{\partial \bar{\zeta}} \frac{\partial \bar{z}}{\partial \zeta}, \quad K = \tfrac{1}{2}(\lambda^2 - 1), \tag{3.6.6}$$

$$\mathscr{I}\left[\frac{\partial W}{\partial I} \frac{\partial^2}{\partial \bar{\zeta}^2}\left(\frac{\partial z}{\partial \zeta} \frac{\partial \bar{z}}{\partial \zeta}\right) + 4\frac{\partial^2 W}{\partial I^2}\left\{\frac{\partial z}{\partial \zeta} \frac{\partial \bar{z}}{\partial \zeta} \frac{\partial^2}{\partial \bar{\zeta}^2}\left(\frac{\partial z}{\partial \bar{\zeta}} \frac{\partial \bar{z}}{\partial \zeta}\right) + 2\frac{\partial}{\partial \bar{\zeta}}\left(\frac{\partial z}{\partial \zeta} \frac{\partial \bar{z}}{\partial \zeta}\right) \frac{\partial}{\partial \bar{\zeta}}\left(\frac{\partial z}{\partial \bar{\zeta}} \frac{\partial \bar{z}}{\partial \zeta}\right)\right\} + \right.$$
$$\left. + 16 \frac{\partial^3 W}{\partial I^3} \frac{\partial z}{\partial \zeta} \frac{\partial \bar{z}}{\partial \zeta}\left\{\frac{\partial}{\partial \bar{\zeta}}\left(\frac{\partial z}{\partial \bar{\zeta}} \frac{\partial \bar{z}}{\partial \zeta}\right)\right\}^2\right] = 0. \tag{3.6.7}$$

The transformation from (3.6.4) to (3.6.7) is lengthy but may be carried out by considering separately the coefficients of the three derivatives of W which occur in the two expressions; these are equivalent to each other apart from the same constant factor. To establish the equivalence of the coefficients of $\partial^2 W/\partial I^2$ in the two cases it is necessary to make use of the equation

$$\frac{\partial z}{\partial \zeta}\frac{\partial^2 \bar{z}}{\partial \bar{\zeta}^2} = \frac{\partial \bar{z}}{\partial \zeta}\frac{\partial^2 z}{\partial \bar{\zeta}^2} + \frac{\partial z}{\partial \bar{\zeta}}\frac{\partial^2 \bar{z}}{\partial \zeta \partial \bar{\zeta}} - \frac{\partial \bar{z}}{\partial \bar{\zeta}}\frac{\partial^2 z}{\partial \zeta \partial \bar{\zeta}}, \qquad (3.6.8)$$

derived by differentiation of (3.6.5), while in transforming the coefficient of $\partial W/\partial I$ the additional relation obtained from (3.6.8) by a further differentiation with respect to $\bar{\zeta}$ must also be employed. Evidently this transformation does not depend upon the form of W as a function of I and K.

The substitutions

$$\zeta' = \lambda^n \zeta, \qquad z' = \lambda^{n+1} z,$$

where n is a real constant, reduce (3.6.5)–(3.6.7) to equations identical with those obtained by replacing z, ζ in (3.6.2)–(3.6.4) by ζ', z' respectively. It follows that if

$$\zeta = f(z, \bar{z}, \lambda) \qquad (3.6.9)$$

defines a solution in the absence of body forces of the finite plane strain equations for an incompressible isotropic or transversely isotropic material with a given form of strain energy function defined by $W = W(I)$ or $W = W(I, K)$, then

$$z' = f(\zeta', \bar{\zeta}', \lambda)$$

or
$$\lambda^{n+1} z = f(\lambda^n \zeta, \lambda^n \bar{\zeta}, \lambda) \qquad (3.6.10)$$

also represents a possible solution for the same material. By writing (3.6.10) as

$$kz = \frac{1}{\lambda} f(k\zeta, k\bar{\zeta}, \lambda) \qquad (k = \lambda^n),$$

we see that these reciprocal solutions, for all values of n, are equivalent apart from the constant scaling factor k. Putting $k = 1$ ($n = 0$), we therefore have

$$z = \frac{1}{\lambda} f(\zeta, \bar{\zeta}, \lambda) \qquad (3.6.11)$$

as the solution associated with (3.6.9). In terms of the real variables x_α, y_α it follows from (3.6.9), (3.6.11) that a solution

$$x_\alpha = f_\alpha(y_1, y_2, \lambda) \qquad (3.6.12)$$

of the plane strain equations automatically implies a related solution of the type

$$y_\alpha = \frac{1}{\lambda} f_\alpha(x_1, x_2, \lambda). \qquad (3.6.13)$$

When the material is isotropic, and the strain energy function W assumes the linear form (1.14.3) suggested by Mooney, the first derivatives take the constant values

$$\left. \begin{array}{c} \dfrac{\partial W}{\partial I_1} = C_1, \qquad \dfrac{\partial W}{\partial I_2} = C_2 \\[6pt] \dfrac{dW(I)}{dI} = C_1 + \lambda^2 C_2 = C \quad (\text{say}) \end{array} \right\}, \qquad (3.6.14)$$

and (3.6.4), (3.6.7) reduce to

$$\left. \begin{array}{c} \left(\dfrac{\partial \zeta}{\partial z}\dfrac{\partial}{\partial \bar{z}} - \dfrac{\partial \zeta}{\partial \bar{z}}\dfrac{\partial}{\partial z}\right)\dfrac{\partial^2 \bar{\zeta}}{\partial z\,\partial \bar{z}} + \left(\dfrac{\partial \bar{\zeta}}{\partial z}\dfrac{\partial}{\partial \bar{z}} - \dfrac{\partial \bar{\zeta}}{\partial \bar{z}}\dfrac{\partial}{\partial z}\right)\dfrac{\partial^2 \zeta}{\partial z\,\partial \bar{z}} = 0 \\[6pt] \left(\dfrac{\partial z}{\partial \zeta}\dfrac{\partial}{\partial \bar{\zeta}} - \dfrac{\partial z}{\partial \bar{\zeta}}\dfrac{\partial}{\partial \zeta}\right)\dfrac{\partial^2 \bar{z}}{\partial \zeta\,\partial \bar{\zeta}} + \left(\dfrac{\partial \bar{z}}{\partial \zeta}\dfrac{\partial}{\partial \bar{\zeta}} - \dfrac{\partial \bar{z}}{\partial \bar{\zeta}}\dfrac{\partial}{\partial \zeta}\right)\dfrac{\partial^2 z}{\partial \zeta\,\partial \bar{\zeta}} = 0 \end{array}\right\}, \qquad (3.6.15)$$

and

respectively. By making use of (3.4.1) and (3.4.3) these equations may be expressed in terms of the real variables x_α, y_α as

$$\left. \begin{array}{c} \dfrac{\partial(x_1, \nabla_1^2 x_1)}{\partial(y_1, y_2)} + \dfrac{\partial(x_2, \nabla_1^2 x_2)}{\partial(y_1, y_2)} = 0 \\[6pt] \dfrac{\partial(y_1, \nabla_2^2 y_1)}{\partial(x_1, x_2)} + \dfrac{\partial(y_2, \nabla_2^2 y_2)}{\partial(x_1, x_2)} = 0 \end{array}\right\}, \qquad (3.6.16)$$

respectively, where

$$\nabla_1^2 \equiv \dfrac{\partial^2}{\partial y_1^2} + \dfrac{\partial^2}{\partial y_2^2}, \qquad \nabla_2^2 \equiv \dfrac{\partial^2}{\partial x_1^2} + \dfrac{\partial^2}{\partial x_2^2},$$

and with these must be coupled the corresponding forms of the incompressibility condition

$$\dfrac{\partial(x_1, x_2)}{\partial(y_1, y_2)} = \lambda, \qquad \dfrac{\partial(y_1, y_2)}{\partial(x_1, x_2)} = \dfrac{1}{\lambda}. \qquad (3.6.17)$$

In addition to the reciprocal property defined by (3.6.12), (3.6.13), we may observe that if (3.6.12) represents a solution of the plane strain equations for a Mooney material, then

$$\left. \begin{array}{c} y_\alpha = f_\alpha\!\left(x_1, x_2, \dfrac{1}{\lambda}\right) \\[6pt] x_\alpha = f_\alpha\!\left(ky_1, ky_2, \dfrac{\lambda}{k^2}\right) \end{array}\right\} \qquad (3.6.18)$$

and

are further possible solutions. The latter expression represents a uniform magnification of the deformation pattern represented by (3.6.12) in the x_1, x_2-plane contingent upon a further uniform extension of the material in the x_3-direction. These further reciprocal solutions do not necessarily hold, in the absence of body forces, for the more general material defined by (3.6.4).

The foregoing results continue to apply if a two-dimensional system of body forces is present, provided that these forces can be represented as the gradient of a potential function U. The original solution (3.6.9) and the reciprocal solutions (3.6.11), (3.6.18) are unaffected but the principal stresses σ_{ii} are now all increased by an amount ρU.†

3.7. Flexure of a cuboid

The flexure problem considered in § 2.15 may evidently be regarded as an example of plane strain. Consider a cuboid of aeolotropic material bounded by the planes $x_1 = A_1$, $x_1 = A_2$, $x_2 = \pm B$, $x_3 = \pm C$.‡ In addition to the symmetry conditions imposed in § 3.3, we shall, for simplicity, suppose the form of the strain energy function for the body to be restricted so that each of the derivatives $\partial W/\partial e_{ij}$ ($i \neq j$) vanishes when $e_{12} = e_{23} = e_{31} = 0$. This implies that W is not linear in any of the components e_{12}, e_{23}, e_{31} alone and excludes from the materials listed in § 3.3 the monoclinic class and the forms (1.10.3), (1.12.11). This restriction is sufficient to ensure the vanishing of each of the stress components τ^{ij} for $i \neq j$ under the deformation being considered. The cuboid is deformed symmetrically with respect to the x_1-axis so that:

(i) each plane initially normal to the x_1-axis becomes, in the deformed state, a portion of a curved surface of a cylinder whose axis is the x_3-axis;

(ii) planes initially normal to the x_2-axis become in the deformed state planes containing the x_3-axis;

(iii) there is a uniform extension λ in the direction of the x_3-axis.

The curvilinear system θ_i may now be identified with a system of cylindrical polar coordinates in the deformed body, so that

$$\left. \begin{array}{l} (\theta_1, \theta_2, \theta_3) = (r, \theta, y_3) \\ y_1 = r\cos\theta, \quad y_2 = r\sin\theta \end{array} \right\} \quad (3.7.1)$$

---

† For details, and illustrations of the reciprocal solutions, reference may be made to the paper by J. E. Adkins, loc. cit., p. 101.

‡ The constant C used in this section is not the same as that defined by (3.6.14).

and the deformation is defined by

$$r = f(x_1), \qquad \theta = x_2/(\lambda k), \qquad y_3 = \lambda x_3, \qquad (3.7.2)$$

where k is a constant. Hence

$$A_{\alpha\beta} = \begin{bmatrix} 1 & 0 \\ 0 & r^2 \end{bmatrix}, \quad A^{\alpha\beta} = \begin{bmatrix} 1 & 0 \\ 0 & \dfrac{1}{r^2} \end{bmatrix}, \quad A = r^2$$

$$a_{\alpha\beta} = \begin{bmatrix} \dfrac{1}{f'^2} & 0 \\ 0 & \lambda^2 k^2 \end{bmatrix}, \quad a^{\alpha\beta} = \begin{bmatrix} f'^2 & 0 \\ 0 & \dfrac{1}{\lambda^2 k^2} \end{bmatrix}, \quad a = \dfrac{\lambda^2 k^2}{f'^2} \qquad (3.7.3)$$

$$I_3 = \frac{r^2 f'^2}{k^2}, \qquad (3.7.4)$$

a prime denoting differentiation with respect to x_1. Also, from (3.7.1) to (3.7.3) it follows that

$$e_{11} = \tfrac{1}{2}(f'^2 - 1), \qquad e_{22} = \frac{1}{2}\left(\frac{r^2}{\lambda^2 k^2} - 1\right), \qquad e_{12} = 0. \qquad (3.7.5)$$

The only non-zero Christoffel symbols formed with the metric tensor $A_{\alpha\beta}$ are now

$$\Gamma^2_{12} = \Gamma^2_{21} = \frac{1}{r}, \qquad \Gamma^1_{22} = -r, \qquad (3.7.6)$$

and if we interpret (3.3.3) or (3.3.8) in polar coordinates and assume that ϕ is independent of θ, we obtain for compressible bodies

$$\phi\|_{11} = \frac{d^2\phi}{dr^2} = \frac{r}{k\lambda^2 f'}\frac{\partial W}{\partial e_{22}}$$

$$\phi\|_{22} = r\frac{d\phi}{dr} = krf'\frac{\partial W}{\partial e_{11}} \qquad (3.7.7)$$

the relation for $\alpha = 1$, $\beta = 2$ vanishing identically by virtue of the restriction on the form of W. Since

$$\frac{de_{11}}{dx_1} = f'f'', \qquad \frac{de_{22}}{dx_1} = \frac{ff'}{k^2\lambda^2},$$

it follows from (3.7.7) that

$$\frac{f'^2}{k}\frac{d^2\phi}{dr^2} + \frac{f''}{k}\frac{d\phi}{dr} = \frac{ff'}{k^2\lambda^2}\frac{\partial W}{\partial e_{22}} + f'f''\frac{\partial W}{\partial e_{11}} = \frac{dW}{dx_1},$$

or

$$\frac{f'}{k}\frac{d}{dr}\left(f'\frac{d\phi}{dr}\right) = \frac{1}{k}\frac{d^2\phi}{dx_1^2} = \frac{dW}{dx_1},$$

so that
$$\frac{f'd\phi}{k\,dr} = \frac{1}{k}\frac{d\phi}{dx_1} = W(x_1)+W_0, \qquad (3.7.8)$$

where W_0 is a constant.

From (3.2.16) and (3.2.11) we have
$$\mathbf{P} = \epsilon^{\rho\beta}\phi_{,\rho}\mathbf{A}_\beta = \frac{1}{r}\frac{d\phi}{dr}\mathbf{A}_2 = \tau^{11}\mathbf{A}_2, \qquad (3.7.9)$$

so that the resultant force per unit length acting on each of the surfaces initially at $x_2 = \pm B$ is the difference in $d\phi/dr$ or $r\tau^{11}$ as r changes from r_1 to r_2. When the applied stresses over the curved surfaces of the deformed cylinder are zero, $d\phi/dr$, and hence also $d\phi/dx_1$, vanishes on these surfaces,

and
$$\left.\begin{array}{c} W_0 = -W(A_1) = -W(A_2) \\ \mathbf{P} = 0 \end{array}\right\}. \qquad (3.7.10)$$

The resultant of the stresses acting on the surfaces initially at $x_2 = \pm B$ then reduces to a couple M, which from (3.2.19) is given by
$$M = 2\lambda C[\phi(A_2)-\phi(A_1)]$$
$$= 2\lambda kC \int_{A_1}^{A_2} [W(x_1)+W_0]\,dx_1$$
$$= 2\lambda kC\left[\int_{A_1}^{A_2} W(x_1)\,dx_1 - (A_1-A_2)W_0\right]. \qquad (3.7.11)$$

From (3.7.7) and (3.7.8)
$$W+W_0 = f'^2\frac{\partial W}{\partial_{11}}, \qquad (3.7.12)$$

which is an equation for f' when the form of W is known. The stresses over the ends of the cuboid may be found from the last of equations (3.3.1). Alternatively, if the resultant force on the ends of the cylinder is given, this relation yields an equation for the determination of λ. For, remembering (3.7.2) and (3.3.1),
$$N = \iint r\tau^{33}\,d\theta\,dr = 2B\lambda\int_{A_2}^{A_1}\frac{\partial W}{\partial e_{33}}\,dx_1. \qquad (3.7.13)$$

When the body is incompressible,
$$f' = k/r = k/f, \qquad r^2 = 2kx_1+K, \qquad (3.7.14)$$

where K is a constant. If r_1, r_2 ($r_1 > r_2$) are the radii of the curved surfaces of the deformed body, then

$$k = \frac{r_1^2 - r_2^2}{2(A_1 - A_2)}, \quad K = \frac{A_1 r_2^2 - A_2 r_1^2}{A_1 - A_2}, \qquad (3.7.15)$$

and (3.7.11) becomes

$$M = \lambda C \left[2 \int_{r_1}^{r_2} r W(r) \, dr - (r_1^2 - r_2^2) W_0 \right]. \qquad (3.7.16)$$

To obtain a detailed examination of the stress distribution for an incompressible cuboid we replace (3.7.7) by

$$\left. \begin{array}{c} \dfrac{d^2\phi}{dr^2} = \dfrac{r^2}{k^2 \lambda^2} \dfrac{\partial W}{\partial e_{22}} + p \\[6pt] r\dfrac{d\phi}{dr} = k^2 \dfrac{\partial W}{\partial e_{11}} + r^2 p \end{array} \right\}, \qquad (3.7.17)$$

where, from (3.7.5) and (3.7.14),

$$e_{11} = \frac{1}{2}\left(\frac{k^2}{r^2} - 1\right), \quad e_{22} = \frac{1}{2}\left(\frac{r^2}{\lambda^2 k^2} - 1\right). \qquad (3.7.18)$$

The elimination of p from (3.7.17) leads to

$$\frac{1}{r}\frac{d\phi}{dr} = W(r) + W_0, \qquad (3.7.19)$$

and thence to (3.7.16) if the conditions (3.7.10) are satisfied. The value of p follows by combining this equation with the second of (3.7.17).

The analysis for isotropic and transversely isotropic bodies follows similar lines. Equations (3.7.8) to (3.7.11) for compressible materials and (3.7.14) to (3.7.16) for the incompressible case are unaffected by the additional symmetry properties possessed by the material; equation (3.7.11) is valid also for incompressible bodies. The equations giving the detailed stress distribution, however, assume different forms, and for isotropic incompressible materials, for example, we see from (3.3.27) that equations (3.7.17) may be replaced by

$$\frac{d}{dr}\left(\frac{1}{r}\frac{d\phi}{dr}\right) = \left(\frac{r}{k^2\lambda^2} - \frac{k^2}{r^3}\right)\mathscr{H} \quad \left[\mathscr{H} = 2\frac{dW(I)}{dI}\right],$$

where

$$I = \lambda^2 + \frac{r^2}{k^2\lambda^2} + \frac{k^2}{r^2},$$

a result which again leads immediately to (3.7.19).

3.8. Generalized shear

We consider the deformation of an isotropic incompressible cuboid bounded initially by the faces $x_1 = \pm A_0$, $x_2 = \pm B_0$, $x_3 = \pm C_0$, where A_0, B_0, C_0 are constants, and examine the possibility of a generalized shear in which each point moves parallel to the x_1-direction. This may be described by the relations

$$x_1 = y_1 + f(y_2), \qquad x_2 = y_2, \qquad x_3 = y_3, \qquad (3.8.1)$$

where f is a function of y_2 to be determined. Identifying the curvilinear system θ_i with the coordinates y_i, we have

$$\left. A_{\alpha\beta} = A^{\alpha\beta} = \delta_{\alpha\beta}, \quad A = 1 \atop a_{\alpha\beta} = \begin{bmatrix} 1 & f' \\ f' & 1+f'^2 \end{bmatrix}, \quad a^{\alpha\beta} = \begin{bmatrix} 1+f'^2 & -f' \\ -f' & 1 \end{bmatrix}, \quad a = 1 \right\}, \qquad (3.8.2)$$

a prime denoting differentiation with respect to y_2. The incompressibility condition is evidently satisfied, and from (3.3.12), with $\lambda = 1$,

$$I = 1 + a^{11} + a^{22} = 3 + f'^2. \qquad (3.8.3)$$

With the help of (3.8.1) and (3.8.2) equations (3.3.27) yield

$$\frac{\partial^2 \phi}{\partial y_1^2} - \frac{\partial^2 \phi}{\partial y_2^2} = -f'^2 \mathscr{H}, \qquad \frac{\partial^2 \phi}{\partial y_1 \partial y_2} = f' \mathscr{H}, \qquad (3.8.4)$$

where we recall that $\mathscr{H}$ is given by (3.3.34) and is a function of $f'(y_2)$. Equations (3.8.4) are compatible if

$$\frac{d^2}{dy_2^2}(f'\mathscr{H}) = 0,$$

or
$$f'\mathscr{H} = K + Ly_2, \qquad (3.8.5)$$

where K and L are constants. This is a differential equation for f which may be solved in the usual manner when the form of W, and hence $\mathscr{H}$ is known.

From (3.8.4) and (3.8.5) we see that ϕ has the form

$$\phi = y_1(Ky_2 + \tfrac{1}{2}Ly_2^2) + \tfrac{1}{6}Ly_1^3 + \psi(y_2) + \tfrac{1}{2}a'(y_1^2 + y_2^2) + b'y_1 + c'y_2 + d', \qquad (3.8.6)$$

† For an examination of the isotropic case, reference may be made to the paper by J. E. Adkins, A. E. Green, and R. T. Shield, loc. cit., p. 101.

where a', b', c', d' are constants and $\psi(y_2)$ is a function of y_2 only which is determined by the equation

$$\frac{d^2\psi}{dy_2^2} = (K+Ly_2)f'. \tag{3.8.7}$$

From (3.2.11) with $\theta_i = y_i$, we obtain for the physical components of stress $\tau^{ij} = \sigma_{ij}$,

$$\left.\begin{aligned}\sigma_{11} &= \frac{\partial^2\phi}{\partial y_2^2} = (K+Ly_2)f' + Ly_1 + a' \\ \sigma_{22} &= \frac{\partial^2\phi}{\partial y_1^2} = Ly_1 + a' \\ \sigma_{12} &= -\frac{\partial^2\phi}{\partial y_1\,\partial y_2} = -(K+Ly_2)\end{aligned}\right\}, \tag{3.8.8}$$

and from (3.3.29)

$$\sigma_{33} = \tfrac{1}{2}f'^2(\mathscr{H}-2\Phi) + \tfrac{1}{2}f'(K+Ly_2) + Ly_1 + a'. \tag{3.8.9}$$

We observe that the terms in (3.8.6) involving a' represent a uniform hydrostatic pressure, while the linear terms contribute nothing to the stresses and could be omitted.

When f' has been determined from (3.8.5), $\psi(y_2)$ may be obtained from (3.8.7) by integration, but this is unnecessary for the evaluation of the stresses. When $f' = $ constant, the deformation becomes a simple shear, I, $\mathscr{H}$, and Φ are constants, and (3.8.5) is satisfied with $L = 0$. The stresses are then uniform throughout the body. This case is considered in detail in § 3.2 of $T.E.$ In the general case it is of interest to note that (3.8.5) may be written

$$(K+Ly_2)f'' = 2f'f''\frac{dW_1}{dI} = \frac{dW_1}{dy_2}, \tag{3.8.10}$$

using (3.8.3), and then, by successive integration,

$$(K+Ly_2)f' - Lf = W_1 + k_1, \tag{3.8.11}$$

and

$$f = (K+Ly_2)\left\{\int \frac{W_1+k_1}{(K+Ly_2)^2}\,dy_2 + k_2\right\}, \tag{3.8.12}$$

where k_1, k_2 are constants. Alternatively, remembering that W_1 is a function of $f'(y_2)$, we may write the differential equation (3.8.11) in Clairaut's form

$$f = y_2 f' + F(f'), \tag{3.8.13}$$

where
$$F(f') = [Kf' - W_1(f') - k_1]/L. \qquad (3.8.14)$$

The general solution
$$f(y_2) = ky_2 + F(k), \qquad (3.8.15)$$

where k is a further arbitrary constant, would imply that f' is a constant, or that in (3.8.10) $f'' = 0$, both sides of the equation vanishing identically.

There exists also, however, the singular solution obtained by eliminating f' between

$$\left. \begin{array}{l} f = y_2 f' + F(f') \\ 0 = y_2 + dF/df' \end{array} \right\}. \qquad (3.8.16)$$

When the strain energy takes the Mooney form (1.14.3),
$$W_1 = C(I-3) = Cf'^2 \quad (C = C_1 + C_2),$$

F becomes a quadratic form in f', and equations (3.8.16) yield the results
$$f' = (K + Ly_2)/(2C),$$
$$f = (K' + Ky_2 + \tfrac{1}{2}Ly_2^2)/(2C),$$

identical with those obtained by a direct integration of (3.8.5). From (3.8.8), (3.8.9) we now obtain
$$\sigma_{11} = (K + Ly_2)^2/(2C) + Ly_1 + a',$$
$$\sigma_{33} = C_2(K + Ly_2)^2/(2C^2) + Ly_1 + a',$$

the expressions for the remaining stresses being unaffected.

3.9. Further special solutions for a Mooney material

The formulae of § 3.6 may be illustrated by a consideration of deformations in which y_1 is a function of x_1 only, or conversely x_1 is a function solely of y_1. Planes normal to the x_1-axis in the unstrained body then become planes normal to the y_1-axis in the deformed body but are displaced relative to each other through a distance which is a function of their x_1-coordinate.

We therefore write
$$x_1 = h(y_1), \qquad x_2 = \lambda y_2 g(y_1) + f(y_1), \qquad (3.9.1)$$

where f, g, and h are functions of y_1. The incompressibility condition (3.6.17) is satisfied if
$$gh' = 1, \qquad (3.9.2)$$

while the first of equations (3.6.16) yields
$$\lambda y_2(g'g'' - gg''') + f'g'' - f'''g = 0.$$

Since this equation must be satisfied for all values of y_2 we have the two independent relations

$$g'g'' - gg''' = 0, \qquad f'g'' - f'''g = 0, \tag{3.9.3}$$

which yield by integration

$$g'' = \pm a^2 g, \qquad f'' = K \pm a^2 f, \tag{3.9.4}$$

where a ($\geqslant 0$) and K are arbitrary real constants. By choosing the positive signs in (3.9.4) and using (3.9.2) we obtain the solutions

$$\left.\begin{aligned} f(y_1) &= A_1 e^{ay_1} + A_2 e^{-ay_1} - K/a^2 \\ g(y_1) &= B_1 e^{ay_1} + B_2 e^{-ay_1} \end{aligned}\right\}, \tag{3.9.5}$$

with

$$h(y_1) = \frac{1}{B_1 ab}\tan^{-1}\!\left(\frac{e^{ay_1}}{b}\right) + K' \qquad (B_1 > 0,\ B_2 = b^2 B_1 > 0), \tag{3.9.6}$$

or $\quad h(y_1) = \dfrac{1}{2B_1 ab}\log\!\left(\dfrac{e^{ay_1}-b}{e^{ay_1}+b}\right) + K'$

$$(B_1 > 0,\ B_2 = -b^2 B_1 < 0,\ e^{ay_1} > |b|), \tag{3.9.7}$$

while the negative signs yield similarly

$$\left.\begin{aligned} f(y_1) &= A_1 \sin(ay_1+c) + K/a^2 \\ g(y_1) &= B_1 \sin(ay_1+d) \\ h(y_1) &= \frac{1}{B_1 a}\log\tan\tfrac{1}{2}(ay_1+d) + K' \\ &\quad [2n\pi < ay_1+d < (2n+1)\pi;\ n \text{ integer}] \end{aligned}\right\}. \tag{3.9.8}$$

The choice $a = 0$ in (3.9.4) yields the further solution

$$\left.\begin{aligned} f(y_1) &= \tfrac{1}{2}Ky_1^2 + cy_1 + d \\ g(y_1) &= A_1 y_1 + B_1 \\ h(y_1) &= \frac{1}{A_1}\log(A_1 y_1 + B_1) + K' \qquad (A_1 y_1 + B_1 > 0) \end{aligned}\right\}. \tag{3.9.9}$$

In (3.9.5)–(3.9.9) A_1, A_2, B_1, B_2, K', c, and d are arbitrary constants of integration. The values of x_1, x_2 are obtained by combining these solutions with (3.9.1). In (3.9.5)–(3.9.8) the constants K/a^2 and K' represent rigid body translations and could be omitted without affecting the stresses; a similar remark applies to the constants d and K' in (3.9.9). For given values of the arbitrary constants, the solutions

(3.9.7)–(3.9.9) represent physically real deformations only within the ranges of the variables x_1, y_1 specified; conversely, for given ranges of these variables, the arbitrary constants must be restricted in the manner indicated.

The stress distributions required to maintain these deformations may be determined by solving equations (3.6.1), with $dW/dI = C$, for ϕ and making use of (3.5.4). Remembering (3.4.1), we have

$$\left. \begin{aligned} \lambda^2 \left(\frac{\partial^2 \phi}{\partial y_1^2} - \frac{\partial^2 \phi}{\partial y_2^2} \right) &= 2C \left[\left(\frac{\partial x_1}{\partial y_1} \right)^2 + \left(\frac{\partial x_2}{\partial y_1} \right)^2 - \left(\frac{\partial x_1}{\partial y_2} \right)^2 - \left(\frac{\partial x_2}{\partial y_2} \right)^2 \right] \\ \lambda^2 \frac{\partial^2 \phi}{\partial y_1 \partial y_2} &= 2C \left[\frac{\partial x_1}{\partial y_1} \frac{\partial x_1}{\partial y_2} + \frac{\partial x_2}{\partial y_1} \frac{\partial x_2}{\partial y_2} \right] \end{aligned} \right\} \quad (3.9.10)$$

which, with (3.9.1), yields

$$\left. \begin{aligned} \lambda^2 \left(\frac{\partial^2 \phi}{\partial y_1^2} - \frac{\partial^2 \phi}{\partial y_2^2} \right) &= 2C[h'^2 + (\lambda y_2 g' + f')^2 - \lambda^2 g^2] \\ \lambda^2 \frac{\partial^2 \phi}{\partial y_1 \partial y_2} &= 2C\lambda g (\lambda y_2 g' + f') \end{aligned} \right\} . \quad (3.9.11)$$

The second of these relations gives, by integration,

$$\lambda^2 \phi = C \left[\tfrac{1}{2} \lambda^2 y_2^2 g^2 + 2\lambda y_2 \int^{y_1} f' g \, dy_1 + \psi_1(y_1) + \psi_2(y_2) \right],$$

or, if we make use of the first equation to determine the arbitrary functions ψ_1, ψ_2,

$$\lambda^2 \phi = C \left\{ \tfrac{1}{2} \lambda^2 y_2^2 g^2 + 2\lambda y_2 \int^{y_1} f' g \, dy_1 + \int^{y_1} dy_1 \int^{y_1} (2h'^2 + 2f'^2 - \lambda^2 g^2) \, dy_1 + \right.$$
$$\left. + k_1 y_2^4 + k_2 y_2^3 + p(y_1^2 + y_2^2) + k_3 y_1 + k_4 y_2 + k_5 \right\}, \quad (3.9.12)$$

where k_1 to k_5 and p are again constants of integration. The terms in p represent a uniform hydrostatic pressure; the linear terms contribute nothing to the stresses. Also, in deriving the expression for ϕ it is necessary to make use of the relations

$$\lambda^2 (gg'' - g'^2) = 12k_1, \quad \lambda(f''g - f'g') = 3k_2,$$

which are satisfied by virtue of (3.9.3), the constants k_1 and k_2 being related to a, A_1, A_2, B_1, B_2, c, and d by means of (3.9.5)–(3.9.9). By combining (3.9.5)–(3.9.9) with (3.9.1) and (3.9.12) we obtain expressions for ϕ, and hence may derive for each case the relevant system of stresses.

For each of the solutions obtained from (3.9.5)–(3.9.9) corresponding to (3.6.12) there exist associated solutions of the types (3.6.13), (3.6.18). Owing to the large number of arbitrary constants which occur in (3.9.5)–(3.9.9), however, not all of these are distinct.

IV
THEORY OF ELASTIC MEMBRANES

THE theory of thin plates and shells for highly elastic materials differs from that for classical elasticity in that, in the former case, the principal extension ratios in the middle surface of the deformed plate or shell are usually appreciably greater than unity. If the variations of these quantities throughout the thickness of the deformed sheet are not too large, shearing stresses and couples may then be neglected in comparison with the stress resultants acting in the tangent plane to the deformed middle surface. Under these circumstances the membrane theory for highly elastic shells becomes of comparably greater importance than is the case for classically small deformations. Furthermore, if attention is confined to a membrane theory, the difficulties are avoided which are experienced in classical elasticity in formulating the equations for the bending of shells.†

In this chapter the theory of plane stress for finite deformations‡ (§§ 4.1–4.4) and the membrane theory for thin shells (§§ 4.5–4.9) are formulated.§ In each case the equations of equilibrium are identical with the corresponding equations of classical elasticity provided all quantities are referred to coordinates in the middle surface of the *deformed* plate or shell. The formulae of Chapter I for finite deformations must, of course, be used in the evaluation of the stress resultants.

In applications attention is confined to axially symmetrical problems.‖ The equations governing the deformation then reduce to a system of ordinary differential equations which may usually be integrated by numerical methods. The application of successive

---

† See *T.E.*, chap. X, for a discussion and further references.
‡ This was developed, for isotropic bodies, by J. E. Adkins, A. E. Green, and G. C. Nicholas, *Phil. Trans. R. Soc.* A, **247** (1954), 279.
§ A theory of stress and strain in shells and rods has also been given by J. L. Ericksen and C. Truesdell, *Arch ration. Mech. Analysis* **1** (1958), 295.
‖ Most of the problems discussed in the present chapter are contained in the following papers:
R. S. Rivlin and A. G. Thomas, *Phil. Trans. R. Soc.* A, **243** (1951), 289.
J. E. Adkins and R. S. Rivlin, ibid. A, **244** (1952), 505.
Other problems have been examined by J. E. Adkins, Ph.D. thesis, University of London, 1951; A. D. Kydoniefs and A. J. M. Spencer, *Int. J. Engng Sci.* **5** (1967), 367; *Q. Jl Mech. appl. Math.* **22** (1969), 87. A. D. Kydoniefs, *Int. J. Engng Sci.* **5** (1967), 477. A. H. Corneliussen and R. T. Shield, *Arch ration. Mech. Analysis* **7** (1961), 273.

approximation methods to non-symmetrical plane problems is considered in Chapter VI. In § 4.12 we have retained the method of numerical integration used by Adkins and Rivlin which was reproduced in the first edition of this book. Since then an alternative procedure has been given by Klingbeil and Shield.†

4.1. Stress resultants and loads: plane sheet

In this section the development of the theory is similar to that given in *T.E.* for the classical theory of plates. We suppose that the unstrained body is a plate of homogeneous elastic material bounded by the plane surfaces $x_3 = \pm h_0$, where h_0 may be a function of x_1, x_2, the body having elastic symmetry about the middle plane $x_3 = 0$. The plate undergoes a finite deformation symmetric about the middle plane $x_3 = 0$, which thus becomes the middle plane $y_3 = 0$ in the deformed state. Here y_i are the coordinates after deformation of a point of the plate which was originally at x_i, referred to the same rectangular cartesian axes. The major surfaces of the plate after deformation are given by $y_3 = \pm h$, where h is, in general, a function of y_1, y_2. We choose the curvilinear coordinate system θ_i so that

$$y_3 = \theta_3, \qquad y_\alpha = y_\alpha(\theta_1, \theta_2), \qquad (4.1.1)$$

Greek indices taking the values 1, 2. It follows that

$$G_{ij} = \begin{bmatrix} A_{11} & A_{12} & 0 \\ A_{12} & A_{22} & 0 \\ 0 & 0 & 1 \end{bmatrix}, \qquad G^{ij} = \begin{bmatrix} A^{11} & A^{12} & 0 \\ A^{12} & A^{22} & 0 \\ 0 & 0 & 1 \end{bmatrix}, \qquad G = A, \qquad (4.1.2)$$

with
$$A = |A_{\alpha\beta}|, \qquad A^{\alpha\rho} A_{\rho\beta} = \delta^\alpha_\beta, \qquad (4.1.3)$$

where $A_{\alpha\beta}$, $A^{\alpha\beta}$ are the covariant and contravariant metric tensors associated with coordinates θ_α in the middle plane $y_3 = 0$ of the deformed plate.

The force acting on an element of area of the coordinate surface $\theta_1 = $ constant in the deformed body is $\mathbf{T}_1\, d\theta^2 d\theta^3$, and the length of the corresponding line element of the middle plane $y_3 = 0$ is

$$\sqrt{(A_{22})}\, d\theta^2 = \sqrt{(AA^{11})}\, d\theta^2.$$

Similar considerations apply for the other surface $\theta_2 = $ constant. The stress across either of the surfaces $\theta_\alpha = $ constant may be replaced by

† W. W. Klingbeil and R. T. Shield, *Z. angew. Math. Phys.* **15** (1964), 608. See also L. J. Hart-Smith and J. D. C. Crisp, *Int. J. Engng Sci.* **5** (1967), 1.

a physical stress resultant $\mathbf{n}_\alpha$, measured per unit length of the curve $\theta_\alpha = $ constant in the plane $y_3 = 0$, where

$$\mathbf{n}_\alpha = \frac{\mathbf{N}_\alpha}{\sqrt{(AA^{\alpha\alpha})}}, \qquad \mathbf{N}_\alpha = \int_{-h}^{h} \mathbf{T}_\alpha \, dy_3, \qquad (4.1.4)$$

and we recall that h is a function of y_1, y_2 or θ_1, θ_2. Since the plate has elastic symmetry and the deformation is symmetric about $y_3 = 0$, the corresponding stress couples are zero. From (1.1.26)

$$\mathbf{T}_\alpha = \sqrt{(A)}\tau^{\alpha j}\mathbf{G}_j,$$

so that, using (4.1.4), we may write

$$\mathbf{n}_\alpha \sqrt{A^{\alpha\alpha}} = n^{\alpha\rho}\mathbf{G}_\rho + q^\alpha \mathbf{G}_3, \qquad \mathbf{N}_\alpha = N^{\alpha\rho}\mathbf{G}_\rho + Q^\alpha \mathbf{G}_3, \qquad (4.1.5)$$

where
$$N^{\alpha\rho} = n^{\alpha\rho}\sqrt{A}, \qquad Q^\alpha = q^\alpha \sqrt{A}, \qquad (4.1.6)$$

and
$$n^{\alpha\rho} = \int_{-h}^{h} \tau^{\alpha\rho} \, dy_3, \qquad q^\alpha = \int_{-h}^{h} \tau^{\alpha 3} \, dy_3. \qquad (4.1.7)$$

Since the stress distribution is symmetric about the plane $y_3 = 0$, it follows that $q^\alpha = Q^\alpha = 0$.

The stress resultant $\mathbf{n}$ per unit length of a line drawn in the middle plane $y_3 = 0$ of the deformed plate, whose unit normal in that plane is

$$\mathbf{u} = u_\alpha \mathbf{G}^\alpha, \qquad (4.1.8)$$

is given by
$$\mathbf{n} = \int_{-h}^{h} \mathbf{t} \, dy_3, \qquad (4.1.9)$$

so that, from (1.1.24), (4.1.4), and (4.1.5),

$$\mathbf{n} = \frac{u_\alpha}{\sqrt{A}} \int_{-h}^{h} \mathbf{T}_\alpha \, dy_3 = \frac{u_\alpha \mathbf{N}_\alpha}{\sqrt{A}} = \sum_{\alpha=1}^{2} u_\alpha \mathbf{n}_\alpha \sqrt{A^{\alpha\alpha}} = u_\alpha(n^{\alpha\rho}\mathbf{G}_\rho + q^\alpha \mathbf{G}_3). \quad (4.1.10)$$

The functions defined in (4.1.7) are (plane) surface tensors. The components of the symmetric contravariant tensor $n^{\alpha\rho}$ and the components of the contravariant tensor q^α are called stress resultants and shearing forces respectively. Mixed and covariant tensors may be formed with the help of the metric tensors $A_{\alpha\beta}$, $A^{\alpha\beta}$. In order to find the physical components of $\mathbf{n}_\alpha$ we express these vectors in terms of unit base vectors along the coordinate curves $\theta_\alpha = $ constant. The physical stress resultants and shearing forces are denoted by $n_{(\alpha\beta)}$, $q_{(\alpha)}$ respectively, the bracket indicating that these quantities are not tensors.

Thus we have
$$\mathbf{n}_\alpha = n_{(\alpha 1)}\frac{\mathbf{G}_1}{\sqrt{A_{11}}} + n_{(\alpha 2)}\frac{\mathbf{G}_2}{\sqrt{A_{22}}} + q_{(\alpha)}\mathbf{G}_3, \qquad (4.1.11)$$

and comparison of this with (4.1.5) yields

$$n_{(\alpha\beta)} = n^{\alpha\beta}\sqrt{(A_{\beta\beta}/A^{\alpha\alpha})}, \qquad q_{(\alpha)} = q^\alpha/\sqrt{A^{\alpha\alpha}}. \qquad (4.1.12)$$

We now consider the external forces acting on the major surfaces of the deformed plate. The covariant components u_i of the unit normal to the surfaces $y_3 = \pm h(\theta_1, \theta_2)$ referred to base vectors $\mathbf{G}^i$ are

$$(u_1, u_2, u_3) = \pm k\left(-\frac{\partial y_3}{\partial \theta^1}, -\frac{\partial y_3}{\partial \theta^2}, 1\right) \qquad (4.1.13)$$

where, remembering (4.1.2),

$$k = (A^{\alpha\beta}h_{,\alpha}h_{,\beta} + 1)^{-\frac{1}{2}}, \qquad (4.1.14)$$

a comma denoting partial differentiation. At these surfaces,

$$\mathbf{t} = \frac{u_i \mathbf{T}_i}{\sqrt{A}} = \pm\frac{k}{\sqrt{A}}(\mathbf{T}_3 - \mathbf{T}_\alpha y_{3,\alpha}) = \pm k(\tau^{3j} - y_{3,\alpha}\tau^{\alpha j})\mathbf{G}_j. \qquad (4.1.15)$$

If the faces of the plate are free from applied stress then

$$\tau^{3j} - \tau^{\alpha j}y_{3,\alpha} = 0 \qquad (y_3 = \pm h). \qquad (4.1.16)$$

4.2. Equations of equilibrium: Airy's stress function

When the body forces are zero the equations (1.1.27) of equilibrium become
$$\mathbf{T}_{i,i} = \mathbf{0}. \qquad (4.2.1)$$

If we integrate these equations through the thickness of the deformed plate we obtain

$$\int_{-h}^{h} \mathbf{T}_{\alpha,\alpha}\, dy_3 + [\mathbf{T}_3]_{-h}^{h} = \mathbf{0}. \qquad (4.2.2)$$

But, from (4.1.4),

$$\mathbf{N}_{\alpha,\alpha} = \frac{\partial}{\partial \theta^\alpha}\int_{-h}^{h}\mathbf{T}_\alpha\, dy_3 = \int_{-h}^{h}\mathbf{T}_{\alpha,\alpha}\, dy_3 + [\mathbf{T}_\alpha y_{3,\alpha}]_{-h}^{h},$$

so that (4.2.2) becomes

$$\mathbf{N}_{\alpha,\alpha} + [\mathbf{T}_3 - \mathbf{T}_\alpha y_{3,\alpha}]_{-h}^{h} = \mathbf{0},$$

and if the surfaces of the plate are free from applied stress this reduces to
$$\mathbf{N}_{\alpha,\alpha} = \mathbf{0}. \qquad (4.2.3)$$

Combining this with (4.1.5) and (4.1.6), with $q^\alpha = 0$, we have

$$n^{\alpha\beta}\|_\alpha = 0, \tag{4.2.4}$$

where the double line denotes covariant differentiation with respect to the plane variables θ_α in the deformed body, using Christoffel symbols formed from the metric tensors $A_{\alpha\beta}$, $A^{\alpha\beta}$.

Equations (4.2.3), (4.2.4), and (4.1.5) are similar in form to the corresponding equations for $\mathbf{T}_\alpha$ and $\tau^{\alpha\beta}$ given in Chapter III for plane strain, when body forces are zero. The results there obtained may therefore be applied to express the stress resultants in terms of an Airy stress function ϕ. Thus

$$\left.\begin{aligned}\mathbf{N}_\alpha &= \sqrt{(A)}\epsilon^{\gamma\alpha}\boldsymbol{\chi}_{,\gamma} = \sqrt{(A)}\epsilon^{\gamma\alpha}\epsilon^{\rho\beta}\phi\|_{\gamma\rho}\mathbf{G}_\beta\\ n^{\alpha\beta} &= \epsilon^{\alpha\gamma}\epsilon^{\beta\rho}\phi\|_{\gamma\rho}\end{aligned}\right\}, \tag{4.2.5}$$

or

$$\phi\|_{\alpha\beta} = \epsilon_{\alpha\gamma}\epsilon_{\beta\rho}n^{\gamma\rho} = (A/a)({}_0\epsilon_{\alpha\gamma})({}_0\epsilon_{\beta\rho})n^{\gamma\rho}. \tag{4.2.6}$$

where $\boldsymbol{\chi}$ is a vector in the plane $y_3 = 0$, ϕ is a scalar function of θ_1, θ_2, and

$$\left.\begin{aligned}{}_0\epsilon^{\alpha\beta}\sqrt{a} &= {}_0\epsilon_{\alpha\beta}/\sqrt{a} = \epsilon^{\alpha\beta}\sqrt{A} = \epsilon_{\alpha\beta}/\sqrt{A} = e_{\alpha\beta3}\\ \epsilon_{\alpha\rho}\epsilon^{\alpha\lambda} &= \delta^\lambda_\rho\end{aligned}\right\}. \tag{4.2.7}$$

In (4.2.7) e_{ijk} is the alternating cartesian tensor [see (3.1.10)]. The double line again indicates covariant differentiation with respect to the plane $y_3 = 0$ of the deformed body, the order of differentiation being immaterial since the Riemann–Christoffel tensor in the plane vanishes.

Let AP be an arc of a curve AB in the middle plane $y_3 = 0$ of the deformed body (Fig. 3.1, p. 106). By an analysis similar to that used for plane strain we may obtain the resultant force across a surface in the deformed body formed by normals to $y_3 = 0$ along AB. Denoting an element of AB by ds and making use of (4.1.10) and (4.2.5), we obtain for the total force $\mathbf{P}$ exerted by the region 1 on the region 2, across the arc AP,

$$\mathbf{P} = -\int_A^P \mathbf{n}\,ds = \boldsymbol{\chi} = \epsilon^{\rho\beta}\phi_{,\rho}\mathbf{G}_\beta \tag{4.2.8}$$

apart from an arbitrary constant which may be absorbed into $\boldsymbol{\chi}$ without loss of generality. Similarly, the total moment about the y_3-axis of the forces exerted by the region 1 on the region 2 is given by

$$\mathbf{M} = \int_A^P [\mathbf{B}\times\boldsymbol{\chi}_{,\beta}]\frac{d\theta^\beta}{ds}\,ds = (B^\alpha\phi_{,\alpha} - \phi)\mathbf{G}_3 \tag{4.2.9}$$

apart from an arbitrary constant vector which may be absorbed into $\phi \mathbf{G}_3$ without affecting the stresses. In (4.2.9)

$$\mathbf{B} = B^\alpha \mathbf{G}_\alpha = B_\alpha \mathbf{G}^\alpha \tag{4.2.10}$$

is the vector defined in (3.1.2). Equation (4.2.9) thus represents a couple of magnitude

$$M = B^\alpha \phi_{,\alpha} - \phi \tag{4.2.11}$$

about the y_3-axis.

The conditions for a single stress-free boundary follow as in (3.2.20) or (3.2.21).

4.3. Values of strains and stresses at middle surface

Since the deformation has been assumed to be completely symmetric about the middle plane of the plate it follows that x_α is an even function of θ_3 ($= y_3$) and x_3 is an odd function of θ_3. Also, on the middle plane $y_3 = 0$,

$$\left. \begin{array}{l} x_\alpha = x_\alpha(\theta_1, \theta_2), \quad \dfrac{\partial x^3}{\partial \theta^3} = \dfrac{1}{\lambda}, \quad \dfrac{\partial \theta^3}{\partial x^3} = \lambda \\[2mm] \dfrac{\partial x^\alpha}{\partial \theta^3} = 0, \quad \dfrac{\partial x^3}{\partial \theta^\alpha} = 0 \\[2mm] \dfrac{\partial \theta^3}{\partial x^\alpha} = 0, \quad \dfrac{\partial \theta^\alpha}{\partial x^3} = 0 \end{array} \right\} (y_3 = 0). \tag{4.3.1}$$

Here λ is a scalar function of θ_1, θ_2 and represents the extension ratio at the middle plane in a direction normal to the plane. We assume that λ is finite and non-zero. The values of the metric tensors g_{ij}, g^{ij} when $y_3 = 0$ are

$$\left. \begin{array}{l} g_{\alpha\beta} = a_{\alpha\beta}, \quad g_{33} = 1/\lambda^2, \quad g_{\alpha 3} = 0 \\ g^{\alpha\beta} = a^{\alpha\beta}, \quad g^{33} = \lambda^2, \quad g^{\alpha 3} = 0 \\ g = a/\lambda^2, \quad a = |a_{\alpha\beta}| \end{array} \right\} (y_3 = 0), \tag{4.3.2}$$

where $a_{\alpha\beta}$, $a^{\alpha\beta}$ are the covariant and contravariant metric tensors associated with curvilinear coordinates θ_α in the plane $x_3 = 0$ of the undeformed plate.

From (1.1.12), (1.1.15), (4.1.2), (4.3.1), and (4.3.2) it follows that

$$\left. \begin{array}{l} \gamma_{\alpha\beta} = \tfrac{1}{2}(A_{\alpha\beta} - a_{\alpha\beta}), \quad \gamma_{\alpha 3} = 0, \quad \gamma_{33} = \tfrac{1}{2}(1 - 1/\lambda^2) \\[2mm] e_{\alpha\beta} = \dfrac{1}{2}\left(\dfrac{\partial y_\lambda}{\partial x_\alpha}\dfrac{\partial y_\lambda}{\partial x_\beta} - \delta_{\alpha\beta}\right), \quad e_{\alpha 3} = 0, \quad e_{33} = \tfrac{1}{2}(\lambda^2 - 1) \end{array} \right\} (y_3 = 0). \tag{4.3.3}$$

Also, the value of the invariants I_r and K_α on $y_3 = 0$ are found from (1.1.18), (1.13.11), (4.1.2), (4.3.2), and (4.3.3). Thus

$$\left.\begin{aligned} I_1 &= \lambda^2 + a^{\alpha\beta} A_{\alpha\beta} \\ I_2 &= \lambda^2 (A/a) a_{\alpha\beta} A^{\alpha\beta} + A/a \\ I_3 &= \lambda^2 A/a \\ K_1 &= \tfrac{1}{2}(\lambda^2 - 1), \quad K_2 = 0 \end{aligned}\right\} (y_3 = 0). \quad (4.3.4)$$

Since $a_{\alpha\beta} A^{\alpha\beta} A = a^{\alpha\beta} A_{\alpha\beta} a$, we have

$$I_3 - \lambda^2 I_2 + \lambda^4 I_1 - \lambda^6 = 0 \qquad (y_3 = 0). \quad (4.3.5)$$

Formulae (4.3.3)–(4.3.5), and many subsequent results in this section, have a formal resemblance to those given in Chapter III for plane strain, but here λ is not constant. The values of the alternative invariants J_r on $y_3 = 0$ can be found from (1.1.20) and (4.3.4). Thus

$$\left.\begin{aligned} J_1 &= I_1 - 3 \\ J_2 &= I_2 - 2I_1 + 3 \\ J_3 &= I_3 - I_2 + I_1 - 1 \end{aligned}\right\} (y_3 = 0), \quad (4.3.6)$$

and in view of (4.3.5) we have

$$J_3 + (1-\lambda^2) J_2 + (1-\lambda^2)^2 J_1 + (1-\lambda^2)^3 = 0 \qquad (y_3 = 0). \quad (4.3.7)$$

We also need values of B^{ij}, M^{ij}, N^{ij} on $y_3 = 0$ and, from (1.15.9), (1.15.10), (4.3.1)–(4.3.4), we obtain

$$\left.\begin{aligned} B^{\alpha\beta} &= \lambda^2 a^{\alpha\beta} + A A^{\alpha\beta}/a \\ B^{33} &= \lambda^2 (I_1 - \lambda^2), \quad B^{\alpha 3} = 0 \\ M^{\alpha\beta} &= 0, \quad M^{\alpha 3} = 0, \quad M^{33} = \lambda^2, \quad N^{ij} = 0 \end{aligned}\right\} (y_3 = 0). \quad (4.3.8)$$

Rhombic system

The stress–strain relations for the compressible rhombic system are given in (1.15.3), or by (1.9.2) and (1.15.1), so that, using (4.3.3), we have

$$\left.\begin{aligned} \tau^{\alpha\beta} \sqrt{I_3} &= \frac{\partial W}{\partial e_{11}} \frac{\partial \theta^\alpha}{\partial x^1} \frac{\partial \theta^\beta}{\partial x^1} + \frac{\partial W}{\partial e_{22}} \frac{\partial \theta^\alpha}{\partial x^2} \frac{\partial \theta^\beta}{\partial x^2} + \\ &\quad + e_{12} \frac{\partial W}{\partial e_{12}^2} \left(\frac{\partial \theta^\alpha}{\partial x^1} \frac{\partial \theta^\beta}{\partial x^2} + \frac{\partial \theta^\alpha}{\partial x^2} \frac{\partial \theta^\beta}{\partial x^1} \right) \\ \tau^{\alpha 3} &= 0, \quad \tau^{33} = \frac{\lambda^2}{\sqrt{I_3}} \frac{\partial W}{\partial e_{33}} \end{aligned}\right\} (y_3 = 0), \quad (4.3.9)$$

where
$$W = W(e_{11}, e_{22}, e_{33}, e_{12}^2) \quad (y_3 = 0). \quad (4.3.10)$$

When the body is incompressible (4.3.9) is replaced by

$$\left.\begin{aligned}\tau^{\alpha\beta} &= \frac{\partial W}{\partial e_{11}}\frac{\partial \theta^{\alpha}}{\partial x^1}\frac{\partial \theta^{\beta}}{\partial x^1}+\frac{\partial W}{\partial e_{22}}\frac{\partial \theta^{\alpha}}{\partial x^2}\frac{\partial \theta^{\beta}}{\partial x^2}+\\ &\quad +e_{12}\frac{\partial W}{\partial e_{12}^2}\left(\frac{\partial \theta^{\alpha}}{\partial x^1}\frac{\partial \theta^{\beta}}{\partial x^2}+\frac{\partial \theta^{\alpha}}{\partial x^2}\frac{\partial \theta^{\beta}}{\partial x^1}\right)+pA^{\alpha\beta}\\ \tau^{\alpha 3} &= 0, \qquad \tau^{33} = \lambda^2\frac{\partial W}{\partial e_{33}}+p\end{aligned}\right\}(y_3=0), \quad (4.3.11)$$

where W is still given by (4.3.10), and p is a scalar function of θ_1, θ_2.

Transverse isotropy

Values of the stresses on $y_3 = 0$ for a body which is transversely isotropic about the x_3-axis may be found in a similar way from (1.15.6)–(1.15.10). The results are identical in form with those given in § 3.3 for plane strain. Thus

$$\left.\begin{aligned}\tau^{\alpha\beta} &= \mathscr{H}a^{\alpha\beta}+\mathscr{L}A^{\alpha\beta}\\ \tau^{\alpha 3} &= 0\\ \tau^{33} &= \lambda^2\Phi+\lambda^2(I_1-\lambda^2)\Psi+p+\mathscr{F}\\ &= \frac{2}{\sqrt{I_3}}\Big\{\lambda^2\frac{\partial W_2}{\partial J_1}+\lambda^2(J_1+1-\lambda^2)\frac{\partial W_2}{\partial J_2}+\\ &\quad +[(1-\lambda^2)(J_1+1-\lambda^2)+J_2+J_3]\frac{\partial W_2}{\partial J_3}\Big\}+\mathscr{F}\end{aligned}\right\}(y_3=0),\quad(4.3.12)$$

where, for a compressible body,

$$\left.\begin{aligned}\mathscr{H} &= \Phi+\lambda^2\Psi = \frac{2}{\sqrt{I_3}}\left(\frac{\partial W}{\partial I_1}+\lambda^2\frac{\partial W}{\partial I_2}\right) = \frac{2}{\sqrt{J}}\frac{\partial W_1}{\partial I}\\ &= \frac{2}{\sqrt{I_3}}\Big\{\frac{\partial W_2}{\partial J_1}+(\lambda^2-2)\frac{\partial W_2}{\partial J_2}-(\lambda^2-1)\frac{\partial W_2}{\partial J_3}\Big\}\end{aligned}\right\}(y_3=0),\quad(4.3.13)$$

$$\left.\begin{aligned}\mathscr{L} &= p+I_3\Psi/\lambda^2 = 2\sqrt{(I_3)}\left(\frac{\partial W}{\partial I_3}+\frac{1}{\lambda^2}\frac{\partial W}{\partial I_2}\right) = 2\sqrt{J}\frac{\partial W_1}{\partial J}\\ &= \frac{2\sqrt{I_3}}{\lambda^2}\Big\{\frac{\partial W_2}{\partial J_2}+(\lambda^2-1)\frac{\partial W_2}{\partial J_3}\Big\}\end{aligned}\right\}(y_3=0),\quad(4.3.14)$$

$$\mathscr{F} = \lambda^2\Theta = \frac{\lambda^2}{\sqrt{I_3}}\frac{\partial W}{\partial K_1} = \frac{\lambda^2}{\sqrt{J}}\frac{\partial W_1}{\partial K} \quad (y_3=0). \quad (4.3.15)$$

In these formulae

$$\left.\begin{array}{c} I = I_1 \qquad J = I_3, \qquad K = K_1 \\ I_2 = \lambda^2(I-\lambda^2)+J/\lambda^2 \\ W(I_1, I_2, I_3, K_1, K_2) = W_1(I, J, K) = W_2(J_1, J_2, J_3, K) \end{array}\right\} (y_3 = 0). $$
(4.3.16)

If the body is incompressible $I_3 = J = 1$ and W becomes a function of I_1, I_2, K_1, K_2 or of I, K when $y_3 = 0$. In this case

$$\left.\begin{array}{c} \mathcal{H} = 2\left(\dfrac{\partial W}{\partial I_1}+\lambda^2\dfrac{\partial W}{\partial I_2}\right) = 2\dfrac{\partial W_1}{\partial I} \\ \mathcal{F} = \lambda^2\dfrac{\partial W}{\partial K_1} = \lambda^2\dfrac{\partial W_1}{\partial K} \\ \mathcal{L} = p+\dfrac{2}{\lambda^2}\dfrac{\partial W}{\partial I_2} \end{array}\right\} (y_3 = 0), \qquad (4.3.17)$$

and p is a scalar function which is determined by the equations of equilibrium and boundary conditions.

Isotropy

Results for an isotropic body can be found immediately from those for a transversely isotropic body by omitting (K_1, K_2) (and hence K) from all formulae and putting $\mathcal{F} \equiv 0$.

4.4. Stress resultants: equations of equilibrium

In the equations derived so far no assumptions have been made about the thickness of the plate. We now confine attention to a plate whose thickness is originally constant and small compared with its linear dimensions. The restriction to constant thickness is, however, only introduced for simplicity and the final results hold for a plate of varying thickness, provided there is symmetry about $x_3 = 0$.

Since the major surfaces of the plate are free from applied tractions we have, from (4.1.16),

$$\tau^{\alpha 3}-\tau^{\alpha\beta}y_{3,\beta} = 0, \qquad \tau^{33}-\tau^{3\alpha}y_{3,\alpha} = 0 \qquad (y_3 = \pm h),$$

so that
$$\tau^{33}-\tau^{\alpha\beta}h_{,\alpha}h_{,\beta} = 0 \qquad (x_3 = \pm h_0). \qquad (4.4.1)$$

In view of (4.3.1),
$$\lim_{h_0 \to 0}\frac{h}{h_0} = \lim_{y_3 \to 0}\frac{\partial y_3}{\partial x_3} = \lambda, \qquad (4.4.2)$$

so that, when $h_0 \to 0$,

$$\tau^{33} \to h_0^2 \tau^{\alpha\beta} \lambda_{,\alpha} \lambda_{,\beta} \qquad (y_3 = \pm h). \tag{4.4.3}$$

We now suppose that $\lambda_{,\alpha}$ are finite and that τ^{33}, $\tau^{\alpha\beta}$ are continuous functions of y_3 in the range $|y_3| \leqslant h$. Also, we assume that $\tau^{\alpha\beta}$ is finite. Then

$$\lim_{h_0 \to 0} [\tau^{33}]_{y_3=0} = 0. \tag{4.4.4}$$

From (4.1.7) we have

$$\frac{n^{\alpha\beta}}{2h_0} = \frac{1}{2h_0} \int_{-h}^{h} \tau^{\alpha\beta} \, dy_3,$$

so that, with the help of (4.4.2) and the assumed continuity of $\tau^{\alpha\beta}$ with respect to y_3, we obtain

$$\lim_{h_0 \to 0} \frac{n^{\alpha\beta}}{2h_0} = \lambda [\tau^{\alpha\beta}]_{y_3=0}. \tag{4.4.5}$$

Equations (4.4.4) and (4.4.5) are exact results for the limiting case $h_0 \to 0$. We now assume that when h_0 is very small compared with the linear dimensions of the plate we may write, approximately,

$$[\tau^{33}]_{y_3=0} = 0, \qquad n^{\alpha\beta} = 2h_0 \lambda [\tau^{\alpha\beta}]_{y_3=0}. \tag{4.4.6}$$

Hence, for a transversely isotropic body, we have, from (4.3.12) and (4.4.6),

$$n^{\alpha\beta} = 2h_0 \lambda (\mathscr{H} a^{\alpha\beta} + \mathscr{L} A^{\alpha\beta}) \tag{4.4.7}$$

and

$$\left.\begin{array}{l} \lambda^2 \Phi + \lambda^2(I_1 - \lambda^2)\Psi + p + \mathscr{F} = 0 \\[6pt] \text{or} \quad \dfrac{2}{\sqrt{I_3}}\left\{\lambda^2 \dfrac{\partial W_2}{\partial J_1} + \lambda^2(J_1 + 1 - \lambda^2)\dfrac{\partial W_2}{\partial J_2} + \right. \\[6pt] \left. +[(1-\lambda^2)(J_1+1-\lambda^2) + J_2 + J_3]\dfrac{\partial W_2}{\partial J_3}\right\} + \mathscr{F} = 0 \end{array}\right\}, \tag{4.4.8}$$

where $\mathscr{H}$, $\mathscr{L}$, $\mathscr{F}$ are given by (4.3.13)–(4.3.15) for a compressible body. When the body is incompressible $\mathscr{H}$, $\mathscr{F}$ are given by (4.3.17) and p is a scalar function.

If we substitute (4.4.7) into (4.2.6) we obtain

$$\phi\|_{\alpha\beta} = 2h_0 \lambda \{(A/a)\mathscr{H} a_{\alpha\beta} + \mathscr{L} A_{\alpha\beta}\}, \tag{4.4.9}$$

where $\mathscr{H}$, $\mathscr{L}$ are given by (4.3.13) and (4.3.14) when the body is compressible. Equations (4.4.9), (4.4.8), and either (4.3.7) or (4.3.5) are then the fundamental equations of the problem.

For an incompressible body

$$I_3 = \lambda^2 A/a = 1$$

and (4.4.9) becomes
$$\phi\|_{\alpha\beta} = 2h_0\lambda\{a_{\alpha\beta}\mathscr{H}/\lambda^2 + \mathscr{L}A_{\alpha\beta}\}, \tag{4.4.10}$$

where $\mathscr{H}$ is given by (4.3.17). Also, if we eliminate p from (4.4.8) and (4.3.17), we have

$$\mathscr{L} = -2\lambda^2\frac{\partial W}{\partial I_1} - 2\left(\lambda^2 I_1 - \lambda^4 - \frac{1}{\lambda^2}\right)\frac{\partial W}{\partial I_2} - \lambda^2\frac{\partial W}{\partial K_1}. \tag{4.4.11}$$

Apart from a factor $2h_0\lambda$, equations (4.4.9) and (4.4.10) are similar to those given in § 3.3 for plane strain. Here, however, λ is not constant and is determined by the extra condition (4.4.8). The equations can be expressed in complex variable notation in a manner similar to that used in §§ 3.4, 3.5 for plane strain but details are omitted.†

4.5. Geometrical relations for a deformed shell

We consider a surface M and we denote the position vector of any point on M by $\mathbf{\bar{R}}$. General curvilinear coordinates on M are denoted by θ_1, θ_2 so that
$$\mathbf{\bar{R}} = \mathbf{\bar{R}}(\theta_1, \theta_2). \tag{4.5.1}$$

At every point of M we erect the unit normal vector $\mathbf{A}_3$ and we denote the perpendicular distance of a point on $\mathbf{A}_3$ from M by θ^3 $(=\theta_3)$ so that $\theta_3 = 0$ is the surface M. Two more surfaces are defined by

$$\theta_3 = \pm h(\theta_1, \theta_2). \tag{4.5.2}$$

These surfaces are taken as the boundaries of a body S which includes the surface M. The body S is called a shell and the boundaries of S are called the faces of the shell, the surface M is the middle surface, and $2h$ is the (variable) thickness of the shell. We imagine that the shell S forms a deformed body which has been obtained from an undeformed shell by a finite deformation, and that S is in equilibrium under the action of surface and edge forces.

The position vector of a point of the shell S may be defined by

$$\mathbf{R} = \mathbf{\bar{R}}(\theta_1, \theta_2) + \theta_3\mathbf{A}_3. \tag{4.5.3}$$

The relations between certain geometrical functions connected with the shell S and the middle surface M may now be obtained from (4.5.3) by a process similar to that used in Chapter X of $T.E.$ for the classical theory of shells. Here, however, the notation of § 1.13 of $T.E.$ for quantities associated with the surface M is replaced by capital letters

† See J. E. Adkins, A. E. Green, and G. C. Nicholas, loc. cit., p. 133.

to denote reference to the deformed body. Thus, using Greek indices to denote the values 1, 2, we have covariant and contravariant base vectors $\mathbf{A}_\alpha$, $\mathbf{A}^\alpha$ respectively of the surface M with metric tensors $A_{\alpha\beta}$, $A^{\alpha\beta}$, where

$$A_{\alpha\beta} = \mathbf{A}_\alpha \cdot \mathbf{A}_\beta, \qquad A^{\alpha\beta} = \mathbf{A}^\alpha \cdot \mathbf{A}^\beta, \qquad A^{\alpha\rho} A_{\rho\beta} = \delta^\alpha_\beta. \qquad (4.5.4)$$

The symmetric tensors associated with the second fundamental form of the surface S are here denoted by $C_{\alpha\beta}$, $C^{\alpha\beta}$ instead of the more natural $B_{\alpha\beta}$, $B^{\alpha\beta}$ to avoid confusion with the tensor B^{ij} defined in Chapter I. Thus

$$\left. \begin{aligned} C_{\alpha\beta} &= -\mathbf{A}_\alpha \cdot \mathbf{A}_{3,\beta} = -\mathbf{A}_\beta \cdot \mathbf{A}_{3,\alpha} = \mathbf{A}_3 \cdot \mathbf{A}_{\alpha,\beta} = \mathbf{A}_3 \cdot \mathbf{A}_{\alpha,\beta} \\ C^\alpha_\beta &= A^{\alpha\rho} C_{\rho\beta}, \qquad C^{\alpha\beta} = A^{\alpha\rho} C^\beta_\rho \end{aligned} \right\}, \qquad (4.5.5)$$

and these are related by the Mainardi–Codazzi equation

$$C_{\alpha 1}\|_2 = C_{\alpha 2}\|_1, \qquad (4.5.6)$$

where a double line denotes covariant differentiation with respect to the θ_α curves on the middle surface M.

The base vectors $\mathbf{G}_i$, $\mathbf{G}^i$ and the corresponding metric tensors G_{ij}, G^{ij} of the deformed shell may now be evaluated using (4.5.3). Hence

$$\mathbf{G}_\alpha = \mathbf{R}_{,\alpha} = (\delta^\rho_\alpha - \theta_3 C^\rho_\alpha) \mathbf{A}_\rho, \qquad \mathbf{G}_3 = \mathbf{R}_{,3} = \mathbf{A}_3 \qquad (4.5.7)$$

and

$$\left. \begin{aligned} G_{\alpha\beta} &= A_{\alpha\beta} - 2\theta_3 C_{\alpha\beta} + \theta_3^2 C^\lambda_\beta C_{\lambda\alpha} \\ G_{\alpha 3} &= 0, \qquad G_{33} = 1 \end{aligned} \right\}. \qquad (4.5.8)$$

Also, since $\quad [\mathbf{G}_1 \, \mathbf{G}_2 \, \mathbf{G}_3] = \sqrt{G}, \quad [\mathbf{A}_1 \, \mathbf{A}_2 \, \mathbf{A}_3] = \sqrt{A},$

it follows from (4.5.7) that

$$\sqrt{(G/A)} = 1 - \theta_3(C^1_1 + C^2_2) + \theta_3^2(C^1_1 C^2_2 - C^1_2 C^2_1). \qquad (4.5.9)$$

4.6. Stress resultants and stress couples

The discussion of the state of stress in the deformed shell in terms of stress resultants and stress couples is identical with that given in § 10.2 of *T.E.* except that all quantities occurring here refer to the deformed surface M. We therefore omit detailed discussion and quote results, referring the reader to *T.E.* for further details.

Stress resultants and stress couples, defined for the surface $\theta_\alpha =$ constant and measured per unit length of the line $\theta_\alpha =$ constant,

4.7 THEORY OF ELASTIC MEMBRANES

$\theta_3 = 0$, are $\mathbf{n}_\alpha$, $\mathbf{m}_\alpha$ respectively, where

$$\left. \begin{array}{cc} \mathbf{n}_\alpha = \dfrac{\mathbf{N}_\alpha}{\sqrt{(AA^{\alpha\alpha})}}, & \mathbf{m}_\alpha = \dfrac{\mathbf{M}_\alpha}{\sqrt{(AA^{\alpha\alpha})}} \\ \mathbf{N}_\alpha = \displaystyle\int_{-h}^{h} \mathbf{T}_\alpha\, d\theta^3, & \mathbf{M}_\alpha = \displaystyle\int_{-h}^{h} (\mathbf{A}_3 \times \mathbf{T}_\alpha)\theta_3\, d\theta^3 \end{array} \right\}, \quad (4.6.1)$$

and†
$$\mathbf{T}_i = (\sigma^{i\rho}\mathbf{A}_\rho + \sigma^{i3}\mathbf{A}_3)\sqrt{A}, \qquad \sigma^{i\rho} = (\delta^\rho_\alpha - \theta_3 C^\rho_\alpha)\tau^{i\alpha}\sqrt{(G/A)},$$
$$\sigma^{i3} = \tau^{i3}\sqrt{(G/A)}. \qquad (4.6.2)$$

The stress resultant $\mathbf{n}$ and stress couple $\mathbf{m}$ per unit length of a line of the middle surface whose unit normal in the surface is

$$\mathbf{u} = u_\alpha \mathbf{A}^\alpha, \qquad (4.6.3)$$

are given by
$$\mathbf{n} = \sum_\alpha u_\alpha \mathbf{n}_\alpha \sqrt{A^{\alpha\alpha}}, \qquad \mathbf{m} = \sum_\alpha u_\alpha \mathbf{m}_\alpha \sqrt{A^{\alpha\alpha}}. \qquad (4.6.4)$$

If we substitute (4.6.2) in (4.6.1) we obtain

$$\mathbf{n}_\alpha \sqrt{A^{\alpha\alpha}} = n^{\alpha\rho}\mathbf{A}_\rho + q^\alpha \mathbf{A}_3, \qquad \mathbf{N}_\alpha = N^{\alpha\rho}\mathbf{A}_\rho + Q^\alpha \mathbf{A}_3, \qquad (4.6.5)$$

$$\mathbf{m}_\alpha \sqrt{A^{\alpha\alpha}} = m^{\alpha\rho}\mathbf{A}_3 \times \mathbf{A}_\rho, \qquad \mathbf{M}_\alpha = M^{\alpha\rho}\mathbf{A}_3 \times \mathbf{A}_\rho, \qquad (4.6.6)$$

where
$$N^{\alpha\rho} = n^{\alpha\rho}\sqrt{A}, \qquad M^{\alpha\rho} = m^{\alpha\rho}\sqrt{A}, \qquad Q^\alpha = q^\alpha\sqrt{A}, \qquad (4.6.7)$$

and
$$n^{\alpha\beta} = \int_{-h}^{h} \sigma^{\alpha\beta}\, d\theta^3, \qquad (4.6.8)$$

$$m^{\alpha\beta} = \int_{-h}^{h} \sigma^{\alpha\beta}\theta_3\, d\theta^3, \qquad (4.6.9)$$

$$q^\alpha = \int_{-h}^{h} \sigma^{\alpha 3}\, d\theta^3. \qquad (4.6.10)$$

Physical components of stress may be obtained as in § 10.2 of *T.E.* but details are omitted.

4.7. Surface conditions

Loads and equations of equilibrium in terms of stress resultants may be obtained by the method used in §§ 10.3, 10.4 of *T.E.* provided we refer all quantities to the deformed surface. Here, however, we

---

† $\sigma^{i\rho}$ need not be confused with physical components of stress $\sigma_{i\rho}$ which are not used in this chapter.

restrict our attention to problems in which the loads on either surface of the shell are entirely normal to these surfaces and we also only consider a membrane theory which is valid for values of h small compared with other dimensions of the shell. In view of this it is convenient to discuss surface conditions, and in the next section equations of equilibrium, by a different method from that used in *T.E.*

The surface $\theta_3 = h$ has the vector equation
$$\mathbf{R} = \bar{\mathbf{R}}(\theta_1, \theta_2) + \mathbf{A}_3 h(\theta_1, \theta_2), \tag{4.7.1}$$
and the unit normal to this surface is
$$\mathbf{u} = \frac{\mathbf{R}_{,1} \times \mathbf{R}_{,2}}{|\mathbf{R}_{,1} \times \mathbf{R}_{,2}|} = \frac{(\mathbf{G}_1 + \mathbf{A}_3 h_{,1}) \times (\mathbf{G}_2 + \mathbf{A}_3 h_{,2})}{|\mathbf{R}_{,1} \times \mathbf{R}_{,2}|}$$
if we use (4.7.1) and (4.5.7). Hence
$$\mathbf{u} = u_i \mathbf{G}^i = \frac{(\mathbf{G}_3 - h_{,\alpha} \mathbf{G}^\alpha)\sqrt{G}}{|\mathbf{R}_{,1} \times \mathbf{R}_{,2}|}. \tag{4.7.2}$$

If p_2 is the normal pressure on this surface, measured per unit area of the middle surface of the shell, then
$$-\frac{p_2 \mathbf{u}\sqrt{A}}{|\mathbf{R}_{,1} \times \mathbf{R}_{,2}|_h} = \left(\frac{u_i \mathbf{T}_i}{\sqrt{G}}\right)_h. \tag{4.7.3}$$

It follows from (4.7.2) and (4.7.3) that
$$-p_2 \mathbf{u}\sqrt{A} = [\mathbf{T}_3 - h_{,\alpha} \mathbf{T}_\alpha]_h. \tag{4.7.4}$$

If $h_{,\alpha}$ is $O(h)$ for small h and if the geometrical quantities C_α^ρ and the stresses τ^{ij} are finite, then, from (4.7.2), (4.7.4), and (4.5.7), we see that†
$$[\mathbf{T}_3 - h_{,\alpha} \mathbf{T}_\alpha]_h \to (\mathbf{T}_3)_h \to -p_2 \mathbf{A}_3 \sqrt{A} \quad (h \to 0). \tag{4.7.5}$$

Hence, using (4.6.2), it follows that
$$(\tau^{33})_h \to -p_2, \quad (\tau^{\alpha 3})_h \to 0 \quad (h \to 0), \tag{4.7.6}$$

Similarly, if there is a normal pressure p_1 on the surface $\theta_3 = -h$ of the deformed shell, measured per unit area of the deformed middle surface M, then
$$[\mathbf{T}_3 + h_{,\alpha} \mathbf{T}_\alpha]_{-h} \to (\mathbf{T}_3)_{-h} \to -p_1 \mathbf{A}_3 \sqrt{A} \quad (h \to 0), \tag{4.7.7}$$

† We interpret $h \to 0$ as the maximum value of $h/l \to 0$ where l represents a linear dimension of the shell, e.g., the length of an edge or a radius of curvature, whichever is smaller.

and it follows that

$$(\tau^{33})_{-h} \to -p_1, \qquad (\tau^{\alpha 3})_{-h} \to 0 \qquad (h \to 0). \tag{4.7.8}$$

Hence, if τ^{ij} are continuous throughout the shell, it is reasonable to assume, in view of (4.7.6) and (4.7.8), that

$$\tau^{\alpha 3} \to 0, \qquad \tau^{33} = O\{|p_1 + p_2|/2\} \qquad (\theta_3 = 0), \tag{4.7.9}$$

when $h \to 0$. If there is zero normal pressure on each surface of the shell then (4.7.9) is replaced by

$$\tau^{\alpha 3} \to 0, \qquad \tau^{33} \to 0 \qquad (\theta_3 = 0; h \to 0). \tag{4.7.10}$$

4.8. Equations of equilibrium

When body forces are zero the equations of equilibrium are

$$\mathbf{T}_{i,i} = \mathbf{0}. \tag{4.8.1}$$

If we integrate these equations through the thickness of the deformed shell we obtain, as in § 4.2,

$$\mathbf{N}_{\alpha,\alpha} + [\mathbf{T}_3 - h_{,\alpha}\mathbf{T}_\alpha]_h - [\mathbf{T}_3 + h_{,\alpha}\mathbf{T}_\alpha]_{-h} = \mathbf{0}. \tag{4.8.2}$$

Since we are confining attention to problems in which the surface loads are normal to each surface, it follows from (4.7.5) and (4.7.7) that

$$\mathbf{N}_{\alpha,\alpha} + (p_1 - p_2)\mathbf{A}_3\sqrt{A} = \mathbf{0} \qquad (h \to 0). \tag{4.8.3}$$

Alternatively,

$$\left.\begin{array}{l} n^{\alpha\beta}\|_\alpha - C^\beta_\alpha q^\alpha = 0 \\ n^{\alpha\beta}C_{\alpha\beta} + q^\alpha\|_\alpha + p_1 - p_2 = 0 \end{array}\right\} \qquad (h \to 0), \tag{4.8.4}$$

the reduction from (4.8.3) to (4.8.4) being the same as that given in § 10.4 of *T.E.* The double line in (4.8.4) denotes covariant differentiation with respect to the deformed surface M so that

$$\left.\begin{array}{l} n^{\alpha\beta}\|_\alpha = n^{\alpha\beta}_{,\alpha} + \Gamma^\beta_{\alpha\rho}n^{\alpha\rho} + \Gamma^\alpha_{\alpha\rho}n^{\rho\beta} \\ q^\alpha\|_\alpha = q^\alpha_{,\alpha} + \Gamma^\alpha_{\alpha\beta}q^\beta \end{array}\right\}, \tag{4.8.5}$$

where $\Gamma^\alpha_{\beta\rho}$ are Christoffel symbols evaluated from the metric tensors $A_{\alpha\beta}$, $A^{\alpha\beta}$. Alternatively,

$$\Gamma^\alpha_{\beta\rho} = \mathbf{A}^\alpha \cdot \mathbf{A}_{\beta,\rho}. \tag{4.8.6}$$

4.9. Stress resultants

We now consider the undeformed position of the shell. The surface M of the deformed shell formed a surface M_0 in the undeformed state and we take its equation to be

$$\bar{\mathbf{r}} = \bar{\mathbf{r}}(\theta_1, \theta_2). \tag{4.9.1}$$

Associated with this surface are covariant and contravariant base vectors $\mathbf{a}_\alpha$, $\mathbf{a}^\alpha$ with metric tensors $a_{\alpha\beta}$, $a^{\alpha\beta}$ of the first fundamental form. For a full discussion of the geometry of this surface the reader is referred again to § 1.13 of $T.E.$, and the notation of that section is used here.

We assume that the deformation from the undeformed to the deformed shell is such that the displacement vector has continuous derivatives at least up to and including the second order with respect to θ_i for all θ_α in M and for $|\theta_3| \leqslant h$. Hence the position vector $\mathbf{r}$ of points in the undeformed shell has the form

$$\mathbf{r} = \mathbf{r}(\theta_1, \theta_2) + \theta_3 \mathbf{k}(\theta_1, \theta_2) + \mathbf{O}(\theta_3^2), \tag{4.9.2}$$

where $\mathbf{k}$ is a vector depending only on θ_α. From (4.9.2) we obtain

$$\left. \begin{array}{ll} \mathbf{g}_\alpha = \mathbf{a}_\alpha + \mathbf{O}(\theta_3), & \mathbf{g}_3 = \mathbf{k} + \mathbf{O}(\theta_3) \\ g_{\alpha\beta} = a_{\alpha\beta} + O(\theta_3), & g_{\alpha 3} = \mathbf{k} \cdot \mathbf{a}_\alpha + O(\theta_3), \quad g_{33} = k^2 + O(\theta_3) \end{array} \right\}, \tag{4.9.3}$$

where $k = |\mathbf{k}|$.

Since $G^{\alpha 3} = 0$ it follows from the expressions (1.15.13) for the stress tensor for an isotropic material that

$$\tau^{\alpha 3} = g^{\alpha 3}\Phi + B^{\alpha 3}\Psi. \tag{4.9.4}$$

Now, from (4.7.9), $\tau^{\alpha 3} \to 0$ when $\theta_3 = 0$ and $h \to 0$, so that using (1.15.9), (4.9.3), and (4.9.4) we see that $g_{\alpha 3} \to 0$ when $\theta_3 = 0$ and $h \to 0$. Therefore

$$\mathbf{k} \cdot \mathbf{a}_\alpha = 0. \tag{4.9.5}$$

Hence $\mathbf{k}$ is a vector perpendicular to the surface $\theta_3 = 0$ in the undeformed shell. If $\mathbf{r}_1$, $\mathbf{r}_2$ are position vectors of points on the surface of the undeformed shell which correspond to $\theta_3 = \pm h$ respectively, then, from (4.9.2), we have

$$\lim_{h \to 0} \frac{(\mathbf{r}_1 - \mathbf{r}_2)}{2h} = \mathbf{k}. \tag{4.9.6}$$

The (small) thickness $2h_0$ of the undeformed shell may now be defined as

$$2h_0 = (\mathbf{r}_1 - \mathbf{r}_2) \cdot (\mathbf{k}/k),$$

so that

$$h = \lambda h_0, \quad \lambda = 1/k \quad (h \to 0). \tag{4.9.7}$$

The thickness h_0 may vary with θ_1, θ_2.

We return now to (4.9.3) and evaluate all quantities on the surface $\theta_3 = 0$. Thus

$$\left.\begin{aligned} g_{\alpha\beta} &= a_{\alpha\beta}, & g_{33} &= 1/\lambda^2, & g_{\alpha 3} &= 0 \\ g^{\alpha\beta} &= a^{\alpha\beta}, & g^{33} &= \lambda^2, & g^{\alpha 3} &= 0 \\ g &= a/\lambda^2, & a &= |a_{\alpha\beta}| & & \end{aligned}\right\} (\theta_3 = 0). \quad (4.9.8)$$

From (1.1.18), (4.5.8), and (4.9.8) it follows that the values of the strain invariants when $\theta_3 = 0$ are

$$\left.\begin{aligned} I_1 &= \lambda^2 + a^{\alpha\beta} A_{\alpha\beta} \\ I_2 &= \lambda^2 (A/a) a_{\alpha\beta} A^{\alpha\beta} + A/a \\ I_3 &= \lambda^2 A/a \end{aligned}\right\} (\theta_3 = 0), \quad (4.9.9)$$

and as in § 4.3 these are related by the equation

$$I_3 - \lambda^2 I_2 + \lambda^4 I_1 - \lambda^6 = 0 \quad (\theta_3 = 0). \quad (4.9.10)$$

Alternatively, the invariants J_1, J_2, J_3 are given by (4.3.6) and (4.9.9) when $\theta_3 = 0$, and these are related by an equation of the form (4.3.7) when $\theta_3 = 0$. Again, from (1.15.9), (4.5.8), and (4.9.8), we see that

$$\left.\begin{aligned} B^{\alpha\beta} &= \lambda^2 a^{\alpha\beta} + A A^{\alpha\beta}/a \\ B^{33} &= \lambda^2 (I_1 - \lambda^2), \quad B^{\alpha 3} = 0 \end{aligned}\right\} (\theta_3 = 0). \quad (4.9.11)$$

The values of the stress tensor τ^{ij} for an isotropic shell, on $\theta_3 = 0$, may be found from (1.15.13), (4.5.8), (4.9.8), and (4.9.11) in the form

$$\left.\begin{aligned} \tau^{\alpha\beta} &= \mathscr{H} a^{\alpha\beta} + \mathscr{L} A^{\alpha\beta}, \quad \tau^{\alpha 3} = 0 \\ \tau^{33} &= \lambda^2 \Phi + \lambda^2 (I_1 - \lambda^2) \Psi + p \\ &= \frac{2}{\sqrt{I_3}} \Big\{ \lambda^2 \frac{\partial W_2}{\partial J_1} + \lambda^2 (J_1 + 1 - \lambda^2) \frac{\partial W_2}{\partial J_2} + \\ & \qquad + [(1-\lambda^2)(J_1 + 1 - \lambda^2) + J_2 + J_3] \frac{\partial W_2}{\partial J_3} \Big\} \end{aligned}\right\} (\theta_3 = 0), \quad (4.9.12)$$

where, for a compressible body,

$$\left.\begin{aligned} \mathscr{H} &= \Phi + \lambda^2 \Psi = \frac{2}{\sqrt{I_3}} \left(\frac{\partial W}{\partial I_1} + \lambda^2 \frac{\partial W}{\partial I_2} \right) \\ &= \frac{2}{\sqrt{I_3}} \left\{ \frac{\partial W_2}{\partial J_1} + (\lambda^2 - 2) \frac{\partial W_2}{\partial J_2} - (\lambda^2 - 1) \frac{\partial W_2}{\partial J_3} \right\} \end{aligned}\right\} (\theta_3 = 0), \quad (4.9.13)$$

$$\left.\begin{aligned} \mathscr{L} &= p + I_3 \Psi/\lambda^2 = 2\sqrt{(I_3)} \left(\frac{\partial W}{\partial I_3} + \frac{1}{\lambda^2} \frac{\partial W}{\partial I_2} \right) \\ &= \frac{2\sqrt{I_3}}{\lambda^2} \left\{ \frac{\partial W_2}{\partial J_2} + (\lambda^2 - 1) \frac{\partial W_2}{\partial J_3} \right\} \end{aligned}\right\} (\theta_3 = 0), \quad (4.9.14)$$

and
$$W(I_1, I_2, I_3) = W_2(J_1, J_2, J_3), \tag{4.9.15}$$

the values of the invariants when $\theta_3 = 0$ being given by (4.9.9) and (4.3.6).

When the shell is incompressible,
$$I_3 = 1 \quad \text{or} \quad \lambda^2 A = a, \tag{4.9.16}$$

$$\left. \begin{aligned} \mathscr{H} &= 2\left(\frac{\partial W}{\partial I_1} + \lambda^2 \frac{\partial W}{\partial I_2}\right) \\ \mathscr{L} &= p + \frac{2}{\lambda^2}\frac{\partial W}{\partial I_2} \\ W &= W(I_1, I_2) \end{aligned} \right\}, \tag{4.9.17}$$

and p is an arbitrary scalar function.

For both the compressible and incompressible shells it is convenient to eliminate p from the expression for $\tau^{\alpha\beta}$ ($\theta_3 = 0$) so that

$$\left. \begin{aligned} \tau^{\alpha\beta} &= \tau^{33}A^{\alpha\beta} + \mathscr{H}a^{\alpha\beta} + \{\mathscr{L} - \lambda^2\Phi - \lambda^2(I_1-\lambda^2)\Psi - p\}A^{\alpha\beta} \\ &= \tau^{33}A^{\alpha\beta} + (\Phi+\lambda^2\Psi)a^{\alpha\beta} + \{(I_3/\lambda^2 + \lambda^4 - \lambda^2 I_1)\Psi - \lambda^2\Phi\}A^{\alpha\beta} \end{aligned} \right\} (\theta_3 = 0). \tag{4.9.18}$$

From (4.5.9), (4.6.2), (4.6.8), (4.6.10), and (4.9.18) we have

$$\lim_{h \to 0} \frac{n^{\alpha\beta}}{2h} = \lim_{h \to 0} \frac{n^{\beta\alpha}}{2h}$$
$$= (\tau^{33})_0 A^{\alpha\beta} + (\Phi+\lambda^2\Psi)a^{\alpha\beta} + \{(I_3/\lambda^2+\lambda^4-\lambda^2 I_1)\Psi - \lambda^2\Phi\}A^{\alpha\beta}, \tag{4.9.19}$$

$$\lim_{h \to 0} \frac{q^\alpha}{2h} = 0, \tag{4.9.20}$$

if we remember the assumption about continuity of τ^{ij}, where $(\tau^{33})_0$ is evaluated at $\theta_3 = 0$.

If the results (4.9.19) and (4.9.20) are substituted in the second equation of (4.8.4) we see that, provided $C_{\alpha\beta}$ are not all zero and $p_1 \neq p_2 \neq 0$, the term $(\tau^{33})_0$ in (4.9.19) may in general be neglected† in view of (4.7.9), and the equations (4.8.4) become

$$\left. \begin{aligned} n^{\alpha\beta}\|_\alpha &= 0 \\ n^{\alpha\beta}C_{\alpha\beta} + p_1 - p_2 &= 0 \end{aligned} \right\} \quad (h \to 0). \tag{4.9.21}$$

Alternatively,
$$\mathbf{N}_{\alpha,\alpha} + (p_1-p_2)\mathbf{A}_3\sqrt{A} = 0 \quad (h \to 0) \tag{4.9.22}$$

† We recall the footnote on p. 146.

where now $\mathbf{N}_\alpha$ is given by (4.6.5) with $q^\alpha = Q^\alpha = 0$. Consistent with (4.9.21) we may now write, from (4.9.19) and (4.9.12),

$$\left.\begin{array}{l}\dfrac{n^{\alpha\beta}}{2h} \to (\Phi+\lambda^2\Psi)a^{\alpha\beta}+\{(I_3/\lambda^2+\lambda^4-\lambda^2 I_1)\Psi-\lambda^2\Phi\}A^{\alpha\beta} \\[2mm] \lambda^2\Phi+\lambda^2(I_1-\lambda^2)\Psi+p = 0 \end{array}\right\} \quad (h \to 0). \quad (4.9.23)$$

We observe that $n^{\alpha\beta}$ in (4.9.23) is now symmetric. If $p_1 = p_2 = 0$ we see from (4.7.10) that (4.9.23) is still valid.

When the shell is thin compared with its other dimensions we shall assume that the above results, which are valid in the limiting case $h \to 0$, are a reasonable approximation. We therefore have equations of equilibrium (4.9.21) or (4.9.22) for small h, together with

$$h = \lambda h_0, \quad (4.9.24)$$

$$n^{\alpha\beta} = 2h_0[\lambda(\Phi+\lambda^2\Psi)a^{\alpha\beta}+\lambda\{(I_3/\lambda^2+\lambda^4-\lambda^2 I_1)\Psi-\lambda^2\Phi\}A^{\alpha\beta}], \quad (4.9.25)$$

$$\lambda^2\Phi+\lambda^2(I_1-\lambda^2)\Psi+p = 0. \quad (4.9.26)$$

For a compressible body,

$$\left.\begin{array}{l}\Phi = \dfrac{2}{\sqrt{I_3}}\dfrac{\partial W}{\partial I_1}, \quad \Psi = \dfrac{2}{\sqrt{I_3}}\dfrac{\partial W}{\partial I_2}, \quad p = 2\sqrt{(I_3)}\dfrac{\partial W}{\partial I_3} \\[2mm] W = W(I_1, I_2, I_3) \end{array}\right\}, \quad (4.9.27)$$

and I_1, I_2, I_3 are given by (4.9.9). For an incompressible body,

$$\left.\begin{array}{l}\Phi = 2\dfrac{\partial W}{\partial I_1}, \quad \Psi = 2\dfrac{\partial W}{\partial I_2} \\[2mm] W = W(I_1, I_2)\end{array}\right\}. \quad (4.9.28)$$

In this case equation (4.9.26) determines the scalar function p and

$$\left.\begin{array}{l}I_1 = \lambda^2+a^{\alpha\beta}A_{\alpha\beta} \\ I_2 = 1/\lambda^2+a_{\alpha\beta}A^{\alpha\beta} \\ I_3 = \lambda^2 A/a = 1\end{array}\right\}. \quad (4.9.29)$$

Equations (4.9.21) are the equations of membrane theory of the deformed shell and may be discussed by methods similar to those used in *T.E.* provided we refer all results to the deformed shell. The complete deformation can, however, only be found with the help of (4.9.24)–(4.9.26).

4.10. Radial deformation of a plane sheet containing a circular hole or inclusion

Consider a thin plane uniform sheet of incompressible material which is transversely isotropic about the x_3-axis and is bounded in the cylindrical polar coordinate system (ρ, ϑ, x_3) by the planes $x_3 = \pm h_0$ and the cylindrical surfaces $\rho = a_1$, $\rho = a_2$ $(a_1 > a_2)$. This sheet is subjected to a system of radial surface tractions acting in its plane and uniformly distributed around its outer edge $\rho = a_1$. A uniformly distributed system of radial surface tractions or displacements is also prescribed along the inner edge $\rho = a_2$; the major surfaces $x_3 = \pm h_0$ are free from applied stresses. Under these circumstances the sheet remains plane and the deformation is symmetric about the ρ-axis, a point initially at (ρ, ϑ) in the plane $x_3 = 0$ being displaced to (r, ϑ) in the same plane, where r is a function of ρ only.

This problem may be examined by using the theory of §§ 4.1–4.4. From symmetry considerations it follows that the principal directions of strain at any point in the middle surface of the deformed sheet coincide with the directions of r, ϑ, x_3. Denoting the principal extension ratios in these directions by λ_1, λ_2, λ_3 respectively we have

$$\lambda_1 = \frac{dr}{d\rho}, \qquad \lambda_2 = \frac{r}{\rho}, \qquad \lambda = \lambda_3 = \frac{1}{\lambda_1 \lambda_2}, \qquad (4.10.1)$$

where r, λ_1, λ_2, λ_3 may be regarded as functions of the single variable ρ. If we allow the curvilinear coordinate system θ_1, θ_2 to coincide with the polar coordinates r, ϑ in the middle plane of the deformed sheet, we obtain, with the help of (4.10.1),

$$A_{\alpha\beta} = \begin{bmatrix} 1 & 0 \\ 0 & r^2 \end{bmatrix}, \qquad A^{\alpha\beta} = \begin{bmatrix} 1 & 0 \\ 0 & \frac{1}{r^2} \end{bmatrix}, \qquad A = r^2, \qquad (4.10.2)$$

$$\left. \begin{array}{c} a_{\alpha\beta} = \begin{bmatrix} \frac{1}{\lambda_1^2} & 0 \\ 0 & \frac{r^2}{\lambda_2^2} \end{bmatrix}, \qquad a^{\alpha\beta} = \begin{bmatrix} \lambda_1^2 & 0 \\ 0 & \frac{\lambda_2^2}{r^2} \end{bmatrix} \\ a = \frac{r^2}{\lambda_1^2 \lambda_2^2} = \lambda_3^2 r^2 \end{array} \right\}. \qquad (4.10.3)$$

Using (4.3.4) with $I_3 = 1$, the invariants I_1, I_2, K_1 then become

$$\left.\begin{array}{c} I_1 = \lambda_1^2+\lambda_2^2+\lambda_3^2, \qquad I_2 = \dfrac{1}{\lambda_1^2}+\dfrac{1}{\lambda_2^2}+\dfrac{1}{\lambda_3^2} \\ K_1 = \tfrac{1}{2}(\lambda_3^2-1) \end{array}\right\}. \qquad (4.10.4)$$

The stress resultant components $n^{\alpha\beta}$ are obtained by combining (4.10.2)–(4.10.4) with (4.4.7) and making use of (4.3.17) and (4.4.11). Thus

$$\left.\begin{array}{l} n^{11} = 2h_0\lambda_3\left[2(\lambda_1^2-\lambda_3^2)\left(\dfrac{\partial W}{\partial I_1}+\lambda_2^2\dfrac{\partial W}{\partial I_2}\right)-\lambda_3^2\dfrac{\partial W}{\partial K_1}\right] = T_1 \text{ (say)} \\ r^2 n^{22} = 2h_0\lambda_3\left[2(\lambda_2^2-\lambda_3^2)\left(\dfrac{\partial W}{\partial I_1}+\lambda_1^2\dfrac{\partial W}{\partial I_2}\right)-\lambda_3^2\dfrac{\partial W}{\partial K_1}\right] = T_2 \text{ (say)} \\ n^{12} = 0 \end{array}\right\}. \qquad (4.10.5)$$

The quantities T_1, T_2 defined by (4.10.5) are the physical stress resultants in the radial and transverse directions respectively in the deformed sheet.

The metric tensors $a_{\alpha\beta}$, $a^{\alpha\beta}$, $A_{\alpha\beta}$, $A^{\alpha\beta}$, the invariants I_1, I_2, K_1, the strain energy function W, and the stress resultants T_1, T_2 are all independent of ϑ and may be regarded as functions of r (or ρ). With the help of (4.10.2) the equations of equilibrium (4.2.4) therefore reduce to the single equation

$$\frac{d}{dr}(T_1 r) = T_2. \qquad (4.10.6)$$

If the expressions (4.10.5) for T_1, T_2 are introduced into (4.10.6) and λ_1, λ_2, λ_3 eliminated by means of (4.10.1), the resulting relation becomes a non-linear second-order differential equation for the determination of r as a function of ρ. In some respects, however, it is more convenient to regard (4.10.5), (4.10.1), and (4.10.6) as a system of six equations for the determination of the six quantities r, λ_1, λ_2, λ_3, T_1, and T_2 as functions of ρ. An approximate numerical method of integration may then be employed. For this purpose it is necessary to assume a specific expression for the strain energy function W, and for convenience we shall restrict attention to isotropic bodies which have the Mooney form (1.14.3). We then have

$$\frac{\partial W}{\partial I_1} = C_1, \qquad \frac{\partial W}{\partial I_2} = C_2, \qquad \frac{\partial W}{\partial K_1} \equiv 0, \qquad (4.10.7)$$

and we write

$$\Gamma = \frac{C_2}{C_1}, \qquad T_1' = \frac{T_1}{4h_0 C_1}, \qquad T_2' = \frac{T_2}{4h_0 C_1}. \qquad (4.10.8)$$

We suppose the deformation at any point P of the deformed sheet initially at (ρ, ϑ) to be specified by means of known values of the extension ratios λ_1, λ_2. From (4.10.1) and (4.10.5) we then derive the scheme

$$r = \lambda_2 \rho, \qquad (4.10.9)$$

$$\lambda_3 = 1/(\lambda_1 \lambda_2), \qquad (4.10.10)$$

$$T_1' = \lambda_3(\lambda_1^2 - \lambda_3^2)(1 + \lambda_2^2 \Gamma), \qquad (4.10.11)$$

$$T_2' = \lambda_3(\lambda_2^2 - \lambda_3^2)(1 + \lambda_1^2 \Gamma), \qquad (4.10.12)$$

for the determination of the values of r, λ_3, T_1', and T_2' at P. From (4.10.1) and (4.10.6) we have

$$\frac{dr}{d\rho} = \lambda_1, \qquad (4.10.13)$$

$$\frac{dT_1'}{d\rho} = -\frac{1}{r}\frac{dr}{d\rho}(T_1' - T_2'), \qquad (4.10.14)$$

and by differentiating (4.10.9), (4.10.11), and (4.10.12) and making use of (4.10.10) we obtain

$$\frac{d\lambda_2}{d\rho} = \frac{1}{\rho}\left(\frac{dr}{d\rho} - \lambda_2\right) = \frac{1}{\rho}(\lambda_1 - \lambda_2), \qquad (4.10.15)$$

$$\frac{d\lambda_1}{d\rho} = \frac{\dfrac{dT_1'}{d\rho} - [(3\lambda_3^2 - \lambda_1^2) + (\lambda_1^2 + \lambda_3^2)\lambda_2^2 \Gamma]\dfrac{\lambda_3}{\lambda_2}\dfrac{d\lambda_2}{d\rho}}{(\lambda_1^2 + 3\lambda_3^2)(1 + \lambda_2^2 \Gamma)\dfrac{\lambda_3}{\lambda_1}}, \qquad (4.10.16)$$

$$\frac{dT_2'}{d\rho} = (\lambda_2^2 + 3\lambda_3^2)(1 + \lambda_1^2 \Gamma)\frac{\lambda_3}{\lambda_2}\frac{d\lambda_2}{d\rho} + [(3\lambda_3^2 - \lambda_2^2) + (\lambda_2^2 + \lambda_3^2)\lambda_1^2 \Gamma]\frac{\lambda_3}{\lambda_1}\frac{d\lambda_1}{d\rho}. \qquad (4.10.17)$$

In each of the relations (4.10.9)–(4.10.17) the left-hand member is expressed as a function of the known quantities ρ, λ_1, λ_2 and quantities which have been determined from them by means of the preceding equations in the scheme. If, therefore, λ_1 and λ_2 have known values at the point P, their first derivatives can also be determined at this point. By successive differentiation of (4.10.13)–(4.10.17) the second and higher order derivatives of r, T_1', λ_2, λ_1, and T_2' can, in principle, be found. If we assume that these quantities are regular functions of ρ expressible as Taylor series in a finite interval containing P, approximate values of λ_1, λ_2 can be found at the point P' $(\rho + \Delta\rho, \vartheta)$ from expansions of the type

$$[\lambda_1]_{\rho + \Delta\rho} = [\lambda_1]_\rho + \left[\frac{d\lambda_1}{d\rho}\right]_\rho \Delta\rho + \frac{1}{2}\left[\frac{d^2\lambda_1}{d\rho^2}\right]_\rho (\Delta\rho)^2 + \dots. \qquad (4.10.18)$$

The values of λ_1, λ_2 thus obtained serve as a basis for a similar calculation at the point P'. We observe that if, at the point P, $\lambda_2 > \lambda_1$, then from (4.10.11), (4.10.12) $T_2' > T_1'$, and from (4.10.14), (4.10.15) $dT_1'/d\rho > 0$ and $d\lambda_2/d\rho < 0$. In the neighbourhood of this point, therefore, λ_2 and T_1' are decreasing and increasing functions of ρ respectively. This suggests that as ρ increases the deformation will tend uniformly towards a state of uniform strain in which $\lambda_1 = \lambda_2$, $T_1' = T_2'$. We should, in fact, expect the effect of the conditions at $\rho = a_2$ to become progressively less important as we proceed outwards away from this boundary.

If values of λ_1, λ_2 can be determined from the boundary conditions at $\rho = a_2$, the deformation throughout the entire sheet can be determined approximately by means of the iterative procedure embodied in equations (4.10.9)–(4.10.17). When the inner boundary is a stress-free hole we have, from (4.10.11), at $\rho = a_2$,

$$T_1' = 0, \quad \lambda_2 = \lambda_0 \text{ (say)}, \quad \lambda_1 = \lambda_3 = \lambda_0^{-\frac{1}{2}}. \quad (4.10.19)$$

If a value is prescribed for λ_0 at $\rho = a_2$, λ_1 and λ_2 are known on this boundary. The value of λ_2 obtained as a result of the numerical integration, for $\rho = a_1$, then gives the value of r at the outer edge and hence the overall extension of the sheet corresponding to a given value of λ_0.

An approximation to the value of r in the neighbourhood of the hole may be obtained by the following method. By Taylor's theorem we can express the value of r for any value of ρ in the form

$$r = \sum_{n=0}^{\infty} \frac{1}{n!}(\rho - a_2)^n \left[\frac{d^n r}{d\rho^n}\right]_{\rho = a_2}, \quad (4.10.20)$$

and since $dr/d\rho = \lambda_0^{-\frac{1}{2}}$ when $\rho = a_2$ we have, as a first approximation,

$$r = \lambda_0 a_2 + \lambda_0^{-\frac{1}{2}}(\rho - a_2). \quad (4.10.21)$$

A closer approximation follows by observing that since $T_1' = 0$ when $\rho = a_2$, we have, from (4.10.11) and (4.10.1),

$$\frac{dr}{d\rho} = \left(\frac{\rho}{r}\right)^{\frac{1}{2}}$$

at the inner boundary. Remembering that $r = \lambda_0 a_2$ when $\rho = a_2$, this yields

$$r^{\frac{3}{2}} = \rho^{\frac{3}{2}} + a_2^{\frac{3}{2}}(\lambda_0^{\frac{3}{2}} - 1), \quad (4.10.22)$$

approximately. For this result to be strictly applicable we should require that $T'_1 = 0$ over the whole range of applicability. We may expect the result to be approximately valid, however, as long as $T'_1 \ll T'_2$.

The problem has been solved numerically for a value of $\Gamma = 0 \cdot 1$ by Rivlin and Thomas.† The theoretical results obtained agree very closely with those obtained experimentally for rubber sheets even for quite large values of λ_0. Since $\partial W/\partial I_2$ is not constant for rubber, this suggests that the deformation is insensitive to the form of the strain energy function. A reason for this emerges from an examination of the variation of I_2 as a function of ρ. For all values of λ_0 in the range $1 \cdot 0 < \lambda_0 < 6 \cdot 0$, calculations based on (4.10.4) show that I_2 does not vary greatly with ρ, and the most marked variation occurs near the hole where the asymptotic formula (4.10.22), which is independent of the form of W, applies. In fact, at no point of the sheet, in the range examined for the extension ratio λ_0, does I_2 attain very large values. The good agreement between theory and experiment for this type of deformation can thus be accounted for by a form of W in which $\partial W/\partial I_1$ and $\partial W/\partial I_2$ are independent of I_1. This form appears to be approximately valid for rubber.‡ For a fuller discussion reference may be made to the original paper.

When the inner boundary is bonded to a rigid circular inclusion of radius a_2, the problem may be solved in a similar manner. In this case the boundary conditions at $\rho = a_2$ are

$$\lambda_2 = 1, \qquad \lambda_1 = \lambda_0 = 1/\lambda_3.$$

4.11. Axially symmetrical problems

A more general class of problems may be examined by using the theory of §§ 4.5–4.9. In these, the middle surface of the sheet forms a surface of revolution before and after deformation, the axis of symmetry being the same in both cases. We restrict attention to the case where the undeformed sheet is uniform and of uniform thickness and is composed of isotropic incompressible material. Under these circumstances the system of deforming forces is also symmetric about the axis of the deformed body. We suppose that this system consists of a uniform distribution acting around each of the edges of the deformed sheet and a continuous distribution of traction normal to

---

† Loc. cit., p. 133. The scheme of integration used in the paper is slightly different from that used in this section.
‡ See Chapter IX.

each of its major surfaces. Singularities are excluded both from the system of forces and from the surfaces of revolution defining the elastic body.

The assumptions of uniformity, isotropy, and incompressibility are evidently not essential for the preservation of the axis of symmetry under an axially symmetric system of deforming forces, and a theory similar to that of the present section could therefore be developed with these postulates replaced by less stringent conditions. In many of the applications, however, the solution is completed by numerical methods and it is then necessary to assume some particular form for the strain energy function W. The Mooney form (1.14.3) for incompressible bodies is convenient for this purpose and has some physical significance in the case of rubber.

We suppose the middle surface S_0 of the undeformed sheet to be generated by the revolution of a plane curve C through an angle 2π about an axis x_3 in its plane. This curve has no double points and does not cut the axis x_3 except possibly at one or both of its end points. The curve C thus becomes a meridian curve on the middle surface of the undeformed sheet; lines of latitude are formed by the family of curves which are orthogonal to the meridians at every point.

We choose a cylindrical polar coordinate system (ρ, ϑ, x_3), the x_3-axis of which coincides with the axis of symmetry of the undeformed sheet. We suppose that any point P_0 (ρ, ϑ, x_3) in the surface S_0 is carried by the deformation to the point P (r, ϑ, y_3) in the deformed middle surface S. The arc length $A_0 P_0$ measured along the meridian through P_0 from a fixed point A_0 is denoted by η; when the sheet is deformed, this distance measured along the corresponding meridian in the deformed sheet is denoted by ξ. From the symmetry of the system it follows that the principal directions of strain at P coincide with the meridians, the lines of latitude and the normal to the deformed middle surface S, and we denote the principal extension ratios in these directions by λ_1, λ_2, λ_3 respectively. We then have

$$\lambda_1 = \frac{d\xi}{d\eta}, \qquad \lambda_2 = \frac{r}{\rho}, \qquad \lambda = \lambda_3 = \frac{1}{\lambda_1 \lambda_2}. \qquad (4.11.1)$$

In these expressions λ_1, λ_2, λ_3, r, ξ, and η are independent of ϑ and may be regarded as functions of the single variable ρ. Alternatively, r, ξ, or η may be chosen as the independent variable. The situation in the deformed sheet is illustrated by Fig. 4.1.

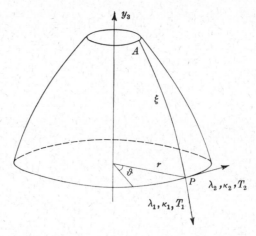

FIG. 4.1. Surface of revolution defining deformed sheet.

In the middle surface of the deformed sheet, an element of length ds is given by
$$ds^2 = d\xi^2 + r^2 \, d\vartheta^2, \tag{4.11.2}$$
the expression for the corresponding element ds_0 in the undeformed middle surface being
$$ds_0^2 = d\eta^2 + \rho^2 \, d\vartheta^2$$
$$= \left(\frac{d\eta}{d\xi}\right)^2 d\xi^2 + \rho^2 \, d\vartheta^2$$
$$= \frac{1}{\lambda_1^2} d\xi^2 + \frac{r^2}{\lambda_2^2} d\vartheta^2, \tag{4.11.3}$$

if we make use of (4.11.1). If we allow the curvilinear coordinates θ_α to coincide with the orthogonal system ξ, ϑ in the deformed body, so that $\theta_1 = \xi$, $\theta_2 = \vartheta$, we obtain from (4.11.2) and (4.11.3)

$$A_{\alpha\beta} = \begin{bmatrix} 1 & 0 \\ 0 & r^2 \end{bmatrix}, \quad A^{\alpha\beta} = \begin{bmatrix} 1 & 0 \\ 0 & \frac{1}{r^2} \end{bmatrix}, \quad A = r^2, \tag{4.11.4}$$

$$\left.\begin{array}{c} a_{\alpha\beta} = \begin{bmatrix} \dfrac{1}{\lambda_1^2} & 0 \\ 0 & \dfrac{r^2}{\lambda_2^2} \end{bmatrix}, \quad a^{\alpha\beta} = \begin{bmatrix} \lambda_1^2 & 0 \\ 0 & \dfrac{\lambda_2^2}{r^2} \end{bmatrix} \\[2em] a = \dfrac{r^2}{\lambda_1^2 \lambda_2^2} = \lambda_3^2 r^2 \end{array}\right\} \tag{4.11.5}$$

Using (4.9.29) the invariants I_1, I_2 then become

$$I_1 = \lambda_1^2+\lambda_2^2+\lambda_3^2, \qquad I_2 = \frac{1}{\lambda_1^2}+\frac{1}{\lambda_2^2}+\frac{1}{\lambda_3^2}. \tag{4.11.6}$$

The stress resultant components $n^{\alpha\beta}$ are obtained by introducing (4.11.4)–(4.11.6) into (4.9.25). Thus

$$\left.\begin{aligned} n^{11} &= 4h_0\lambda_3(\lambda_1^2-\lambda_3^2)\left(\frac{\partial W}{\partial I_1}+\lambda_2^2\frac{\partial W}{\partial I_2}\right) = T_1 \quad\text{(say)} \\ r^2 n^{22} &= 4h_0\lambda_3(\lambda_2^2-\lambda_3^2)\left(\frac{\partial W}{\partial I_1}+\lambda_1^2\frac{\partial W}{\partial I_2}\right) = T_2 \quad\text{(say)} \\ n^{12} &= 0 \end{aligned}\right\}. \tag{4.11.7}$$

Equations (4.11.4)–(4.11.7) may be compared with the corresponding formulae of § 4.10. The quantities defined by (4.11.7) are now the physical stress resultants in the directions of the meridian curves and curves of latitude respectively in the deformed sheet.

The metric tensors $a_{\alpha\beta}$, $a^{\alpha\beta}$, $A_{\alpha\beta}$, $A^{\alpha\beta}$, the invariants I_1, I_2, the strain energy function W, and the stress resultants T_1, T_2 are all independent of ϑ and may be regarded as functions of any one of the variables ρ, r, η, ξ which define the position of P on a meridian curve. The θ_1, θ_2 curves are lines of principal curvature of the deformed middle surface and, denoting the normal curvatures in these directions by κ_1, κ_2 respectively (Fig. 4.1), we have† in (4.5.5)–(4.5.9)

$$C_1^1 = -\kappa_1, \qquad C_2^2 = -\kappa_2, \qquad C_2^1 = C_1^2 = 0. \tag{4.11.8}$$

These quantities are again independent of ϑ. From (4.11.4) it follows that the only non-vanishing components of the Christoffel tensor $\Gamma^{\alpha}_{\beta\gamma}$ are

$$\Gamma^1_{22} = -r\frac{dr}{d\xi}, \qquad \Gamma^2_{12} = \frac{1}{r}\frac{dr}{d\xi}, \tag{4.11.9}$$

and the equations of equilibrium (4.9.21) reduce to

$$\left.\begin{aligned} \frac{d}{d\xi}(T_1 r) &= T_2 \frac{dr}{d\xi} \\ \kappa_1 T_1 + \kappa_2 T_2 &= P \end{aligned}\right\}, \tag{4.11.10}$$

where
$$P = p_1 - p_2 \tag{4.11.11}$$

is the resultant pressure in the direction of the outward normal to the deformed middle surface.

† See, for example, I. S. Sokolnikoff, *Tensor Analysis* (New York, 1951), p. 191.

With the help of (4.11.8) and (4.11.9) the Codazzi equations (4.5.6) with $\alpha = 2$ yield

$$\frac{d}{d\xi}(\kappa_2 r) = \kappa_1 \frac{dr}{d\xi}, \tag{4.11.12}$$

the remaining equation vanishing identically. Also, from the elementary formula for the curvature of a plane curve, we obtain for a meridian curve in the deformed sheet

$$\kappa_1 = -\frac{d^2 r/d\xi^2}{[1-(dr/d\xi)^2]^{\frac{1}{2}}}. \tag{4.11.13}$$

By introducing this value of κ_1 into (4.11.12) and integrating the resulting equation we obtain

$$\kappa_2 r = \left[1-\left(\frac{dr}{d\xi}\right)^2\right]^{\frac{1}{2}}, \tag{4.11.14}$$

if we assume that when $r = 0$, κ_2 is finite and the surface cuts the axis of symmetry orthogonally so that $dr/d\xi = 1$. The condition (4.11.14) may also be obtained by using Meusnier's theorem. From (4.11.13) and (4.11.14) we have the alternative relation

$$\kappa_1 \kappa_2 r = -\frac{d^2 r}{d\xi^2}. \tag{4.11.15}$$

From (4.11.10) and (4.11.12) we obtain

$$T_1 \frac{d}{d\xi}(\kappa_2 r) + \kappa_2 \frac{d}{d\xi}(T_1 r) = P \frac{dr}{d\xi},$$

which integrates to the form

$$\kappa_2 T_1 = \frac{1}{r^2}\int^r Pr\, dr + \frac{L}{r^2}, \tag{4.11.16}$$

where L is a constant. When P is constant this becomes

$$\kappa_2 T_1 = \tfrac{1}{2}P + L/r^2. \tag{4.11.17}$$

In this case, if the deformed sheet cuts the axis of symmetry orthogonally, we have at this point† ($\rho = 0$)

$$\kappa_1 = \kappa_2 = \kappa \text{ (say)}, \quad \lambda_1 = \lambda_2 = \lambda \text{ (say)}, \quad T_1 = T_2 = T \text{ (say)},$$

and (4.11.17) and the second of equations (4.11.10) yield

$$2\kappa T = P, \quad L = 0. \tag{4.11.18}$$

† For the rest of this chapter λ differs from λ defined in § 4.9 and in (4.11.1).

4.12. Inflation of a circular plane sheet

Let the undeformed body be a uniform sheet of isotropic incompressible material bounded in the cylindrical polar coordinate system (ρ, ϑ, x_3) by the planes $x_3 = \pm h_0$. This is subjected to a uniform two-dimensional extension in its own plane of ratio λ_0 so that the point (ρ, ϑ, x_3) is displaced to $(\lambda_0 \rho, \vartheta, x_3/\lambda_0^2)$. The sheet is then clamped in a circular clamp of radius a (Fig. 4.2a) and inflated by a uniform pressure

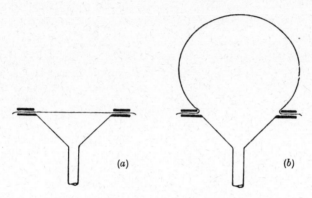

Fig. 4.2. Inflation of a plane circular sheet.

P applied to one face, so that it takes up a cylindrically symmetrical form as shown in Fig. 4.2b.

The deformation thus produced may be examined by using the results of § 4.11. In (4.11.1) we have

$$\eta = \rho, \qquad \lambda_1 = \frac{d\xi}{d\rho}, \qquad (4.12.1)$$

and in the remainder of the equations of § 4.11 the independent variable may be changed from ξ to ρ by making use of (4.12.1). Equations (4.11.1), (4.11.7), (4.11.10), (4.11.12), and (4.11.14) or (4.11.15) then yield eight equations for the determination of the eight unknowns r, λ_1, λ_2, λ_3, T_1, T_2, κ_1, κ_2 as functions of ρ.

These equations may be solved numerically† along the lines indicated for the problems of plane deformation examined in § 4.10. For convenience we again assume the material to have the Mooney form of strain energy function and use the notation of equations (4.10.7) and (4.10.8). If the values of λ_1, λ_2, κ_1, κ_2 are known at the point P initially at (ρ, ϑ) in the middle plane of the undeformed sheet, the

† For an alternative procedure see Klingbeil and Shield, loc. cit., p. 134.

remaining quantities and their first derivatives may be found from the scheme

$$\lambda_3 = \frac{1}{\lambda_1\lambda_2}, \qquad (4.12.2)$$

$$T'_1 = \lambda_3(\lambda_1^2-\lambda_3^2)(1+\lambda_2^2\Gamma), \qquad (4.12.3)$$

$$T'_2 = \lambda_3(\lambda_2^2-\lambda_3^2)(1+\lambda_1^2\Gamma), \qquad (4.12.4)$$

$$r = \lambda_2\rho, \qquad (4.12.5)$$

$$\frac{dr}{d\rho} = \lambda_1(1-\kappa_2^2 r^2)^{\frac{1}{2}}, \qquad (4.12.6)$$

$$\frac{d\lambda_2}{d\rho} = \frac{1}{\rho}\left(\frac{dr}{d\rho}-\lambda_2\right), \qquad (4.12.7)$$

$$\frac{d\kappa_2}{d\rho} = \frac{1}{r}\frac{dr}{d\rho}(\kappa_1-\kappa_2), \qquad (4.12.8)$$

$$\frac{dT'_1}{d\rho} = -\frac{1}{r}\frac{dr}{d\rho}(T'_1-T'_2), \qquad (4.12.9)$$

$$\frac{d\lambda_1}{d\rho} = \frac{\dfrac{dT'_1}{d\rho}-[(3\lambda_3^2-\lambda_1^2)+(\lambda_1^2+\lambda_3^2)\lambda_2^2\Gamma]\dfrac{\lambda_3}{\lambda_2}\dfrac{d\lambda_2}{d\rho}}{(\lambda_1^2+3\lambda_3^2)(1+\lambda_2^2\Gamma)\dfrac{\lambda_3}{\lambda_1}}, \qquad (4.12.10)$$

$$\frac{dT'_2}{d\rho} = (\lambda_2^2+3\lambda_3^2)(1+\lambda_1^2\Gamma)\frac{\lambda_3}{\lambda_2}\frac{d\lambda_2}{d\rho}+[(3\lambda_3^2-\lambda_2^2)+(\lambda_2^2+\lambda_3^2)\lambda_1^2\Gamma]\frac{\lambda_3}{\lambda_1}\frac{d\lambda_1}{d\rho}, \qquad (4.12.11)$$

$$\frac{d\kappa_1}{d\rho} = -\frac{1}{T'_1}\left(\kappa_1\frac{dT'_1}{d\rho}+\kappa_2\frac{dT'_2}{d\rho}+T'_2\frac{d\kappa_2}{d\rho}\right). \qquad (4.12.12)$$

In each of these equations the left-hand member is expressed as a function of the known quantities λ_1, λ_2, κ_1, κ_2 or quantities which can be determined from them by means of the preceding relations in the scheme. In this system of equations, (4.12.2) and (4.12.5) are obtained from (4.11.1) and (4.12.3), and (4.12.4) from (4.11.7). The relations (4.12.6), (4.12.8), (4.12.9) follow from (4.11.14), (4.11.12), and the first of (4.11.10) respectively by making use of (4.12.1) to effect a change of independent variable; (4.12.7), (4.12.10), (4.12.11), and (4.12.12) are obtained from (4.12.5), (4.12.3), (4.12.4), and the second of (4.11.10) respectively by differentiation.

Expressions for the second and higher-order derivatives of r, λ_1, λ_2, κ_1, κ_2, T'_1, and T'_2 in terms of quantities previously determined may be obtained by successive differentiation of (4.12.6)–(4.12.12). Alternatively, for $d^2r/d\rho^2$ we obtain, from (4.11.15) and (4.12.1),

$$\frac{d^2r}{d\rho^2} = \frac{1}{\lambda_1}\left(\frac{dr}{d\rho}\frac{d\lambda_1}{d\rho} - \kappa_1\kappa_2 r\lambda_1^3\right). \qquad (4.12.13)$$

At the centre of the inflated sheet where $r = 0$ we have the symmetry conditions

$$\left.\begin{array}{ll} \kappa_1 = \kappa_2 = \kappa \text{ (say)}, & \lambda_1 = \lambda_2 = \lambda \text{ (say)} \\ T'_1 = T'_2 = T' \text{ (say)}, & \dfrac{dr}{d\rho} = \lambda, \quad \lambda_3 = \dfrac{1}{\lambda^2} \end{array}\right\}, \qquad (4.12.14)$$

it being assumed that κ, λ, and T' are finite and non-zero. Equations (4.11.6), (4.12.3), and the second of (4.11.10) yield

$$I_1 = 2\lambda^2 + \frac{1}{\lambda^4}, \qquad I_2 = \frac{2}{\lambda^2} + \lambda^4, \qquad (4.12.15)$$

$$T' = \left(1 - \frac{1}{\lambda^6}\right)(1 + \lambda^2 \Gamma), \qquad (4.12.16)$$

$$2\kappa T' = P/(4h_0 C_1) = P' \quad \text{(say)}, \qquad (4.12.17)$$

respectively, when $r = 0$.

If we differentiate (4.12.7)–(4.12.9), and assume $d^2\lambda_2/d\rho^2$, $d^2\kappa_2/d\rho^2$, $d^2T'_1/d\rho^2$ to be finite, the resulting equations, with (4.12.10)–(4.12.13) yield, at $\rho = 0$, seven linear homogeneous equations for $d^2r/d\rho^2$ and the first derivatives of λ_1, λ_2, κ_1, κ_2, T'_1, T'_2. It therefore follows from the symmetry conditions (4.12.14) that at the pole of the inflated sheet

$$\frac{d^2r}{d\rho^2} = \frac{d\lambda_1}{d\rho} = \frac{d\lambda_2}{d\rho} = \frac{d\kappa_1}{d\rho} = \frac{d\kappa_2}{d\rho} = \frac{dT'_1}{d\rho} = \frac{dT'_2}{d\rho} = 0 \quad (\rho = 0), \qquad (4.12.18)$$

except possibly for special values of λ, κ, and T.

By writing (4.12.9) in the form

$$\frac{d}{d\rho}(T'_1 r) = T'_2 \frac{dr}{d\rho},$$

differentiating twice, and inserting the conditions (4.12.14), (4.12.18) at $\rho = r = 0$, we obtain

$$3\frac{d^2 T'_1}{d\rho^2} = \frac{d^2 T'_2}{d\rho^2} \quad (\rho = 0). \qquad (4.12.19)$$

Similarly, from (4.12.8) we have

$$3\frac{d^2 \kappa_2}{d\rho^2} = \frac{d^2 \kappa_1}{d\rho^2} \quad (\rho = 0), \qquad (4.12.20)$$

while with the help of (4.12.18)–(4.12.20) the second of equations (4.11.10) yields

$$\kappa \frac{d^2 T_1'}{d\rho^2} + T' \frac{d^2 \kappa_2}{d\rho^2} = 0 \quad (\rho = 0). \tag{4.12.21}$$

Again, by differentiating (4.12.13) and using the conditions at $\rho = 0$ we have

$$\frac{d^3 r}{d\rho^3} - \frac{d^2 \lambda_1}{d\rho^2} = -\kappa^2 \lambda^3 \quad (\rho = 0), \tag{4.12.22}$$

and by differentiating (4.12.5) three times

$$\frac{d^3 r}{d\rho^3} = 3 \frac{d^2 \lambda_2}{d\rho^2} \quad (\rho = 0). \tag{4.12.23}$$

This equation, with (4.12.22), yields

$$3 \frac{d^2 \lambda_2}{d\rho^2} - \frac{d^2 \lambda_1}{d\rho^2} = -\kappa^2 \lambda^3 \quad (\rho = 0). \tag{4.12.24}$$

Two further relations are obtained by differentiating (4.12.3), (4.12.4) twice and introducing the symmetry conditions at $\rho = 0$. Thus

$$\left. \begin{aligned}
\frac{d^2 T_1'}{d\rho^2} &= \left(\lambda^2 + \frac{3}{\lambda^4}\right)(1+\lambda^2 \Gamma)\frac{1}{\lambda^3}\frac{d^2 \lambda_1}{d\rho^2} + \\
&\quad + \left[\left(\frac{3}{\lambda^4} - \lambda^2\right) + \left(\lambda^4 + \frac{1}{\lambda^2}\right)\Gamma\right]\frac{1}{\lambda^3}\frac{d^2 \lambda_2}{d\rho^2} \\
\frac{d^2 T_2'}{d\rho^2} &= \left(\lambda^2 + \frac{3}{\lambda^4}\right)(1+\lambda^2 \Gamma)\frac{1}{\lambda^3}\frac{d^2 \lambda_2}{d\rho^2} + \\
&\quad + \left[\left(\frac{3}{\lambda^4} - \lambda^2\right) + \left(\lambda^4 + \frac{1}{\lambda^2}\right)\Gamma\right]\frac{1}{\lambda^3}\frac{d^2 \lambda_1}{d\rho^2} \quad (\rho = 0)
\end{aligned} \right\}. \tag{4.12.25}$$

The relations (4.12.19)–(4.12.21), (4.12.24), (4.12.25) are six linear equations for the determination of the second derivatives at $\rho = 0$ of $\lambda_1, \lambda_2, \kappa_1, \kappa_2, T_1', T_2'$. From (4.12.19), (4.12.24), and (4.12.25) we obtain the solutions

$$\left. \begin{aligned}
\frac{d^2 \lambda_1}{d\rho^2} &= -\frac{\kappa^2 \lambda^3 \left[\left(2 - \frac{3}{\lambda^6}\right) - \lambda^2 \Gamma\right]}{4\left(1 + \frac{3}{\lambda^6}\right)(1 + \lambda^2 \Gamma)} \\
\frac{d^2 \lambda_2}{d\rho^2} &= -\frac{\kappa^2 \lambda^3 \left[\left(2 + \frac{3}{\lambda^6}\right) + \lambda^2 \left(1 + \frac{4}{\lambda^6}\right)\Gamma\right]}{4\left(1 + \frac{3}{\lambda^6}\right)(1 + \lambda^2 \Gamma)}
\end{aligned} \right\}, \tag{4.12.26}$$

$$\frac{d^2T_1'}{d\rho^2} = -\frac{\kappa^2\lambda^2\left[\dfrac{3}{\lambda^6}+\left(1+\dfrac{2}{\lambda^6}+\dfrac{3}{\lambda^{12}}\right)\lambda^2\Gamma+\dfrac{1}{\lambda^2}\left(1+\dfrac{2}{\lambda^6}\right)\Gamma^2\right]}{2\left(1+\dfrac{3}{\lambda^6}\right)(1+\lambda^2\Gamma)} \qquad (\rho=0), \quad (4.12.27)$$

and these with (4.12.19)–(4.12.21) yield expressions for $d^2T_2'/d\rho^2$, $d^2\kappa_1/d\rho^2$, and $d^2\kappa_2/d\rho^2$. A similar calculation may be carried out for a general form of strain energy function, and it is then found that for large values of λ, higher order derivatives of W with respect to I_1, I_2 can exert an important influence on the values of $d^2\lambda_1/d\rho^2$, $d^2\lambda_2/d\rho^2$ at $\rho=0$ even if these higher derivatives are quite small compared with $\partial W/\partial I_1$ and $\partial W/\partial I_2$.† When W has the Mooney form, we see from (4.12.26) that whereas $d^2\lambda_2/d\rho^2$ is always negative, $d^2\lambda_1/d\rho^2$ is positive both when $\lambda^6 < 3/2$ and for sufficiently large values of λ, but may be negative in an intermediate range of values of λ if Γ is sufficiently small. When $\Gamma=0$ (or $C_2=0$) we have

$$\frac{d^2\lambda_1}{d\rho^2} = -\frac{\kappa^2\lambda^3\left(2-\dfrac{3}{\lambda^6}\right)}{4\left(1+\dfrac{3}{\lambda^6}\right)}, \qquad \frac{d^2\lambda_2}{d\rho^2} = -\frac{\kappa^2\lambda^3\left(2+\dfrac{3}{\lambda^6}\right)}{4\left(1+\dfrac{3}{\lambda^6}\right)}, \qquad (4.12.28)$$

and $d^2\lambda_1/d\rho^2$ is then negative for all values of λ such that $\lambda^6 > 3/2$.

By continued differentiation of (4.12.5)–(4.12.13) it may be shown that all odd order derivatives of λ_1, λ_2, κ_1, κ_2, T_1', T_2', and $dr/d\rho$ vanish at $\rho=0$.‡ The fourth and higher order even derivatives of these quantities may be evaluated (in principle) by a procedure similar to the foregoing, but the resulting expressions increase greatly in complexity.§

If given numerical values are assumed for λ and κ at the centre of the sheet, the appropriate values of T' and P' may be calculated from (4.12.16) and (4.12.17) and the values of the second derivatives of λ_1, λ_2, κ_1, κ_2, T_1', T_2' at $\rho=0$ may be evaluated by means of (4.12.26), (4.12.27), and (4.12.19)–(4.12.21). Approximate values of λ_1, λ_2, κ_1, κ_2,

---

† For a fuller discussion the reader is referred to the paper by J. E. Adkins and R. S. Rivlin, loc. cit., p. 133.

‡ We may observe that if all odd-order derivatives of λ_1, λ_2, κ_1, κ_2, T_1', T_2', and $dr/d\rho$ up to and including the $(2n-1)$th vanish at $\rho=0$, then (4.12.7)–(4.12.13) yield, by continued differentiation, seven linear homogeneous equations for the $(2n+1)$th derivatives of these quantities. The latter therefore also vanish at $\rho=0$ except possibly for special values of λ, κ, and T'.

§ The calculation has been carried out by J. E. Adkins for the case $\Gamma=0$; thesis, loc. cit., p. 133.

T'_1 and T'_2 at the point $(\Delta\rho, \vartheta)$ may then be calculated from formulae of the type

$$[\lambda_1]_{\Delta\rho} = \lambda + \frac{1}{2}\left[\frac{d^2\lambda_1}{d\rho^2}\right]_0 (\Delta\rho)^2, \qquad (4.12.29)$$

assuming the Taylor expansions to be valid and $\Delta\rho$ to be small enough to ensure sufficiently rapid convergence for higher-order terms in (4.12.29) to be neglected. From the system of equations (4.12.2)–(4.12.12) the values of the first derivatives of λ_1, λ_2, κ_1, and κ_2 may then be determined at the point $\Delta\rho$ and the numerical step-by-step integration carried out, as for the plane problems (§ 4.10), by using the first and second terms of the expansion (4.10.18). Greater accuracy may be achieved for a given value of the interval $\Delta\rho$ by calculating the second derivatives of λ_1, λ_2, κ_1, κ_2 and employing the first three terms of the series (4.10.18).†

Errors introduced due to neglect of higher-order terms in the series (4.10.18) are cumulative and an accurate assessment of the total error in the values of λ_1, λ_2, κ_1, κ_2 near the edges of the sheet is difficult to carry out. An indication of the magnitude of the error introduced at any stage of the integration may be obtained by calculating the value of $[\lambda_1]_{\rho+2\Delta\rho}$ from $[\lambda_1]_\rho$ in a single stage of numerical integration and comparing it with the value obtained by two stages of numerical integration using intervals $\Delta\rho$. Further checks on the computations are provided by the second of equations (4.11.10) and (4.11.17) with $L = 0$. These equations do not appear explicitly in the scheme of integration, and the values of P' calculated from them should not vary greatly throughout the sheet.

The foregoing process of numerical integration is continued until λ_2 attains the boundary value λ_0; for a sheet which is unstretched before the inflating pressure is applied, $\lambda_0 = 1$. In general, for given values of λ and κ we find that $\lambda_2 = \lambda_0$ when $\rho = A$ where $A \neq a$. To adapt the solution to a sheet of given initial radius a it is necessary to employ a scaling process. This involves multiplying all linear dimensions by the factor $\mu = a/A$. Thus, the equations of §§ 4.11 and 4.12 remain unchanged in form if we replace ρ, r, ξ by

$$\rho' = \mu\rho, \qquad r' = \mu r, \qquad \xi' = \mu\xi,$$

respectively, κ_1, κ_2, κ by

$$\kappa'_1 = \kappa_1/\mu, \qquad \kappa'_2 = \kappa_2/\mu, \qquad \kappa' = \kappa/\mu,$$

respectively, and P' by $P'_0 = P'/\mu$. The principal extension ratios and the stress resultants are unchanged. The primed quantities are those appropriate to a sheet which before inflation has radius a.

† This is the method used by J. E. Adkins and R. S. Rivlin, loc. cit., p. 133.

THEORY OF ELASTIC MEMBRANES

To determine the shape of the profile of the inflated sheet, we observe that

$$\left(\frac{d\xi}{d\rho}\right)^2 = \left(\frac{dr}{d\rho}\right)^2 + \left(\frac{dy_3}{d\rho}\right)^2,$$

and this relation, with (4.12.1) and (4.12.6), yields

$$\frac{dy_3}{d\rho} = -\kappa_2\lambda_1 r, \qquad (4.12.30)$$

the sign being chosen from the consideration that y_3 is a decreasing function of ρ. This relation, and those obtained from it by differentiation with respect to ρ, enable the derivatives of y_3 to be calculated

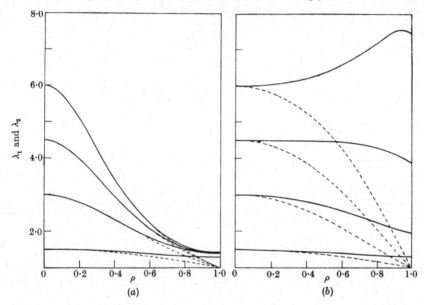

FIG. 4.3. Calculated (λ_1, ρ) curves (full lines) and (λ_2, ρ) curves (broken lines) (a) for $\Gamma = 0$; (b) for $\Gamma = 0.1$.

from the previously determined values of κ_2, λ_1, and r and their derivatives. The successive determination of y_3 at all points of the sheet may then proceed with the help of Taylor's theorem. The value of y_3 at $\rho = 0$ is adjusted so that $y_3 = 0$ when $\rho = A$. Actual values of y_3 for a sheet of radius a are obtained by multiplying by the scaling factor μ. The profile of the inflated sheet is then obtained by plotting the curve of y_3 as a function of r.

Curves of λ_1 and λ_2 against ρ calculated for the values $\Gamma = 0$, $\Gamma = 0.1$ and scaled for a sheet of unit radius, are shown in Figs.

4.3a, 4.3b respectively. These may be compared with experimental curves obtained for rubber sheets by Treloar† (Fig. 4.4). It will be observed that the deformation, for large values of the inflating pressure, is very sensitive to the form of W, the Mooney form with $\Gamma = 0\cdot 1$ giving a much better agreement with experiment than that for $\Gamma = 0$. Fig. 4.5 gives a comparison of calculated and experimental

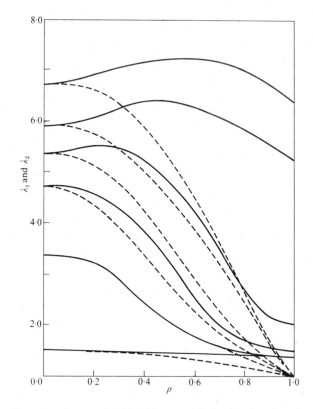

FIG. 4.4. Curves of λ_1 (full lines) and λ_2 (broken lines) against ρ obtained experimentally by Treloar.

profiles.‡ For a further discussion of the theoretical and experimental results, reference may be made to the paper by Adkins and Rivlin.§

† L. R. G. Treloar, *I.R.I. Trans.* **19** (1944), 201.

‡ Apparently these experimental profiles, shown by Adkins and Rivlin, are not those of Treloar's paper as intended but have been reproduced from a different set of data. For further discussion see Klingbeil and Shield, loc. cit., p. 134.

§ Loc. cit., p. 133.

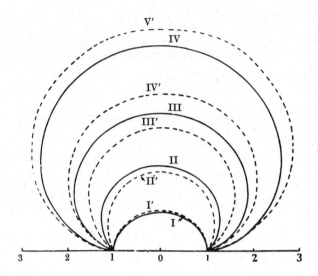

FIG. 4.5. Profiles of inflated sheet calculated for $\Gamma = 0.1$ (full lines) and obtained by Treloar (broken lines).

I, $\lambda = 1\cdot 5$; II, $\lambda = 3\cdot 0$; III, $\lambda = 4\cdot 5$; IV, $\lambda = 6\cdot 0$;
I′, $\lambda = 1\cdot 49$; II′, $\lambda = 3\cdot 36$; III′, $\lambda = 4\cdot 7$; IV′, $\lambda = 5\cdot 35$; V′, $\lambda = 5\cdot 9$.

4.13. Inflation of a spherical shell†

The theory of § 4.11 provides an immediate solution to the problem of a thin spherical shell deformed by means of a uniform inflating pressure P. We assume the shell to be initially of radius a, this radius being changed to A by the deformation. From symmetry considerations, we may write in the formulae of § 4.11

$$\left. \begin{array}{l} \lambda_1 = \lambda_2 = \dfrac{r}{\rho} = \dfrac{A}{a}, \quad \lambda_3 = 1/\lambda_1^2 \\[2mm] \kappa_1 = \kappa_2 = \dfrac{1}{A} = \dfrac{1}{\lambda_1 a} = \kappa \text{ (say)} \end{array} \right\} \qquad (4.13.1)$$

From (4.11.6) we then have

$$I_1 = 2\lambda_1^2 + 1/\lambda_1^4, \qquad I_2 = \lambda_1^4 + 2/\lambda_1^2, \qquad (4.13.2)$$

while from (4.11.7)

$$T_1 = T_2 = 4h_0\left(1 - \frac{1}{\lambda_1^6}\right)\left(\frac{\partial W}{\partial I_1} + \lambda_1^2 \frac{\partial W}{\partial I_2}\right) = T \text{ (say)}. \qquad (4.13.3)$$

† This problem has been discussed in T.E. (§ 3.10) as a special case of the inflation of a thick spherical shell.

These relations with the second of (4.11.10) yield the equation

$$P = 2\kappa T = \frac{8h_0}{a\lambda_1}\left(1-\frac{1}{\lambda_1^6}\right)\left(\frac{\partial W}{\partial I_1}+\lambda_1^2\frac{\partial W}{\partial I_2}\right),\qquad(4.13.4)$$

between the inflating pressure P and the extension ratio λ_1 representing the degree of inflation.

If the elastic material has the Mooney form of strain energy function (1.14.3), we obtain from (4.13.4)

$$\frac{dP}{d\lambda_1} = \frac{8h_0}{a}\left[-\left(\frac{1}{\lambda_1^2}-\frac{7}{\lambda_1^8}\right)C_1+\left(1+\frac{5}{\lambda_1^6}\right)C_2\right].\qquad(4.13.5)$$

From this it follows that if $C_1 = 0$, P increases monotonically with λ_1. If $C_2 = 0$, on the other hand, the pressure increases to a maximum value when $\lambda_1^6 = 7$ and then decreases monotonically with λ_1. When neither C_1 nor C_2 is zero, $dP/d\lambda_1 = 0$ when

$$\Gamma x^4 - x^3 + 5\Gamma x + 7 = 0,$$

where $\qquad\Gamma = C_2/C_1 \quad \text{and} \quad x = \lambda_1^2.$

By Descartes's rule of signs, this equation cannot have more than two real positive roots, so that there are at most two positive values of λ_1 for which P has a stationary value. Furthermore, from (4.13.5) it follows that if C_1 and C_2 are both positive, $dP/d\lambda_1$ is positive for $\lambda_1 = 1$ and also for sufficiently large values of λ_1. In order that P may rise to a maximum and then fall as λ_1 increases, $dP/d\lambda_1$ must become negative for some value of $\lambda_1 > 1$. This implies that for some value of λ_1 we must have

$$f(x) > \Gamma \quad \text{where} \quad f(x) = \frac{x^3-7}{x^4+5x}.$$

This is possible only if Γ is less than the maximum value which $f(x)$ can attain. Solving the equation $df/dx = 0$ we find that $f(x)$ has a maximum value of 0·21 (approximately) when $x = 3\cdot39$, that is, when $\lambda_1 = 1\cdot84$. Hence if the material of the shell has a strain energy of the Mooney form with $\Gamma < 0\cdot21$ an initial rise in pressure will be followed by a fall as inflation proceeds before the pressure commences to rise again at high degrees of inflation. This phenomenon is readily observed during the inflation of a spherical rubber balloon. This is to be expected, since although the ratio $(\partial W/\partial I_2)/(\partial W/\partial I_1)$ for rubber is not independent of the deformation, it remains less than the critical value of 0·21 except possibly at comparatively small degrees of inflation. The variation of inflating pressure with λ_1 is illustrated by the calculated curves of Fig. 4.6.

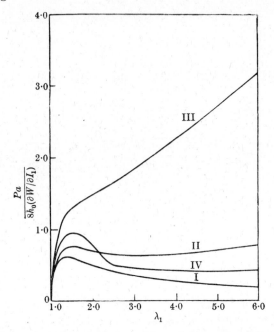

Fig. 4.6. Variation of inflating pressure with degree of inflation for spherical balloon: I, $\Gamma = 0$; II, $\Gamma = 0.1$; III, $\Gamma = 0.5$; IV, Γ varying in manner obtained experimentally by Rivlin and Saunders (see § 9.5).

4.14. Deformation of surface of revolution into one of similar shape

We consider now the circumstances under which a sheet in the form of a surface of revolution can be deformed under the influence of edge tractions and a uniform inflating pressure P into one of related shape which is such that the principal extension ratios λ_1, λ_2 are constant throughout the sheet. With this assumption it follows from (4.11.7) that T_1 and T_2 are also constants and the first of equations (4.11.10) implies that $T_1 = T_2$. We therefore write

$$\left.\begin{aligned} \lambda_1 &= \lambda_2, \quad \lambda_3 = \frac{1}{\lambda_1^2} \\ \xi &= \lambda_1 \eta, \quad r = \lambda_1 \rho \\ T_1 = T_2 &= 4h_0\left(1-\frac{1}{\lambda_1^6}\right)\left(\frac{\partial W}{\partial I_1}+\lambda_1^2\frac{\partial W}{\partial I_2}\right) = T \quad \text{(say)} \end{aligned}\right\} \quad (4.14.1)$$

From (4.11.17) it follows that
$$\kappa_2 = A + B/r^2, \qquad (4.14.2)$$
where A and B are constants defined by
$$A = P/(2T), \qquad B = L/T. \qquad (4.14.3)$$
From (4.11.14) we now have
$$\frac{dr}{d\xi} = (1-\kappa_2^2 r^2)^{\frac{1}{2}}, \qquad (4.14.4)$$
if we assume r and ξ to be measured so that they increase together. This equation, with (4.14.2), yields
$$\frac{dr}{d\xi} = \frac{A}{r}[L_1^2 - (r^2 - L_2)^2]^{\frac{1}{2}}, \qquad (4.14.5)$$
where
$$L_1^2 = \frac{1-4AB}{4A^4}, \qquad L_2 = \frac{1-2AB}{2A^2}, \qquad (4.14.6)$$
and for a real solution we must have $4AB < 1$, that is, from (4.14.3), $L < T^2/(2P)$. By integration of (4.14.5) we obtain
$$r^2 = L_2 + L_1 \sin(2A\xi + \alpha), \qquad (4.14.7)$$
where α is a constant of integration. This relation between ξ and r defines a meridian curve in the deformed sheet. The corresponding curve for the undeformed body is obtained from (4.14.7) as a relation between η and ρ by replacing ξ, r by $\lambda_1 \eta$, $\lambda_1 \rho$ respectively.

By choosing $B = 0$, $L_1 = L_2 = 1/(2A^2)$ we recover the problem of § 4.13, the meridian curves (4.14.7) in the deformed body becoming circles of radius $1/A$.

If there is no inflating pressure we have $P = 0$, $A = 0$, and (4.14.2), (4.14.4) yield
$$\left.\begin{aligned}\kappa_2 &= B/r^2 \\ \frac{dr}{d\xi} &= \frac{1}{r}(r^2 - B^2)^{\frac{1}{2}}\end{aligned}\right\}, \qquad (4.14.8)$$
or, by integration,
$$r^2 = \xi^2 + B^2, \qquad (4.14.9)$$
if ξ is measured from the point where r has its minimum value B. Equation (4.14.9) represents a catenary of parameter B with its directrix coinciding with the axis of symmetry $r = 0$ of the deformed sheet. From (4.14.9), (4.14.1) the corresponding curve on the undeformed sheet is obtained as a catenary
$$\rho^2 = \eta^2 + B^2/\lambda_1^2$$
of parameter B/λ_1.

The results of the present section are independent of the form of the strain energy function W of the material of the sheet. The analysis leading to the equations (4.14.7), (4.14.9) for the shape of the deformed sheet depends only upon the production of stress resultants T_1, T_2 which are equal to each other and are constant throughout the sheet. The assumptions of isotropy and incompressibility are not essential for this, but how far these conditions can be relaxed and the corresponding shape which the undeformed sheet would then need to assume are problems requiring further investigation.

V
THEORY OF SUCCESSIVE APPROXIMATIONS

Owing to the non-linearity of the differential equations for large deformations the problems discussed so far have been restricted to those in which marked simplifying features can be introduced in the deformation. In Chapter II the differential equations are effectively reduced to ordinary non-linear equations which can be integrated when the body is incompressible. Other simplifications are considered in Chapters III and IV. When we are concerned with compressible bodies or with general types of deformations in both incompressible and compressible bodies it is natural to consider methods of solution by successive approximation.† If we examine the known exact solutions of elastic problems presented in $T.E.$ and also in Chapters II and III of the present book, we see that the displacement vector (denoted here by **d** instead of **v**) is often an analytic function of some parameter ϵ which represents the magnitude of the deformation. For example, when a solid incompressible circular cylinder is twisted about its axis through an angle ϵ per unit length, the displacement referred to rectangular cartesian axes (x, y, z) has components

$$-x(1-\cos \epsilon z)-y \sin \epsilon z, \qquad -y(1-\cos \epsilon z)+x \sin \epsilon z, \qquad 0,$$

and these are analytic functions of ϵ for all finite values of x, y, z in the cylinder. Moreover, these components can be expanded as absolutely convergent power series in ϵ which are twice differentiable with respect to x, y, and z. In this example the series converge for all values of ϵ but in other examples the convergence is for a limited range of values of ϵ.

In the present chapter we shall assume that the displacement vector **d** can be expanded as an absolutely convergent series in ϵ with non-zero radius of convergence, and that **d** is twice differentiable term by

---

† Methods of successive approximation have been suggested by:
A. Signorini, *Atti 24th Riun. Soc. ital. Prozi. Sci.* **3** (1936), 6.
M. Misicu, *Studii Cerc. Mec. Metal.* **4** (1953), 31.
R. S. Rivlin, *J. ration. Mech. Analysis* **2** (1953), 53.
R. S. Rivlin and C. Topakoglu, ibid. **3** (1954), 581.
A. E. Green and E. B. Spratt, *Proc. R. Soc.* A, **224** (1954), 347.
A. J. M. Spencer (*Q. Jl Mech. appl. Math.* **12** (1959), 129) has considered approximations based on perturbations of the strain energy function but these problems are not examined here.

term with respect to θ, the resulting series being absolutely convergent with the same radius of convergence. Signorini† has shown that when the applied tractions are specified over the boundary, such that the total applied load is equipollent to zero but does not possess an axis of equilibrium, then a series expansion of the elastic equations is unique if it exists. Further, when the applied tractions and body forces both contain a multiplying parameter ϵ, Stoppelli‡ has given a proof of existence and uniqueness of solution of the general elastic equations and has shown that the displacement can be expanded as an absolutely convergent power series in ϵ with non-zero radius of convergence, provided ϵ is sufficiently small, and provided sufficiently smooth solutions of the classical linear equations of elasticity exist. Suitable assumptions have, of course, to be made about the displacement function and its first and second derivatives with respect to θ. The reader is referred to Stoppelli's papers for further details but, as already stated, we shall present here a theory which makes more restrictive initial assumptions, although these assumptions appear to be reasonable in the light of the behaviour of known exact solutions. There may, however, be solutions of the elastic equations which do not satisfy these conditions.

Before setting up a general analysis which can be used for finding successive approximations for isotropic materials, we discuss the question of the uniqueness of the assumed type of solution, supposing that such a solution exists, the discussion being valid for isotropic or aeolotropic bodies.

5.1. Uniqueness of series solution

Here it is convenient to take rectangular cartesian axes such that the coordinates of corresponding points in the body before and after deformation are x_i and y_i respectively and we put $y_i = x_i + u_i$. From (1.1.16) the strain components are

$$e_{ij} = \frac{1}{2}\left(\frac{\partial u_i}{\partial x_j} + \frac{\partial u_j}{\partial x_i} + \frac{\partial u_r}{\partial x_i}\frac{\partial u_r}{\partial x_j}\right) \tag{5.1.1}$$

and as in Chapter I we suppose that the strain energy function W is a polynomial in e_{ij} with coefficients which are absolute constants so that the body is initially homogeneous. From (1.1.43) and (1.1.46) we

---

† A. Signorini, *Annali Mat. pure appl.* **30** (1949), 10.
‡ F. Stoppelli, *Ric. Mat.* **3** (1954), 247; **4** (1955), 58.

see that the stress components t^{ij} become

$$t^{ij} = \frac{1}{2}\left(\delta_r^j + \frac{\partial u_j}{\partial x_r}\right)\left(\frac{\partial W}{\partial e_{ir}} + \frac{\partial W}{\partial e_{ri}}\right), \tag{5.1.2}$$

and this can be written in the form

$$t^{ij} = E^{ijrs}e_{rs} + H^{ij}, \tag{5.1.3}$$

where H^{ij} is a polynomial in the displacement gradients $\partial u_r/\partial x_s$ containing no constant or linear terms. The constants E^{ijrs} are the usual elastic constants of the classical infinitesimal theory. Although there is no difficulty in retaining body forces we shall, for simplicity, suppose they are zero, so that the equations of equilibrium (1.1.47) become

$$\frac{\partial t^{ij}}{\partial x_i} = 0. \tag{5.1.4}$$

Finally, the stress vector $_0\mathbf{t}$, per unit area of the undeformed body, acting over an area in the deformed body, whose unit normal in its undeformed state is $_0\mathbf{n}$, is, from (1.1.45), given by

$$_0\mathbf{t} = {_0n_i}\, t^{ij}\mathbf{i}_j, \qquad {_0\mathbf{n}} = {_0n_r}\mathbf{i}_r, \tag{5.1.5}$$

where $\mathbf{i}_r$ are unit vectors along the rectangular cartesian axes.

We suppose that the displacement components u_i are prescribed over a (non-zero) part of the surface (S_1) of the undeformed body, and that the surface tractions, per unit area of the undeformed body, are prescribed over the remainder (S_2) of the boundary. Thus

$$\left.\begin{aligned} u_i &= \sum_{n=1}^{\infty} \epsilon^n \overset{(n)}{h_i} \quad \text{on } S_1 \\ t^{ij} &= \sum_{n=1}^{\infty} \epsilon^n \overset{(n)}{p^{ij}} \quad \text{on } S_2 \end{aligned}\right\} \tag{5.1.6}$$

where both series are absolutely convergent for values of ϵ in the range $0 \leqslant \epsilon \leqslant \epsilon_0$ ($\epsilon_0 > 0$), for each value of x_i. Also $\overset{(n)}{h_i}$, for each n, are continuous functions of x_i on S_1, and $\overset{(n)}{p^{ij}}$, for each n, are sectionally continuous functions of x_i on S_2.

In view of the discussion at the beginning of this chapter we now assume that the displacement u_i at points inside the body, together with their first and second partial derivatives, can be expanded as absolutely convergent power series in ϵ in the range $0 \leqslant \epsilon \leqslant \epsilon_0$, in the form

$$\left.\begin{aligned}u_i &= \sum_{n=1}^{\infty} \epsilon^n \overset{(n)}{u_i} \\ \frac{\partial u_i}{\partial x_j} &= \sum_{n=1}^{\infty} \epsilon^n \frac{\partial \overset{(n)}{u_i}}{\partial x_j} \\ \frac{\partial^2 u_i}{\partial x_j \partial x_k} &= \sum_{n=1}^{\infty} \epsilon^n \frac{\partial^2 \overset{(n)}{u_i}}{\partial x_j \partial x_k}\end{aligned}\right\} \quad (5.1.7)$$

Further, we assume that each coefficient in these series is a continuous function of x_i in the body. Since absolutely convergent power series can be multiplied together and rearranged as an absolutely convergent series in ϵ, and since H^{ij} in (5.1.3) is a polynomial in $\partial u_r/\partial x_s$, we have

$$t^{ij} = \sum_{n=1}^{\infty} \epsilon^n \overset{(n)}{t}{}^{ij}, \qquad (5.1.8)$$

where

$$\left.\begin{aligned}\overset{(1)}{t}{}_{ij} &= \tfrac{1}{2} E^{ijrs}\left(\frac{\partial \overset{(1)}{u_r}}{\partial x_s} + \frac{\partial \overset{(1)}{u_s}}{\partial x_r}\right) \\ \overset{(n)}{t}{}^{ij} &= \tfrac{1}{2} E^{ijrs}\left(\frac{\partial \overset{(n)}{u_r}}{\partial x_s} + \frac{\partial \overset{(n)}{u_s}}{\partial x_r}\right) + \overset{(n)}{H}{}^{ij} \quad (n \geqslant 2)\end{aligned}\right\}, \quad (5.1.9)$$

and $\overset{(n)}{H}{}^{ij}$ is a polynomial of degree $\geqslant 2$ in $\partial \overset{(\alpha)}{u_r}/\partial x_s$ for $\alpha = 1, 2, \ldots, n-1$. Also, $\partial t^{ij}/\partial x_i$ can be obtained by term-by-term differentiation of (5.1.8) so that, in view of (5.1.4),

$$\frac{\partial \overset{(n)}{t}{}^{ij}}{\partial x_i} = 0. \qquad (5.1.10)$$

If the displacements (5.1.7) and stresses (5.1.8) take the values (5.1.6) on the surface then

$$\left.\begin{aligned}\overset{(n)}{u_i} &= \overset{(n)}{h_i} \quad (n = 1, 2, \ldots) \quad \text{on } S_1 \\ \overset{(n)}{t}{}^{ij} &= \overset{(n)}{p}{}^{ij} \quad (n = 1, 2, \ldots) \quad \text{on } S_2\end{aligned}\right\}. \quad (5.1.11)$$

Equations (5.1.9), (5.1.10), and (5.1.11) for $n = 1$ are the equations of classical infinitesimal elasticity with zero body force, and provided the surface S_1 is non-zero, the solution $\overset{(1)}{u_i}$ of these equations, if it exists, is unique. Hence $\overset{(2)}{H}{}^{ij}$ is uniquely determined and the equations

(5.1.9), (5.1.10), and (5.1.11) for $n = 2$ are the equations of classical elasticity with body forces $\partial \overset{(2)}{H}{}^{ij}/\partial x_j$. The solution $\overset{(2)}{u}_i$, if it exists, is unique. It follows that $\overset{(3)}{H}{}^{ij}$ is uniquely determined and so, successively, we can show that each coefficient $\overset{(n)}{u}_i$ in (5.1.7), if it exists, can be found uniquely.

In the next sections we obtain equations in general tensor notation for the first two terms in the series expansions when the body is isotropic. The only operations involved are multiplication of absolutely convergent power series and will be carried out formally without further comment.

5.2. Expansion of displacement, strain, and invariants

We develop a general theory† of solution in series form using the notations of Chapter I except that the displacement vector is now denoted by $\mathbf{d}(\theta_1, \theta_2, \theta_3, t)$. We assume that $\mathbf{d}$ is also a function of a parameter ϵ which can be expanded as an absolutely and uniformly convergent series

$$\mathbf{d} = \epsilon \mathbf{v}(\theta_1, \theta_2, \theta_3, t) + \epsilon^2 \mathbf{w}(\theta_1, \theta_2, \theta_3, t) + \ldots \qquad (5.2.1)$$

in the range $0 \leqslant \epsilon \leqslant \epsilon_0$, and for relevant values of θ_i where ϵ_0 is a value of the parameter which depends on the particular problem under discussion. We also assume that $\partial \mathbf{d}/\partial \theta^i$, $\partial^2 \mathbf{d}/\partial \theta^j \partial \theta^k$ can be obtained by term-by-term differentiation of the series (5.2.1) and that the resulting series are absolutely and uniformly convergent in the range $0 \leqslant \epsilon \leqslant \epsilon_0$ and for the relevant values of θ_i. In the subsequent work the only operations involved are multiplications of absolutely and uniformly convergent power series which are justified under the above assumptions and will be carried out without further comment. The actual computations will, however, be limited to two terms in the series, the first term corresponding to the usual classical infinitesimal theory of elasticity.

The covariant base vectors $\mathbf{G}_i$ of the strained body are given by

$$\mathbf{G}_i = \mathbf{g}_i + \mathbf{d}_{,i} \qquad (5.2.2)$$

so that, from (5.2.1),

$$\mathbf{G}_i = \mathbf{g}_i + \epsilon \mathbf{v}_{,i} + \epsilon^2 \mathbf{w}_{,i} + \ldots. \qquad (5.2.3)$$

† The theory is similar to that presented by A. E. Green and E. B. Spratt, loc. cit. p. 174. See also p. 199.

5.2 THEORY OF SUCCESSIVE APPROXIMATIONS

The displacement vectors, **v**, **w** may be expressed in terms of components in various ways, the most convenient for our purpose being

$$\mathbf{v} = v_m \mathbf{\mathring{g}}^m = v^m \mathbf{\mathring{g}}_m, \qquad \mathbf{w} = w_m \mathbf{\mathring{g}}^m = w^m \mathbf{\mathring{g}}_m, \qquad (5.2.4)$$

so that

$$\mathbf{v}_{,i} = v_m|_i \mathbf{\mathring{g}}^m = v^m|_i \mathbf{\mathring{g}}_m, \qquad \mathbf{w}_{,i} = w_m|_i \mathbf{\mathring{g}}^m = w^m|_i \mathbf{\mathring{g}}_m. \qquad (5.2.5)$$

The covariant metric tensor G_{ij} of the strained body, given by $G_{ij} = \mathbf{G}_i \cdot \mathbf{G}_j$, can be expressed as a power series in ϵ with the help of (5.2.3). Thus

$$G_{ij} = (\mathbf{\mathring{g}}_i + \epsilon \mathbf{v}_{,i} + \epsilon^2 \mathbf{w}_{,i} + \ldots) \cdot (\mathbf{\mathring{g}}_j + \epsilon \mathbf{v}_{,j} + \epsilon^2 \mathbf{w}_{,j} + \ldots)$$

and using (5.2.5) this becomes

$$G_{ij} = g_{ij} + \epsilon(v_i|_j + v_j|_i) + \epsilon^2(w_i|_j + w_j|_i + v_m|_i v^m|_j) + \ldots. \qquad (5.2.6)$$

Also,
$$\gamma_{ij} = \tfrac{1}{2}(G_{ij} - g_{ij})$$
$$= \tfrac{1}{2}\epsilon(v_i|_j + v_j|_i) + \tfrac{1}{2}\epsilon^2(w_i|_j + w_j|_i + v_m|_i v^m|_j) + \ldots, \qquad (5.2.7)$$

$$\gamma^i_j = \tfrac{1}{2}\epsilon(v^i|_j + v_j|^i) + \tfrac{1}{2}\epsilon^2(w^i|_j + w_j|^i + v_m|^i v^m|_j) + \ldots, \qquad (5.2.8)$$

and, since $G = |G_{ij}|$, we have, from (5.2.6),

$$I_3 = G/g = 1 + 2\epsilon v^i|_i + \epsilon^2(2w^i|_i + 2v^i|_i v^j|_j - v^i|_j v^j|_i). \qquad (5.2.9)$$

In addition, from (1.1.18) and (5.2.8),

$$I_1 = 3 + 2\epsilon v^i|_i + \epsilon^2(2w^i|_i + v^i|_j v_i|^j) + \ldots, \qquad (5.2.10)$$

$$I_2 = 3 + 4\epsilon v^i|_i + \epsilon^2(4w^i|_i + 2v^i|_i v^j|_j - v^i|_j v^j|_i + v^i|_j v_i|^j) + \ldots. \qquad (5.2.11)$$

Hence, using (1.1.20) and (5.2.9)–(5.2.11), we have

$$\left.\begin{array}{l} J_1 = 2\epsilon v^i|_i + \epsilon^2(2w^i|_i + v^i|_j v_i|^j) + \ldots \\ J_2 = \epsilon^2(2v^i|_i v^j|_j - v^i|_j v^j|_i - v^i|_j v_i|^j) + \ldots \\ J_3 = O(\epsilon^3) \end{array}\right\}. \qquad (5.2.12)$$

We also need the expansion† of $I_3 G^{ij}$. Now

$$I_3 = \frac{G}{g} = \frac{1}{3!} \delta^{imn}_{rst} g^{ru} G_{iu} g^{sv} G_{vm} g^{tw} G_{wn}, \qquad (5.2.13)$$

and

$$I_3 G^{ij} = \frac{\partial I_3}{\partial G_{ij}}, \qquad (5.2.14)$$

so that

$$I_3 G^{ij} = \tfrac{1}{2} \delta^{imn}_{rst} g^{rj} g^{sv} g^{tw} G_{vm} G_{wn}. \qquad (5.2.15)$$

† The expansion of $I_3 G^{ij}$ can be found by evaluating G^{ij} as a quotient from $G^{ir}G_{rj} = \delta^i_j$ and multiplying the result by I_3 which is given by (5.2.9). The above method, however, only depends on the multiplication of absolutely convergent power series.

Hence, with the help of (5.2.6), we find that

$$I_3 G^{ij} = g^{ij} + \epsilon(2g^{ij}v^m|_m - v^i|^j - v^j|^i) + \\
+ \epsilon^2[g^{ij}(2w^m|_m + 2v^m|_m v^n|_n - v^m|_n v^n|_m) - \\
- w^i|^j - w^j|^i - 2v^m|_m(v^i|^j + v^j|^i) + \\
+ v^i|_m v^m|^j + v^i|_m v^j|^m + v^m|^i v^j|_m] + \ldots \quad (5.2.16)$$

5.3. Expansion of stress: compressible body

The most convenient stress tensors to use here are the stresses s^{ij} and t^{ij} (see § 1.1). For a compressible isotropic body we have, from (1.15.13) and (1.15.7),

$$s^{ij} = \tau^{ij}\sqrt{I_3} = 2g^{ij}\frac{\partial W}{\partial I_1} + 2B^{ij}\frac{\partial W}{\partial I_2} + 2G^{ij}I_3\frac{\partial W}{\partial I_3}, \quad (5.3.1)$$

where

$$B^{ij} = g^{ij}I_1 - g^{ir}g^{js}G_{rs}, \quad (5.3.2)$$

and

$$W = W(I_1, I_2, I_3). \quad (5.3.3)$$

For compressible bodies it is more convenient to use strain invariants J_1, J_2, J_3 defined in terms of I_1, I_2, I_3 by (1.1.20). If

$$W(I_1, I_2, I_3) = W'(J_1, J_2, J_3), \quad (5.3.4)$$

then (5.3.1) becomes

$$s^{ij} = 2g^{ij}\frac{\partial W'}{\partial J_1} + 2(B^{ij} - 2g^{ij})\frac{\partial W'}{\partial J_2} + 2(g^{ij} - B^{ij} + G^{ij}I_3)\frac{\partial W'}{\partial J_3}. \quad (5.3.5)$$

From (5.2.6), (5.2.10), (5.2.16), and (5.3.2) we have

$$B^{ij} - 2g^{ij} = \epsilon(2g^{ij}v^m|_m - v^i|^j - v^j|^i) + \epsilon^2[g^{ij}(2w^m|_m + v^m|_n v_m|^n) - \\
- w^i|^j - w^j|^i - v^m|^i v_m|^j] + \ldots, \quad (5.3.6)$$

$$g^{ij} - B^{ij} + G^{ij}I_3 = \epsilon^2[g^{ij}(2v^m|_m v^n|_n - v^m|_n v^n|_m - v^m|^n v_m|_n) + \\
+ v^i|_m v^m|^j + v^i|_m v^j|^m + v^m|^i v^j|_m + v^m|^i v_m|^j - \\
- 2v^m|_m(v^i|^j + v^j|^i)] + \ldots \quad (5.3.7)$$

If W is a polynomial in I_1, I_2, I_3, then W' is a polynomial in J_1, J_2, J_3 and hence a power series in ϵ. Also, $\partial W'/\partial J_i$ are polynomials in J_1, J_2, J_3 and therefore power series in ϵ. From (5.3.5)–(5.3.7) we see that the only term in s^{ij} independent of ϵ arises from $\partial W'/\partial J_1$, and if initial stresses are zero $\partial W'/\partial J_1$ must vanish when $\epsilon = 0$. Hence, using

(5.2.12), we may write

$$\frac{\partial W'}{\partial J_1} = 2A\epsilon v^m|_m + \epsilon^2[A(2w^m|_m + v^m|^n v_m|_n) + 2Hv^m|_m v^n|_n +$$
$$+ C(2v^m|_m v^n|_n - v^m|_n v^n|_m - v^m|_n v_m|^n)] + \ldots$$

$$\frac{\partial W'}{\partial J_2} = B + 2C\epsilon v^m|_m + \ldots, \qquad \frac{\partial W'}{\partial J_3} = F + 2D\epsilon v^m|_m + \ldots$$

$$\left. \right\} \quad (5.3.8)$$

where†
$$A = \frac{\partial^2 W'}{\partial J_1^2}, \qquad B = \frac{\partial W'}{\partial J_2}, \qquad C = \frac{\partial^2 W'}{\partial J_1 \partial J_2}$$
$$F = \frac{\partial W'}{\partial J_3}, \qquad H = \frac{\partial^3 W'}{\partial J_1^3}, \qquad D = \frac{\partial^2 W'}{\partial J_1 \partial J_3}$$

$$\left. \right\} , \quad (5.3.9)$$

and the derivatives in (5.3.9) are evaluated at $J_1 = J_2 = J_3 = 0$ (or $\epsilon = 0$), so that A, B, C, D, F, H are constants.

Using (5.3.6)–(5.3.9) the stress (5.3.5) can be expanded in the form

$$s^{ij} = \epsilon s'^{ij} + \epsilon^2 s''^{ij} + \ldots, \qquad (5.3.10)$$

where
$$s'^{ij} = 4(A+B)g^{ij}v^m|_m - 2B(v^i|^j + v^j|^i), \qquad (5.3.11)$$

$$s''^{ij} = 2Ag^{ij}(2w^m|_m - v^m|^n v_n|_m) +$$
$$+ 2(A-C)g^{ij}(v^m|^n v_m|_n + v^m|^n v_n|_m) +$$
$$+ 2Bg^{ij}(2w^m|_m + v^m|^n v_m|_n) - 2Fg^{ij}(v^m|^n v_m|_n + v^m|^n v_n|_m) +$$
$$+ 4(H+F+3C)g^{ij}v^m|_m v^n|_n - 2B(w^i|^j + w^j|^i + v^m|^i v_m|^j) +$$
$$+ 2F(v^i|^m + v^m|^i)(v^j|_m + v_m|^j) - 4(C+F)(v^i|^j + v^j|^i)v^m|_m. \qquad (5.3.12)$$

The usual Lamé constants of classical elasticity λ, μ, and Poisson's ratio η and Young's modulus E, are given in terms of A, B by the relations

$$\lambda = 4(A+B), \qquad \mu = -2B, \qquad \eta = \frac{A+B}{2A+B}, \qquad E = -\frac{4B(3A+2B)}{2A+B}.$$
$$(5.3.13)$$

If the stress tensor t^{ij} is also expanded in the form

$$t^{ij} = \epsilon t'^{ij} + \epsilon^2 t''^{ij} + \ldots \qquad (5.3.14)$$

then, from (1.1.46) and (5.3.10),

$$t'^{ij} = s'^{ij}, \qquad t''^{ij} = s''^{ij} + s'^{im}v^j|_m. \qquad (5.3.15)$$

† The constant D is for use in § 5.4 in the passage to the incompressible body, but does not appear in the second-order stresses.

Finally, equations (5.3.11), (5.3.12), and (5.3.15) yield

$$t'^{ij} = 4(A+B)g^{ij}v^m|_m - 2B(v^i|^j + v^j|^i), \qquad (5.3.16)$$

$$\begin{aligned}
t''^{ij} = {}& 2Ag^{ij}(2w^m|_m - v^m|^n v_n|_m) + \\
& + 2(A-C)g^{ij}(v^m|^n v_m|_n + v^m|^n v_n|_m) + \\
& + 2Bg^{ij}(2w^m|_m + v^m|^n v_m|_n) - 2Fg^{ij}(v^m|^n v_m|_n + v^m|^n v_n|_m) + \\
& + 4(H+F+3C)g^{ij}v^m|_m v^n|_n + 4(A+B)v^m|_m v^j|^i - \\
& - 2B(w^i|^j + w^j|^i + v^m|^i v_m|^j + v^m|^i v^j|_m + v^i|^m v^j|_m) + \\
& + 2F(v^i|^m + v^m|^i)(v^j|_m + v_m|^j) - 4(C+F)(v^i|^j + v^j|^i)v^m|_m.
\end{aligned} \qquad (5.3.17)$$

We remind the reader that the stress tensor s^{ij}, given by (5.3.10)–(5.3.12), is referred to coordinate surfaces $\theta_i = \text{constant}$ in the deformed body. In special problems we may require stresses referred to other coordinate surfaces but these may readily be obtained by tensor transformations.

From (1.1.45) and (5.3.14) we see that the stress vector across any surface in the deformed body, whose unit normal $_0\mathbf{n}$ in its undeformed state is $_0n_i\mathbf{g}^i$, is given by

$$_0\mathbf{t} = {}_0n_i(\epsilon t'^{ij} + \epsilon^2 t''^{ij} + \ldots)\mathbf{g}_j. \qquad (5.3.18)$$

The stress is measured per unit area of the undeformed body and physical components of stress may be found immediately from (5.3.18).

5.4. Expansion of stress: incompressible body

When the body is incompressible $I_3 = 1$ for all values of ϵ so that, from (5.2.9),

$$v^m|_m = 0, \qquad 2w^m|_m - v^m|^n v_n|_m = 0, \qquad (5.4.1)$$

and

$$\left.\begin{aligned} I_1 &= 3 + \epsilon^2(v^m|^n v_n|_m + v^m|^n v_m|_n) + \ldots \\ I_2 &= 3 + \epsilon^2(v^m|^n v_n|_m + v^m|^n v_m|_n) + \ldots \end{aligned}\right\}. \qquad (5.4.2)$$

The components of the stress tensor s^{ij} (which in the case of incompressibility is equal to τ^{ij}) may be expanded in terms of ϵ by starting from the expression

$$s^{ij} = 2g^{ij}\frac{\partial W}{\partial I_1} + 2B^{ij}\frac{\partial W}{\partial I_2} + pG^{ij}, \qquad (5.4.3)$$

where p is a scalar.† It is, however, convenient to deduce the

---

† See A. E. Green and E. B. Spratt, loc. cit., p. 174.

appropriate results from those of the previous section. Using (1.1.20) we have

$$\frac{\partial W}{\partial I_1} = \frac{\partial W'}{\partial J_1} - 2\frac{\partial W'}{\partial J_2} + \frac{\partial W'}{\partial J_3},$$

$$\frac{\partial W}{\partial I_2} = \frac{\partial W'}{\partial J_2} - \frac{\partial W'}{\partial J_3},$$

$$\frac{\partial W}{\partial I_3} = \frac{\partial W'}{\partial J_3},$$

and with the help of (5.3.8) these become

$$\left.\begin{aligned}\frac{\partial W}{\partial I_1} &= F - 2B + 2(A - 2C + D)\epsilon v^m|_m + \cdots \\ \frac{\partial W}{\partial I_2} &= B - F + 2(C - D)\epsilon v^m|_m + \cdots \\ \frac{\partial W}{\partial I_3} &= F + 2D\epsilon v^m|_m + \cdots\end{aligned}\right\}. \quad (5.4.4)$$

Since $v^m|_m \to 0$ when the body is incompressible and $\partial W/\partial I_3$ becomes an unknown hydrostatic pressure, the constant $D \to \infty$ in such a way that

$$4Dv^m|_m \to p' \quad \text{(say)}, \quad (5.4.5)$$

where p' is a finite scalar. Further, since I_1, I_2, as given by (5.4.2), contain no terms in ϵ, it follows from (5.4.4) that the constants A, C must both tend to infinity in such a way that

$$4Av^m|_m \to p', \quad 4Cv^m|_m \to p'. \quad (5.4.6)$$

Also F and B remain finite. Now consider a deformation of a *compressible* body which is such that volume changes are zero. The second-order stresses are given by (5.3.12) subject to the conditions (5.4.1) so that

$$s''^{ij} = 2(A - C + B - F)g^{ij}(v^m|^n v_m|_n + v^m|^n v_n|_m) - \\ -2B(w^i|^j + w^j|^i + v^m|^i v_m|^j) + 2F(v^i|^m + v^m|^i)(v^j|_m + v_m|^j). \quad (5.4.7)$$

If we now pass to the limiting case of an incompressible body we see that although A and C both tend to infinity, $A - C$ is finite if s''^{ij} remains finite. Hence we have

$$F, B, A - C \quad \text{finite}. \quad (5.4.8)$$

Again, if the stress s''^{ij} in (5.3.12) for a general deformation is to remain finite as we pass to the limiting case of an incompressible body, then, remembering (5.4.1), we have

$$\left.\begin{array}{r}2A(2w^m|_m - v^m|^n v_n|_m) \to p_0 \\ Hv^m|_m\, v^n|_n \to q\end{array}\right\}, \qquad (5.4.9)$$

where p_0, q are finite scalars.

For incompressible bodies it is convenient to denote the values of $\partial W/\partial I_1$ and $\partial W/\partial I_2$ by C_1 and C_2 respectively,† when $\epsilon = 0$, so that

$$F - 2B = C_1, \qquad B - F = C_2, \qquad 2(C_1 + C_2) = \mu. \qquad (5.4.10)$$

Then, using (5.4.6) and (5.4.8)–(5.4.10), together with the incompressibility condition (5.4.1), we see that the stresses (5.3.11) and (5.3.12) become

$$s'^{ij} = p'g^{ij} + \mu(v^i|^j + v^j|^i), \qquad (5.4.11)$$

$$s''^{ij} = p''g^{ij} - p'(v^i|^j + v^j|^i) - 2C_2(v^i|^m + v^m|^i)(v^j|_m + v_m|^j) + \\ + \mu(w^i|^j + w^j|^i - v^i|^m v_m|^j - v^i|^m v^j|_m - v^m|^i v^j|_m), \qquad (5.4.12)$$

where p'' is a combination of p_0, q and invariants of $v^i|_j$ and is therefore a scalar. Further, from (5.3.15), (5.4.11), and (5.4.12), we have

$$t'^{ij} = p'g^{ij} + \mu(v^i|^j + v^j|^i), \qquad (5.4.13)$$

$$t''^{ij} = p''g^{ij} - p'v^i|^j - 2C_2(v^i|^m + v^m|^i)(v^j|_m + v_m|^j) + \\ + \mu(w^i|^j + w^j|^i - v^i|^m v_m|^j). \qquad (5.4.14)$$

5.5. Equations of equilibrium: surface stresses

From (1.1.45) and (5.3.14) we have

$$\mathbf{T}_i = \sqrt{(g)}(\epsilon t'^{ij} + \epsilon^2 t''^{ij} + \ldots)\mathbf{g}_j. \qquad (5.5.1)$$

We assume that the body is in equilibrium and that the body-force vector can be expanded in the form

$$\mathbf{F} = \epsilon \mathbf{F}' + \epsilon^2 \mathbf{F}'' + \ldots, \qquad (5.5.2)$$

and that
$$\mathbf{F}' = {}_0F'^i\, \mathbf{g}_i, \qquad \mathbf{F}'' = {}_0F''^i\, \mathbf{g}_i. \qquad (5.5.3)$$

The vector equation of equilibrium, obtained from (1.1.40), is

$$\mathbf{T}_{i,i} + \rho_0 \mathbf{F}\sqrt{g} = 0. \qquad (5.5.4)$$

Hence, substituting in this equation from (5.5.1), (5.5.2), and (5.5.3),

† These are the constants for the Mooney material with strain energy given by (1.14.3).

and equating coefficients of ϵ to zero, we have
$$t'^{ij}|_i + \rho_0({}_0F'^j) = 0, \tag{5.5.5}$$
$$t''^{ij}|_i + \rho_0({}_0F''^j) = 0. \tag{5.5.6}$$

Alternatively these equations follow directly from (1.1.47) and (5.3.14).

If ${}_0\mathbf{P}$ is the stress vector, measured per unit area of the undeformed body, and associated with a surface in the deformed body whose unit normal in its undeformed state is ${}_0\mathbf{n}$, then, from (5.3.18),

$${}_0\mathbf{P} = {}_0n_i(\epsilon t'^{im} + \epsilon^2 t''^{im} + \dots)\mathbf{g}_m. \tag{5.5.7}$$

If the applied stresses at the surface of the deformed body vanish, then
$${}_0n_i t'^{ij} = 0, \qquad {}_0n_i t''^{ij} = 0. \tag{5.5.8}$$

In the remaining sections of this chapter we consider some applications of the above theory.†

5.6. Hydrostatic pressure

We consider a unit cube of compressible isotropic material bounded by the planes $x_i = 0$, $x_i = 1$, where x_i are rectangular cartesian axes, and study the deformation of the cube under a slowly increasing hydrostatic pressure. We therefore consider the deformation

$$\left. \begin{array}{lll} v^1 = -x_1, & v^2 = -x_2, & v^3 = -x_3 \\ w^1 = \alpha x_1, & w^2 = \alpha x_2, & w^3 = \alpha x_3 \end{array} \right\}, \tag{5.6.1}$$

where α is a constant. We choose moving coordinates θ_i so that

$$\theta_i = x_i, \qquad g^{ij} = g_{ij} = \delta_{ij}, \tag{5.6.2}$$

and observe that the physical components of stress over planes $x_i = $ constant in the deformed cube measured per unit area of the undeformed cube are t^{ij} since the x_i-planes are not rotated by the deformation. We assume that the cube is deformed by equal and opposite

---

† Other work on second-order effects may be found in the following book and papers:
F. D. Murnaghan, *Finite Deformation of an Elastic Solid* (New York, 1951).
R. S. Rivlin, loc. cit., p. 174.
A. E. Green and R. T. Shield, *Phil. Trans. R. Soc.* A, **244** (1951), 47.
A. E. Green and E. B. Spratt, loc. cit., p. 174. See Corrigenda, p. 199.
A. E. Green, *Proc. Camb. phil. Soc. math. phys. Sci.* **50** (1954), 488.
W. S. Blackburn and A. E. Green, *Proc. R. Soc.* A, **240** (1957), 408. See p. 199.
W. S. Blackburn, *Proc. Camb. phil. Soc. math. phys. Sci.* **53** (1957), 907; *Q. Jl Mech. appl. Math.* **11** (1958), 142.
P. L. Sheng, *Secondary Elasticity* (Taipei, 1955).
A. D. Kydoniefs and A. J. M. Spencer, *Int. J. Engng Sci.* **3** (1965), 173.
A. D. Kydoniefs, *Int. J. Engng Sci.* **4** (1966), 125.
Second-order effects in two-dimensional elasticity are considered in Chapter VI and further references are given there.

resultant pressures p uniformly distributed over all the faces of the cube. Then, from (5.3.16), (5.3.17), and (5.6.1), we obtain

$$-p = \epsilon t'^{11} = \epsilon t'^{22} = \epsilon t'^{33} = -\frac{E\epsilon}{1-2\eta}, \qquad t'^{ij} = 0 \quad (i \neq j), \quad (5.6.3)$$

and t''^{ij} vanishes if

$$-\frac{E\alpha}{1-2\eta} = k = 18A + 12B + 72C + 8F + 36H. \qquad (5.6.4)$$

The total displacement, to the second order in ϵ, is

$$-\frac{1-2\eta}{E}\left[p + \frac{k(1-2\eta)^2 p^2}{E^2}\right]x_i. \qquad (5.6.5)$$

5.7. Simple extension

We again consider the unit cube described in the previous section and suppose that it is slowly extended by a resultant force T uniformly distributed parallel to the x_3-axis over the plane ends $x_3 = 0$ and $x_3 = 1$, the remaining faces being free from applied stresses. Choosing the same coordinate system θ_i as in (5.6.2), we assume that

$$\left.\begin{array}{lll} v^1 = -\eta x_1, & v^2 = -\eta x_2, & v^3 = x_3 \\ w^1 = \beta x_1, & w^2 = \beta x_2, & w^3 = \alpha x_3 \end{array}\right\}, \qquad (5.7.1)$$

where α, β are constants and η is the classical Poisson ratio. Since the x_i-planes are not rotated, the physical components of stress over the x_i-planes in the deformed body, measured per unit area of the undeformed body, are again t^{ij}. We now attempt to choose α, β so that

$$T = \epsilon t'^{33}, \qquad t'^{ij} = 0 \quad (i \neq 3 \text{ or } j \neq 3), \qquad t''^{ij} = 0. \qquad (5.7.2)$$

From (5.3.16), (5.3.17), and (5.7.1) we see that this is possible provided

$$\epsilon = T/E, \qquad (5.7.3)$$

and

$$\left.\begin{array}{l} \dfrac{2\eta B\alpha}{1-2\eta} + \dfrac{2B\beta}{1-2\eta} + \dfrac{B\eta(1+\eta)}{1-2\eta} - \\ \qquad\qquad -4C(1-5\eta+3\eta^2) + 4\eta F - 2(1-2\eta)^2 H = 0 \\[6pt] \dfrac{2(1-\eta)B\alpha}{1-2\eta} + \dfrac{4\eta B\beta}{1-2\eta} + \dfrac{B(3-3\eta-4\eta^2+2\eta^3)}{1-2\eta} + \\ \qquad\qquad +4\eta(4-5\eta)C - 4\eta^2 F - 2(1-2\eta)^2 H = 0 \end{array}\right\}. \quad (5.7.4)$$

These equations may be solved for α, β, and the total displacement then has components

$$\frac{T}{E}\left(-\eta+\frac{\beta T}{E}\right)x_1, \quad \frac{T}{E}\left(-\eta+\frac{\beta T}{E}\right)x_2, \quad \frac{T}{E}\left(1+\frac{\alpha T}{E}\right)x_3. \quad (5.7.5)$$

5.8. Simple shear

Consider a cuboid of compressible isotropic material whose faces are given by
$$x_1 = \pm a, \quad x_2 = \pm b, \quad x_3 = \pm c,$$
where a, b, c are constants and take θ_i coordinates as in (5.6.2). If the cuboid is subject to a simple shearing deformation in which each point of the material moves parallel to the x_1-axis by an amount which is proportional to x_2, then the final coordinates of the point which is initially at x_i are y_i in the deformed state, referred to the same rectangular cartesian axes, where

$$y_1 = x_1 + \epsilon x_2, \quad y_2 = x_2, \quad y_3 = x_3, \quad (5.8.1)$$

and ϵ is a constant. If we now identify ϵ with the parameter used throughout this chapter we see that the first- and second-order displacements which correspond to (5.8.1) are

$$v^1 = x_2, \quad v^2 = v^3 = 0, \quad w^i = 0. \quad (5.8.2)$$

From (5.3.11) and (5.3.16) we see that the only non-zero components of stress s'^{ij} (or t'^{ij}) are

$$s'^{12} = t'^{12} = -2B = \mu. \quad (5.8.3)$$

Also, from (5.3.12), (5.3.17), and (5.8.2), we have

$$\left.\begin{array}{ll} s''^{11} = 2(A+B-C), & s''^{22} = 2(A-C) \\ s''^{33} = 2(A+B-C-F), & s''^{ij} = 0 \quad (i \neq j) \\ t''^{11} = 2(A-C), & t''^{22} = 2(A-C) \\ t''^{33} = 2(A+B-C-F), & t''^{ij} = 0 \quad (i \neq j) \end{array}\right\}. \quad (5.8.4)$$

The stresses required over the deformed surfaces of the cuboid to maintain the deformation (5.8.2) may be found in a number of ways. We use (5.3.18) and obtain stress vectors ${}_0\mathbf{t}_i$ over surfaces $x_1 = a$, $x_2 = b$, $x_3 = c$ in the deformed cuboid, measured per unit area of the undeformed cuboid. Thus, for the surface $x_1 = a$ we have ${}_0n_i = (1, 0, 0)$, so that

$${}_0\mathbf{t}_1 = \mu\epsilon\mathbf{g}_2 + 2(A-C)\epsilon^2\mathbf{g}_1 \quad (x_1 = a). \quad (5.8.5)$$

Similarly,
$$_0t_2 = \mu\epsilon g_1 + 2(A-C)\epsilon^2 g_2 \quad (x_2 = b), \qquad (5.8.6)$$
$$_0t_3 = 2(A+B-C-F)\epsilon^2 g_3 \quad (x_3 = c). \qquad (5.8.7)$$

The interpretation of (5.8.6) and (5.8.7) is straightforward, showing that a simple shear cannot be maintained only by shearing forces but requires second-order normal stresses as well over the faces $x_2 = \pm b$, $x_3 = \pm c$ of the deformed cuboid. The stresses (5.8.5) over $x_1 = a$ can also be resolved into a shearing stress $\mu\epsilon$ along the deformed face and a normal stress of $O(\epsilon^2)$.

The examples given in §§ 5.5–5.7 are the simplest applications of the general second-order theory. In the next section we consider some general results for the deformation of a cylinder and then discuss in later sections the problem of torsion. It is convenient to introduce complex coordinates which simplify the subsequent algebra.

5.9. Deformation of a cylinder†

We take the unstrained body to be a homogeneous isotropic cylinder of constant cross-section R whose generators are parallel to the z-axis and the plane ends of which are $z = 0, z = l$, where (x, y, z) is a fixed rectangular cartesian system of axes. We take the curvilinear coordinates θ_i to be defined by

$$\theta_1 = x+iy = \zeta, \qquad \theta_2 = x-iy = \bar{\zeta}, \qquad \theta_3 = z. \qquad (5.9.1)$$

If **i**, **j**, **k** are unit vectors along the (x, y, z) axes then

$$\left.\begin{array}{l}\mathbf{r} = \tfrac{1}{2}(\zeta+\bar{\zeta})\mathbf{i} - \tfrac{1}{2}i(\zeta-\bar{\zeta})\mathbf{j} + z\mathbf{k} \\ = \zeta\mathbf{g}_1 + \bar{\zeta}\mathbf{g}_2 + z\mathbf{g}_3\end{array}\right\}, \qquad (5.9.2)$$

so that
$$\mathbf{g}_1 = \tfrac{1}{2}(\mathbf{i}-i\mathbf{j}), \qquad \mathbf{g}_2 = \bar{\mathbf{g}}_1, \qquad \mathbf{g}_3 = \mathbf{k}, \qquad (5.9.3)$$

where a bar over a quantity denotes the complex conjugate of that quantity. Hence the only non-zero components of the metric tensors are

$$g_{12} = \tfrac{1}{2}, \qquad g_{33} = 1, \qquad g^{12} = 2, \qquad g^{33} = 1, \qquad \sqrt{g} = \tfrac{1}{2}i. \qquad (5.9.4)$$

We note that taking the complex conjugate of any tensor in the coordinate system $(\zeta, \bar{\zeta}, z)$ is equivalent to interchanging the indices 1 and 2. Since the metric tensor has constant components covariant differentiation reduces to partial differentiation.

---
† §§ 5.9–5.11 follow the paper of W. S. Blackburn and A. E. Green, loc. cit., p. 185. The constant D of that paper is denoted by F here. See also p. 199.

The position vector **R** of a point in the deformed body is given by
$$\mathbf{R} = \mathbf{r} + \epsilon v^i \mathbf{g}_i + \epsilon^2 w^i \mathbf{g}_i + \dots. \tag{5.9.5}$$
Hence if
$$\mathbf{R} = y_1 \mathbf{i} + y_2 \mathbf{j} + y_3 \mathbf{k} = Y^i \mathbf{g}_i, \tag{5.9.6}$$
where
$$Y^1 = y_1 + iy_2, \qquad Y^2 = y_1 - iy_2, \qquad Y^3 = y_3, \tag{5.9.7}$$
we have
$$Y^1 = \zeta + \epsilon v^1 + \epsilon^2 w^1 + \dots, \qquad Y^2 = \bar{\zeta} + \epsilon v^2 + \epsilon^2 w^2 + \dots,$$
$$Y^3 = z + \epsilon v^3 + \epsilon^2 w^3 + \dots. \tag{5.9.8}$$

We assume that the cylinder is deformed by the action of forces over the ends $z = 0, l$, the curved surfaces of the cylinder being free from applied traction, and that the cylinder is held so that†

$$d^1 = d^3 = \frac{\partial d^1}{\partial z} = \frac{\partial d^1}{\partial \zeta} - \frac{\partial d^2}{\partial \bar{\zeta}} = 0 \quad (\zeta = z = 0), \tag{5.9.9}$$

where
$$\mathbf{d} = d^m \mathbf{g}_m. \tag{5.9.10}$$

We suppose that each cross-section $z = $ constant of the original cylinder is bounded internally by a number of non-intersecting closed curves $L_1, L_2, \dots, L_n$ and externally by a closed curve L_0 which entirely surrounds the other closed curves. The equation of the entire boundary is
$$G(\zeta, \bar{\zeta}) = 0, \tag{5.9.11}$$
where G is a real function. The normal to this surface has components proportional to $\partial G/\partial x_i$ referred to axes $x_i = (x, y, z)$, and by applying a tensor transformation it is found that

$$({}_0 n_1, {}_0 n_2, {}_0 n_3) \propto (\partial G/\partial \zeta, \partial G/\partial \bar{\zeta}, 0). \tag{5.9.12}$$

Since there is no external stress on this surface, the boundary conditions (5.5.8) may be applied to yield

$$t'^{1j} \frac{\partial G}{\partial \zeta} + t'^{2j} \frac{\partial G}{\partial \bar{\zeta}} = 0, \qquad t''^{1j} \frac{\partial G}{\partial \zeta} + t''^{2j} \frac{\partial G}{\partial \bar{\zeta}} = 0, \tag{5.9.13}$$

when $G(\zeta, \bar{\zeta}) = 0$. On using the result that when $G = 0$

$$\frac{\partial G}{\partial \zeta} d\zeta + \frac{\partial G}{\partial \bar{\zeta}} d\bar{\zeta} = 0, \tag{5.9.14}$$

---

† We assume that the origin is a point of the cylinder. This condition makes the displacement and rotation zero at the origin.

the equations (5.9.13) may be written in the alternative forms

$$t'^{1j}\, d\bar{\zeta} - t'^{2j}\, d\zeta = 0, \qquad t''^{1j}\, d\bar{\zeta} - t''^{2j}\, d\zeta = 0 \quad (G=0). \qquad (5.9.15)$$

At every point of the surface $z = l$

$$({}_0n_1,\ {}_0n_2,\ {}_0n_3) = (0, 0, 1), \qquad (5.9.16)$$

and the element of area of this surface is dR. The surface tractions on the surface $z = l$ in the deformed body are equivalent to a resultant force

$$\mathbf{N} = (\epsilon N'^i + \epsilon^2 N''^i + \ldots)\mathbf{g}_i, \qquad (5.9.17)$$

at the centroid ζ_G, $\bar{\zeta}_G$, l of the surface of the undeformed body, and a resultant couple

$$\mathbf{M} = (\epsilon M'^i + \epsilon^2 M''^i + \ldots)\mathbf{g}_i, \qquad (5.9.18)$$

where

$$\mathbf{N} = \iint_R {}_0\mathbf{P}\, dR, \qquad (5.9.19)$$

$$\mathbf{M} = \iint_R (\mathbf{r} - \mathbf{r}_G + \mathbf{d}) \times {}_0\mathbf{P}\, dR. \qquad (5.9.20)$$

Here $\mathbf{r}_G$ is the position vector of the centre of gravity of the section $z = l$ and ${}_0\mathbf{P}$ is given by (5.5.7). Hence, from (5.9.2), (5.9.4), (5.5.7), and (5.9.16)–(5.9.20), we have

$$\left.\begin{aligned} N'^1 &= \iint_R t'^{31}\, dR, & M'^1 &= -i\iint_R (\zeta - \zeta_G) t'^{33}\, dR \\ N'^3 &= \iint_R t'^{33}\, dR, & M'^3 &= \mathscr{I}\iint_R (\bar{\zeta} - \bar{\zeta}_G) t'^{31}\, dR \end{aligned}\right\}, \qquad (5.9.21)$$

$$N''^1 = \iint_R t''^{31}\, dR, \qquad N''^3 = \iint_R t''^{33}\, dR, \qquad (5.9.22)$$

$$M''^1 = -i\iint_R \{v^1 t'^{33} - v^3 t'^{31} + (\zeta - \zeta_G) t''^{33}\}\, dR, \qquad (5.9.23)$$

$$M''^3 = \mathscr{I}\iint_R \{v^2 t'^{31} + (\bar{\zeta} - \bar{\zeta}_G) t''^{31}\}\, dR, \qquad (5.9.24)$$

where $\mathscr{I}$ denotes the imaginary part of the following complex quantity.

We recall here for future use Stokes's theorem, that if χ is a function of ζ and $\bar{\zeta}$, satisfying suitable conditions,

$$\oint_{G=0} \chi\, d\zeta = 2i \iint_R \frac{\partial \chi}{\partial \bar{\zeta}}\, dR. \qquad (5.9.25)$$

5.10. Torsion of a compressible cylinder: classical theory

We now consider a cylinder of compressible isotropic material which is twisted about the z-axis by the action of forces over the ends $z = 0$, $z = l$. We take ϵ to be the angle of twist per unit length of the undeformed cylinder and seek for a solution as far as terms of $O(\epsilon^2)$. The solution of the first-order problem when the body forces are zero is well known,† and we recapitulate the main results in the notation of § 5.9. Thus

$$\begin{aligned}\mathbf{v} &= -yz\mathbf{i}+xz\mathbf{j}+\phi(x,y)\mathbf{k}\\ &= i\zeta z\mathbf{g}_1-i\bar{\zeta}z\mathbf{g}_2+\tfrac{1}{2}\{f(\zeta)+\bar{f}(\bar{\zeta})\}\mathbf{g}_3\end{aligned}, \quad (5.10.1)$$

on using (5.9.3), so that

$$v^1 = i\zeta z, \quad v^2 = -i\bar{\zeta}z, \quad v^3 = \tfrac{1}{2}(f+\bar{f}), \quad (5.10.2)$$

where $\phi = \tfrac{1}{2}(f+\bar{f})$ is the harmonic torsion function.‡ From (5.3.11) and (5.9.4) we have

$$s'^{11} = s'^{12} = s'^{21} = s'^{22} = s'^{33} = 0, \quad s'^{13} = \overline{s'^{23}} = \mu(\bar{f}'+i\zeta), \quad (5.10.3)$$

and the boundary conditions (5.9.15) on the surface $G = 0$ reduce to

$$(\bar{f}'+i\zeta)\,d\bar{\zeta}-(f'-i\bar{\zeta})\,d\zeta = 0 \quad (G=0) \quad (5.10.4)$$

or
$$f-\bar{f}-i\zeta\bar{\zeta} = iC_r, \quad (5.10.5)$$

where the real constants $C_0, C_1,..., C_n$ correspond to each of the closed curves $L_0, L_1,..., L_n$. The value of f is undetermined to the extent of an arbitrary additive real constant, so in view of (5.9.9) we choose

$$f(0)+\bar{f}(0) = 0. \quad (5.10.6)$$

We assume that $f'(\zeta)$ is regular in the open region R of a cross-section of the cylinder and that $f(\zeta)$ and $f'(\zeta)$ are continuous on to the boundary except possibly at an isolated singular point. Also $f+\bar{f}$ must be single-valued.

From (5.9.21) and (5.10.3) we have

$$\left.\begin{aligned}N'^1 = N'^3 = M'^1 &= 0\\ M'^3 &= \mu S\end{aligned}\right\} \quad (5.10.7)$$

if we use (5.10.4) and (5.9.25), where

$$S = I+i\iint_R \zeta f'\,dR = I-\iint_R f'\bar{f}'\,dR, \quad (5.10.8)$$

$$I = \iint_R \zeta\bar{\zeta}\,dR. \quad (5.10.9)$$

---

† See, for example, I. S. Sokolnikoff, *Mathematical Theory of Elasticity* (edn. 2, New York, 1956).

‡ The arguments ζ, $\bar{\zeta}$ will be omitted from $f(\zeta)$, $\bar{f}(\bar{\zeta})$ respectively for brevity. A similar notation will be used later for other functions of the complex variables ζ, $\bar{\zeta}$.

5.11. Torsion: second order

If we remember that covariant differentiation reduces to partial differentiation, we find, from (5.10.2) and (5.9.4),

$$\left.\begin{aligned}&v^1|_1 = iz, \quad v^1|_2 = 0, \quad\; v^1|_3 = i\zeta, \quad\; v^3|_1 = \tfrac{1}{2}f', \quad v^3|_3 = 0\\&v_1|_1 = 0, \quad\; v_1|_2 = -\tfrac{1}{2}iz, \quad v_1|_3 = -\tfrac{1}{2}i\bar{\zeta}, \quad v_3|_1 = \tfrac{1}{2}f', \quad v_3|_3 = 0\\&v^1|^1 = 0, \quad\; v^1|^2 = 2iz, \quad\; v^1|^3 = i\zeta, \quad\; v^3|^1 = \bar{f}', \quad\; v^3|^3 = 0\end{aligned}\right\}. \quad (5.11.1)$$

From (5.3.12), (5.9.4), and (5.11.1) we deduce that

$$\left.\begin{aligned}s''^{11} &= -2B\left(4\frac{\partial w^1}{\partial \bar{\zeta}} + \bar{f}'^2\right) + 2F(\bar{f}' + i\zeta)^2\\s''^{12} &= 4A(2w^i|_i + 2z^2 + \zeta\bar{\zeta} + f'\bar{f}') - 2(2C+F)(\bar{f}' + i\zeta)(f' - i\bar{\zeta}) +\\&\quad + 2B\left(2w^i|_i + 2\frac{\partial w^3}{\partial z} + 2z^2 + 2\zeta\bar{\zeta} + f'\bar{f}'\right)\\s''^{13} &= -2B\left(\frac{\partial w^1}{\partial z} + 2\frac{\partial w^3}{\partial \bar{\zeta}} + \zeta z\right)\\s''^{33} &= 2A(2w^i|_i + 2z^2 + \zeta\bar{\zeta} + f'\bar{f}') - 2C(\bar{f}' + i\zeta)(f' - i\bar{\zeta}) +\\&\quad + 2B\left(2w^i|_i - 2\frac{\partial w^3}{\partial z} + 2z^2 + f'\bar{f}'\right)\end{aligned}\right\}. \quad (5.11.2)$$

Also, from (5.3.15), (5.10.3), (5.11.1) and (5.11.2), we have

$$\left.\begin{aligned}t''^{11} &= -2B\left(4\frac{\partial w^1}{\partial \bar{\zeta}} + \bar{f}'^2 + i\zeta\bar{f}' - \zeta^2\right) + 2F(\bar{f}' + i\zeta)^2\\t''^{12} &= 4A(2w^i|_i + 2z^2 + \zeta\bar{\zeta} + f'\bar{f}') - 2(2C+F)(\bar{f}' + i\zeta)(f' - i\bar{\zeta}) +\\&\quad + 2B\left(2w^i|_i + 2\frac{\partial w^3}{\partial z} + 2z^2 + f'\bar{f}' + i\bar{\zeta}\bar{f}' + \zeta\bar{\zeta}\right)\\t''^{13} &= -2B\left(\frac{\partial w^1}{\partial z} + 2\frac{\partial w^3}{\partial \bar{\zeta}} + \zeta z\right)\\t''^{31} &= -2B\left(\frac{\partial w^1}{\partial z} + 2\frac{\partial w^3}{\partial \bar{\zeta}} + iz\bar{f}'\right)\\t''^{33} &= 2A(2w^i|_i + 2z^2 + \zeta\bar{\zeta} + f'\bar{f}') - 2C(\bar{f}' + i\zeta)(f' - i\bar{\zeta}) +\\&\quad + B\left(4w^i|_i - 4\frac{\partial w^3}{\partial z} + 4z^2 - i\zeta f' + i\bar{\zeta}\bar{f}'\right)\\t''^{21} &= \bar{t}''^{12},\; t''^{22} = \bar{t}''^{11},\; t''^{23} = \bar{t}''^{13},\; t''^{32} = \bar{t}''^{31}\end{aligned}\right\}. \quad (5.11.3)$$

If these values of the stresses are substituted in the equations of equilibrium (5.5.6), in which body forces are zero, and if we also use

(5.10.2) and (5.10.3), we have

$$(2A+B)\frac{\partial}{\partial z}(w^i|_i+z^2)-B\,\nabla^2 w^3 = 0, \tag{5.11.4}$$

$$(2A+B)\left(2\frac{\partial}{\partial \bar\zeta}w^i|_i+\zeta+f'\bar f''\right)-B(\nabla^2 w^1+2i\bar f'-2\bar\zeta)$$
$$= (2C+F)(f'\bar f''-i\bar\zeta \bar f'')+(2C+3F)(\zeta-i\bar f'). \tag{5.11.5}$$

We observe that we also have the complex conjugate of (5.11.5) so that (5.11.5) represents two equations. We look for a solution of (5.11.4) and (5.11.5) which is such that

$$w^3 = \{h - \tfrac{1}{2}k(\bar\zeta - \bar\zeta_G) - \tfrac{1}{2}\bar k(\zeta - \zeta_G)\}z, \tag{5.11.6}$$

where k is a complex constant and h is a real constant which represents the average extension of each cross-section, parallel to the z-axis. The value of k is chosen later to make the second-order couple about the x- and y-axes zero. Also, we look for a deformation in which the torsional couple is an odd function of ϵ, so we take $\bar\zeta t''^{31}+v^2 t'^{31}$ in (5.9.24) to be zero, and therefore

$$\frac{\partial w^1}{\partial z}+2\frac{\partial w^3}{\partial \bar\zeta}+\zeta z = 0. \tag{5.11.7}$$

The remaining term in (5.9.24) is proportional to N''^1 and we see later that $N''^1 = 0$. Hence, from (5.11.6) and (5.11.7),

$$w^1 = \tfrac{1}{2}kz^2 - \tfrac{1}{2}\zeta z^2 + H(\zeta, \bar\zeta), \tag{5.11.8}$$

where H is an arbitrary function of ζ and $\bar\zeta$. If we substitute (5.11.6) and (5.11.8) in (5.11.5) we see that

$$2(2A-B)\frac{\partial^2 H}{\partial \zeta\,\partial \bar\zeta}+2(2A+B)\frac{\partial^2 \bar H}{\partial \bar\zeta^2}$$
$$= 2(A+B)k-3B\bar\zeta+2iB\bar f'-(2A+B)(\zeta+f'\bar f'')+$$
$$+(2C+F)(f'\bar f''-i\bar\zeta \bar f'')+(2C+3F)(\zeta-i\bar f'), \tag{5.11.9}$$

while (5.11.4) is satisfied identically.

This equation may be integrated to give

$$2(2A-B)\frac{\partial H}{\partial \zeta}+2(2A+B)\frac{\partial \bar H}{\partial \bar\zeta}$$
$$= 4Br'(\zeta)+g'(\zeta)+2(A+B)k(\bar\zeta-\bar\zeta_G)+$$
$$+(2C+F-2A-B)f'\bar f'-2i(F-B)\bar f+$$
$$+(2C+3F-2A-4B)\zeta\bar\zeta-i(2C+F)\bar\zeta \bar f', \tag{5.11.10}$$

where $r'(\zeta)$ is an arbitrary function of ζ and $g'(\zeta)$ will be chosen later. From (5.11.10) and its complex conjugate we may solve for $\partial H/\partial \zeta$ to give

$$\frac{\partial H}{\partial \zeta} = \{(2A+B)(4B\bar{r}'+\bar{g}')-(2A-B)(4Br'+g')\}/(16AB)+$$
$$+(A+B)\{(2A+B)\bar{k}(\zeta-\zeta_G)-(2A-B)k(\bar{\zeta}-\bar{\zeta}_G)\}/(8AB)+$$
$$+i(2C+F)\{(2A+B)\zeta f'+(2A-B)\bar{\zeta}\bar{f}'\}/(16AB)+$$
$$+i(F-B)\{(2A+B)f+(2A-B)\bar{f}\}/(8AB)+$$
$$+(2C+F-2A-B)f'\bar{f}'/(8A)+$$
$$+(2C+3F-2A-4B)\zeta\bar{\zeta}/(8A). \tag{5.11.11}$$

We now choose $g'(\zeta)$ so that

$$(2A-B)g'(\zeta) = i(2A+B)\{2(F-B)f+(2C+F)\zeta f'\}+$$
$$+2(A+B)(2A-B)^2\bar{k}(\zeta-\zeta_G)/(2A+B)-$$
$$-2(A+B)(2A-B)(4Ah-Bk\bar{\zeta}_G+B\bar{k}\zeta_G)/(2A+B), \tag{5.11.12}$$

and then (5.11.11) may be integrated in the form

$$H(\zeta, \bar{\zeta}) = \bar{q}(\bar{\zeta})+\{(2A+B)\zeta\bar{r}'-(2A-B)r\}/(4A)+$$
$$+(2C+F-2A-B)f\bar{f}'/(8A)+$$
$$+(2C+3F-2A-4B)\zeta^2\bar{\zeta}/(16A)-$$
$$-\tfrac{1}{2}i\{2(F-B)\zeta\bar{f}+(2C+F)\zeta\bar{\zeta}\bar{f}'\}/(2A-B)-$$
$$-(A+B)\{(h+\tfrac{1}{2}k\bar{\zeta}_G+\tfrac{1}{2}\bar{k}\zeta_G)\zeta-\tfrac{1}{2}\bar{k}\zeta^2\}/(2A+B), \tag{5.11.13}$$

where $q(\zeta)$ is an arbitrary function of ζ.

In order to simplify the equations derived from the boundary conditions we put

$$\left.\begin{aligned}16ABq &= (BF-2BC+2AB-2B^2+4AF)\omega(\zeta)-\\&\quad -4A(B-F)\int_0^\zeta \{f'(t)\}^2\, dt-\\&\quad -2B(2C+F-2A-B)ff'\\4B(4A^2-B^2)r &= (2A-B)(BF-2BC+2AB-2B^2+4AF)\times\\&\quad \times\Omega(\zeta)-16iA(AF-BC-B^2)\int_0^\zeta f(t)\, dt-\\&\quad -2B(2A+B)(2A-B+2C+F)i\zeta f\end{aligned}\right\} \tag{5.11.14}$$

where ω and Ω are functions of ζ. Substitution in (5.11.8) and (5.11.13) then yields†

$$w^1 = \tfrac{1}{2}(k-\zeta)z^2 + \frac{BF-2BC+2AB-2B^2+4AF}{16AB(2A+B)} \times$$
$$\times \{(2A+B)(\bar\omega+\zeta\bar\Omega')-(2A-B)\Omega\} -$$
$$- \frac{A+B}{2A+B}\{(h+\tfrac{1}{2}k\bar\zeta_G+\tfrac{1}{2}\bar k\zeta_G)\zeta - \tfrac{1}{2}\bar k\zeta^2\} +$$
$$+ \frac{2C+3F-2A-4B}{16A}\zeta^2\bar\zeta -$$
$$- \frac{2C+F-2A-B}{8A}\left(\bar f f' - f\bar f' + i\zeta\bar f + i\zeta\bar\zeta\bar f' + \frac{2A-B}{2A+B}i\zeta f\right) +$$
$$+ \frac{iF\zeta\bar f}{2B} + \frac{i(2C+F)\zeta f}{2(2A+B)} +$$
$$+ \frac{i(AF-B^2-BC)}{B(2A+B)}\int_0^\zeta f(t)\,dt + \frac{F-B}{4B}\int_0^\zeta\{\bar f'(\bar t)\}^2\,d\bar t. \qquad (5.11.15)$$

From (5.11.2), (5.11.3), (5.11.6), and (5.11.15) we obtain the second-order stresses s''^{ij} and t''^{ij} in the form

$$\left.\begin{aligned}
s''^{11} &= (2BC+2B^2-4AF-BF-2AB)(\bar\omega'+\zeta\bar\Omega'')/(2A) - \\
&\quad - (2BC+3BF-2AB-4B^2+4AF)\zeta^2/(2A) + \\
&\quad + B(2C+F-2A-B)(\bar f f'' - f\bar f'' + \bar f'^2 + 2i\zeta\bar f' + i\zeta\bar\zeta\bar f'')/A \\
s''^{12} &= (4AF+BF+2AB-2BC-2B^2)(\Omega'+\bar\Omega')/(2A) + \\
&\quad + B(2C+F-2A-B)(f'\bar f'+i\zeta f'+if-i\bar\zeta\bar f'-i\bar f)/A - \\
&\quad - 4iB(f-\bar f) + (2BC+3BF+4AF-6AB-4B^2)\zeta\bar\zeta/A \\
s''^{13} &= 0 \\
s''^{33} &= (A+B)(BF+2AB+4AF-2BC-2B^2) \times \\
&\quad \times (\Omega'+\bar\Omega')/\{2A(2A+B)\} + \\
&\quad + i(AF-AB+2BC+BF-B^2)(\zeta f'-\bar\zeta\bar f')/A + \\
&\quad + iB(A+B)(2C+F-6A-B)(f-\bar f)/\{A(2A+B)\} + \\
&\quad + (2BC+BF-AB+AF-B^2)f'\bar f'/A + \\
&\quad + (2BC+3BF-4B^2+3AF-6AB)\zeta\bar\zeta/A - \\
&\quad - 2B(3A+2B)(2h+k\bar\zeta_G+\bar k\zeta_G - k\bar\zeta - \bar k\zeta)/(2A+B)
\end{aligned}\right\}, \quad (5.11.16)$$

† The displacement w^1 in (5.11.15) is undetermined to the extent of terms $iK\zeta + K'$, where K, K' are real and complex constants respectively. Terms of this type may be added and the end conditions (5.9.9) satisfied by a suitable choice of K and K'.

$$\left.\begin{aligned}
t''^{11} &= (2BC+2B^2-4AF-BF-2AB)(\bar{\omega}'+\zeta\bar{\Omega}'')/(2A)- \\
&\quad -(2BC+3BF-2AB-4B^2+4AF)\zeta^2/(2A)+ \\
&\quad +B(2C+F-2A-B)(\bar{f}\bar{f}''-f\bar{f}''+\bar{f}'^2+2i\zeta\bar{f}'+i\zeta\bar{\zeta}\bar{f}'')/A- \\
&\quad -2iB\zeta(\bar{f}'+i\zeta) \\
t''^{12} &= (4AF+BF+2AB-2BC-2B^2)(\Omega'+\bar{\Omega}')/(2A)+ \\
&\quad +B(2C+F-2A-B)(f'\bar{f}'+i\zeta f'+if-i\bar{\zeta}\bar{f}'-i\bar{f})/A- \\
&\quad -4iB(f-\bar{f})+(2BC+3BF+4AF-6AB-4B^2)\zeta\bar{\zeta}/A+ \\
&\quad +2iB\bar{\zeta}(\bar{f}'+i\zeta) \\
t''^{33} &= (A+B)(BF+2AB+4AF-2BC-2B^2)\times \\
&\qquad\qquad \times(\Omega'+\bar{\Omega}')/\{2A(2A+B)\}+ \\
&\quad +i(AF-2AB+2BC+BF-B^2)(\zeta f'-\bar{\zeta}\bar{f}')/A+ \\
&\quad +iB(A+B)(2C+F-6A-B)(f-\bar{f})/\{A(2A+B)\}+ \\
&\quad +(2BC+BF-3AB+AF-B^2)f'\bar{f}'/A+ \\
&\quad +(2BC+3BF-4B^2+3AF-6AB)\zeta\bar{\zeta}/A- \\
&\quad -2B(3A+2B)(2h+k\bar{\zeta}_G+\bar{k}\zeta_G-k\bar{\zeta}-\bar{k}\zeta)/(2A+B) \\
t''^{31} &= -2iBz(\bar{f}'+i\zeta), \quad t''^{13}=0, \quad t''^{21}=\bar{t}''^{12} \\
t''^{22} &= \bar{t}''^{11}, \quad t''^{23}=0, \quad t''^{32}=\bar{t}''^{31}
\end{aligned}\right\} \quad (5.11.17)$$

On substituting from (5.11.17) in the boundary conditions (5.9.15), and using (5.10.4), we find, after some reduction,

$$(2BC+2B^2-2AB-BF-4AF)\,d(\Omega+\zeta\bar{\Omega}'+\bar{\omega}+\zeta^2\bar{\zeta})+ \\
+2B(2C+F-2A-B)\,d\{\bar{f}'(\bar{f}-f+i\zeta\bar{\zeta})\}+ \\
+2iB(2C+F-6A-B)\,d\{\zeta(\bar{f}-f+i\zeta\bar{\zeta})\} = 0 \quad (G=0). \tag{5.11.18}$$

If we multiply (5.11.18) by $\bar{\zeta}$, integrate around the complete boundary, and transform by Stokes's theorem (5.9.25), we have

$$(BF+2AB+4AF-2BC-2B^2)\iint_R (\Omega'+\bar{\Omega}')\,dR+ \\
+2iB(2C+F-6A-B)\iint_R (f-\bar{f})\,dR \\
= 2B(6A+B-2C-F)S+2(4AB+3B^2-2BF-4AF)I, \tag{5.11.19}$$

provided Ω, ω satisfy suitable conditions. We are now able to evaluate the second-order resultant forces and couples, acting over the end

$z = l$ of the cylinder, by substituting from (5.11.17) in (5.9.22)–(5.9.24). If we also make use of the results (5.10.4), (5.10.8), (5.10.9), and (5.11.19), we find that

$$N''^1 = M''^3 = 0, \qquad (5.11.20)$$

$$(2A+B)N''^3 = 2(AF+BC+BF+2AB+2B^2)S - 2B(3A+2B)(I+2hR). \quad (5.11.21)$$

We may also evaluate M''^1 and, in general, choose the complex constant k so that $M''^1 = 0$, but we do not record here the resulting values of k. Further, if the resultant force $N''^3 = 0$, we have†

$$h = \frac{(AF+BC+BF+2AB+2B^2)S}{2BR(3A+2B)} - \frac{I}{2R}. \qquad (5.11.22)$$

The stresses s^{ij} in (5.10.3) and (5.11.16) are referred to θ_i coordinates in the deformed body. Components referred to cartesian coordinates y_i in the deformed body (see (5.9.6)) may be found by tensor transformations. Thus, if σ'_{ij} represent the components of stresses referred to y_i-axes in the deformed body, and T^{ij} are the corresponding components referred to Y^i-axes, all per unit area of the undeformed body, then

$$T^{ij} = \frac{\partial Y^i}{\partial y^r}\frac{\partial Y^j}{\partial y^s}\sigma'_{rs}, \qquad (5.11.23)$$

and

$$T^{ij} = \frac{\partial Y^i}{\partial \theta^m}\frac{\partial Y^j}{\partial \theta^n}s^{mn}. \qquad (5.11.24)$$

From (5.9.7) and (5.11.23)

$$\left.\begin{array}{ll} T^{11} = \sigma'_{11}-\sigma'_{22}+2i\sigma'_{12}, & T^{22} = \sigma'_{11}-\sigma'_{22}-2i\sigma'_{12} \\ T^{12} = \sigma'_{11}+\sigma'_{22}, & T^{13} = \sigma'_{13}+i\sigma'_{23} \\ T^{23} = \sigma'_{13}-i\sigma'_{23}, & T^{33} = \sigma'_{33} \end{array}\right\}, \quad (5.11.25)$$

and T^{ij} can be evaluated from (5.11.24) and (5.9.8). The results are not, however, recorded here.

When the body is incompressible, a corresponding theory may be developed using the formulae of § 5.4. Alternatively, results may be obtained from those found for a compressible body by using the limiting process (5.4.6) and (5.4.8), that is,

$$A \to \infty, \qquad C \to \infty, \qquad A-C,\ B,\ F \text{ finite}, \qquad (5.11.26)$$

† This formula was first found by R. S. Rivlin, loc. cit., p. 174, by a different method.

together with the formulae (5.4.10). Thus the boundary condition (5.11.18) reduces to

$$(C_1+2C_2)\,d(\Omega+\zeta\bar{\Omega}'+\bar{\omega}+\zeta^2\bar{\zeta})+$$
$$+2i(C_1+C_2)\,d\{\zeta(\bar{f}-f+i\zeta\bar{\zeta})\} = 0 \quad (G=0), \quad (5.11.27)$$

while (5.11.22) becomes†

$$h = \frac{(4C_1+5C_2)S}{6(C_1+C_2)R} - \frac{I}{2R}. \quad (5.11.28)$$

Further discussion is required about the potential functions Ω, ω in order that the displacements and stresses should be single-valued. In the next section we consider this for a cross-section which is simply-connected.

5.12. Simply-connected cross-section

When the cross-section is bounded by a single closed curve L_0 we may take the first-order boundary condition (5.10.5) to be

$$f-\bar{f}-i\zeta\bar{\zeta} = 0, \quad (5.12.1)$$

without loss of generality. Also, we assume that $f(\zeta)$ is regular in the cross-section R so that the corresponding first-order displacements and stresses are single-valued. With (5.12.1) the second-order boundary condition (5.11.18) reduces to

$$\Omega+\zeta\bar{\Omega}'+\bar{\omega}+\zeta^2\bar{\zeta} = 0, \quad (5.12.2)$$

where the arbitrary constant of integration can be taken to be zero without loss of generality. We now assume that Ω and ω are regular functions in the cross-section R and continuous on to the boundary so that, from (5.11.15) and (5.11.16), we see that the resulting displacements and stresses are single-valued.

When the cross-section can be mapped on a unit circle, the interior of the circle corresponding to the region R, it is possible to solve (5.12.2) by the methods of Muskhelishvili and obtain integral equations for Ω and ω.‡

Other three-dimensional problems can be discussed using the general theory of this chapter, for example, the inflation of a circular cylindrical tube or spherical shell of isotropic compressible material, but details

---

† See R. S. Rivlin, loc. cit., p. 174, and A. E. Green, loc. cit., p. 185.
‡ A. E. Green and R. T. Shield, loc. cit., p. 185.

are omitted.† The torsion of an arbitrary solid of revolution has been studied by Green and Spratt.‡

Corrigenda. In the paper by A. E. Green and E. B. Spratt, *Proc. R. Soc.* A **224** (1954), 347, a term

$$-4Ag^{ik}v^m|_m \, v^n|_n$$

should be added to the expression (4.14) for τ''^{ik}. A resulting correction is needed in a paper by W. S. Blackburn and A. E. Green, *Proc. R. Soc.* A **240** (1957) 408. Add the above term to τ''^{ik} in (2.8). Replace J, K from $p.$ 418 onwards by J', K', where

$$J' = J - \frac{2AB^2}{(2A+B)^2}, \qquad K' = K - \frac{AB^2}{(2A+B)^2},$$

respectively, and J, K take the values defined after (5.7) on p. 418 of the paper.

† See also F. D. Murnaghan, loc. cit., p. 185, and papers quoted on pp. 174 and 185.
‡ Loc. cit., p. 174.

VI

APPROXIMATE SOLUTION OF TWO-DIMENSIONAL PROBLEMS

THE approximation methods of Chapter V find a ready application to two-dimensional problems. In the consideration of these, we again assume that the displacement vector can be expanded as a power series in a real parameter ϵ and that the restrictions upon this vector indicated in Chapter V and the general discussion of § 5.1 continue to apply. To avoid undue complication of the analysis the theory is developed in full only for the finite plane strain of isotropic incompressible materials; the results for isotropic compressible bodies and for plane stress are quoted without proof.†

Following the procedure of the preceding chapter, we expand the equations for incompressible plane strain in terms of the parameter ϵ and by considering separately the coefficients of corresponding powers of ϵ obtain a series of relations for the determination of successive terms in the expansions of the stress and displacement functions. We again restrict attention to the determination of the first- and second-order terms. For two-dimensional problems it is convenient to use complex variable notation. Explicit expressions can then be obtained for the stress and displacement functions in terms of complex potential functions, two such functions being introduced at each succeeding stage of the approximation process. The procedure forms a natural extension of the corresponding methods of classical elasticity theory.‡ Expressions may also be derived for the force and couple resultants on a simple closed contour and, by making use of these, the forms at infinity of the complex potential functions may be deduced for an infinite body subject to a uniform distribution of stress and strain at infinity. For convenience the theory is developed initially using the complex coordinates in the deformed body as the independent

---

† The theory of plane strain for incompressible bodies was developed by J. E. Adkins, A. E. Green, and R. T. Shield, loc. cit., p. 101. For further details of the general two-dimensional theory reference may be made to the following papers:
J. E. Adkins, A. E. Green, and G. C. Nicholas, *Phil. Trans. R. Soc.* A, **247** (1954), 279.
J. E. Adkins and A. E. Green, *Proc. R. Soc.* A, **239** (1957), 557.
D. E. Carlson and R. T. Shield, *Arch ration. Mech. Analysis* **19** (1965), 189.
‡ See, for example, *T.E.*, chap. VI.

variables. The corresponding formulae for coordinates in the undeformed body have a similar structure and may be deduced by a simple change of independent variable.

Approximation theories for the plane strain of compressible bodies and for plane stress may be developed using the appropriate formulae of Chapters III and IV. The resulting formulae are similar in structure to the corresponding relations for incompressible plane strain. In each case, the expressions in terms of the complex potentials for the stress and displacement functions may be put in forms which are similar to each other, the only differences arising from the different values of the constant coefficients. These similarities of structure which exist between the stress and displacement functions, between the formulae for plane stress and plane strain, between the relations for coordinates in the deformed and undeformed bodies, and between the formulae for compressible and incompressible materials, make it possible to solve a large number of problems simultaneously. It should be noted, however, that the solutions obtained for plane stress are approximations to results which are themselves approximations for thin sheets subject to the limitations indicated in § 4.4.

The solution of particular problems is facilitated, as in the infinitesimal theory, by the use of the Hilbert problem. Owing to the complexity of the expressions involved, attention is confined to some simple problems for an infinite body containing a single internal circular boundary. In principle, however, the method can be extended to other boundaries, by the use, if appropriate, of conformal transformations.†

6.1. Plane strain: isotropic incompressible bodies

It is convenient to use the complex variable formulation of §§ 3.4 and 3.5 and to choose initially the complex coordinates z, $\bar{z}$ in the deformed body as the independent variables. We further restrict attention to plane strain for which $\lambda = 1$. For isotropic, incompressible bodies we then have from (3.5.18), (3.3.22), and (3.4.21)

$$\frac{\partial^2 \phi}{\partial z^2} = \frac{\partial \bar{D}}{\partial z}\left(\frac{\partial D}{\partial z} - 1\right)\mathscr{H}, \tag{6.1.1}$$

with
$$\mathscr{H} = 2\frac{dW_1(I)}{dI}, \tag{6.1.2}$$

† A number of such problems for the first-order classical theory are considered in T.E., chap. VIII.

and
$$I = 3 + 4\frac{\partial D}{\partial \bar{z}}\frac{\partial \bar{D}}{\partial z}. \tag{6.1.3}$$

With these equations we must couple the incompressibility condition, which, from (3.4.21) and (3.4.8), may be written

$$\frac{\partial D}{\partial z} + \frac{\partial \bar{D}}{\partial \bar{z}} - \frac{\partial D}{\partial z}\frac{\partial \bar{D}}{\partial \bar{z}} + \frac{\partial D}{\partial \bar{z}}\frac{\partial \bar{D}}{\partial z} = 0. \tag{6.1.4}$$

Following the method of § 5.2, we assume that the complex displacement function D is a function of a real parameter ϵ which can be expanded as an absolutely convergent series

$$D = \epsilon[^0D(z,\bar{z})] + \epsilon^2[^1D(z,\bar{z})] + \ldots, \tag{6.1.5}$$

in the range $0 \leqslant \epsilon \leqslant \epsilon_0$ for relevant values of z, ϵ_0 being a value of the parameter which depends upon the problem under consideration. We assume that the first and second partial derivatives of D with respect to z and $\bar{z}$ exist and can be obtained by term-by-term differentiation of the series (6.1.5), and that the resulting series are absolutely convergent in the range $0 \leqslant \epsilon \leqslant \epsilon_0$ and for relevant values of z. The only operations then involved in subsequent work are multiplications of absolutely convergent power series, which are justified under the above assumptions. As in Chapter V, the calculations are limited to the first two terms of the power series expansion in terms of ϵ, the first term again corresponding to the classical infinitesimal theory of elasticity.

From (6.1.5) and (6.1.3),

$$I = 3 + 4\epsilon^2 \frac{\partial {}^0D}{\partial \bar{z}}\frac{\partial {}^0\bar{D}}{\partial z} + O(\epsilon^3), \tag{6.1.6}$$

so that
$$\mathscr{H} = {}^0\mathscr{H} + 4\epsilon^2({}^2\mathscr{H})\frac{\partial {}^0D}{\partial \bar{z}}\frac{\partial {}^0\bar{D}}{\partial z} + O(\epsilon^3), \tag{6.1.7}$$

where ${}^0\mathscr{H}$, ${}^2\mathscr{H}$ are constants defined by

$$^0\mathscr{H} = 2\left[\frac{dW_1(I)}{dI}\right]_{I=3}, \quad {}^2\mathscr{H} = 2\left[\frac{d^2W_1(I)}{dI^2}\right]_{I=3}. \tag{6.1.8}$$

We observe that ${}^0\mathscr{H} = \tfrac{1}{3}E$, where E is the value of Young's modulus for strains corresponding to classical theory. If we now put

$$\phi = {}^0\mathscr{H}\,\epsilon\{{}^0\phi(z,\bar{z}) + \epsilon[{}^1\phi(z,\bar{z})] + \ldots\}, \tag{6.1.9}$$

and introduce the expansions (6.1.5), (6.1.7), and (6.1.9) into (6.1.1) and (6.1.4), we obtain, by equating the coefficients of ϵ, ϵ^2 separately to zero,

$$\frac{\partial^2(^0\phi)}{\partial z^2}+\frac{\partial ^0D}{\partial z}=0, \quad \frac{\partial ^0D}{\partial z}+\frac{\partial ^0\bar{D}}{\partial \bar{z}}=0, \qquad (6.1.10)$$

$$\left.\begin{array}{l}\dfrac{\partial^2(^1\phi)}{\partial z^2}+\dfrac{\partial ^1D}{\partial z}=\dfrac{\partial ^0D}{\partial z}\dfrac{\partial ^0D}{\partial z}\\[2mm] \dfrac{\partial ^1D}{\partial z}+\dfrac{\partial ^1\bar{D}}{\partial \bar{z}}=\dfrac{\partial ^0D}{\partial z}\dfrac{\partial ^0\bar{D}}{\partial \bar{z}}-\dfrac{\partial ^0\bar{D}}{\partial z}\dfrac{\partial ^0D}{\partial \bar{z}}=-\left[\dfrac{\partial ^0D}{\partial z}\right]^2-\dfrac{\partial ^0\bar{D}}{\partial z}\dfrac{\partial ^0D}{\partial \bar{z}}\end{array}\right\}. \qquad (6.1.11)$$

Similar equations may be obtained from the coefficients of higher powers of ϵ but we restrict attention to the first and second approximations represented by equations (6.1.10) and (6.1.11) respectively. The equations for the first approximation, corresponding to the classical theory of elasticity, may be integrated in terms of complex potential functions $\Omega(z)$, $\omega(z)$. Thus

$$\left.\begin{array}{l}^0\phi(z,\bar{z})=z\bar{\Omega}(\bar{z})+\bar{z}\Omega(z)+\omega(z)+\bar{\omega}(\bar{z})\\ ^0D(z,\bar{z})=\Omega(z)-z\bar{\Omega}'(\bar{z})-\bar{\omega}'(\bar{z})\end{array}\right\}. \qquad (6.1.12)$$

Using (6.1.12), equations (6.1.11) now become

$$\frac{\partial^2(^1\phi)}{\partial z^2}+\frac{\partial ^1D}{\partial z}=-[\Omega'(z)-\bar{\Omega}'(\bar{z})][\bar{z}\Omega''(z)+\omega''(z)], \qquad (6.1.13)$$

$$\frac{\partial ^1D}{\partial z}+\frac{\partial ^1\bar{D}}{\partial \bar{z}}=-[\Omega'(z)-\bar{\Omega}'(\bar{z})]^2-[z\bar{\Omega}''(\bar{z})+\bar{\omega}''(\bar{z})][\bar{z}\Omega''(z)+\omega''(z)]. \qquad (6.1.14)$$

The first of these equations may be integrated to give

$$\frac{\partial ^1\phi}{\partial z}+{}^1D=2\bar{\Delta}(\bar{z})+\tfrac{1}{2}\int^{\bar{z}}[\bar{\Omega}'(\bar{z})]^2\,d\bar{z}-\int^z\Omega'(z)\omega''(z)\,dz+$$
$$+\bar{\Omega}'(\bar{z})[\bar{z}\Omega'(z)+\omega'(z)-\bar{\Omega}(\bar{z})]-\tfrac{1}{2}\bar{z}[\Omega'(z)]^2, \qquad (6.1.15)$$

where $\bar{\Delta}(\bar{z})$ is an arbitrary function of $\bar{z}$ and the integral term, which is a function of $\bar{z}$ only, has been added for convenience. From (6.1.14) and (6.1.15) it follows that

$$2\frac{\partial^2(^1\phi)}{\partial z\,\partial \bar{z}}=2[\Delta'(z)+\bar{\Delta}'(\bar{z})]+\bar{\Omega}''(\bar{z})[\bar{z}\Omega'(z)+\omega'(z)-\bar{\Omega}(\bar{z})]+$$
$$+\Omega''(z)[z\bar{\Omega}'(\bar{z})+\bar{\omega}'(\bar{z})-\Omega(z)]+[z\bar{\Omega}''(\bar{z})+\bar{\omega}''(\bar{z})][\bar{z}\Omega''(z)+\omega''(z)],$$

and hence

$$\frac{\partial ^1\phi(z,\bar{z})}{\partial \bar{z}}=\Delta(z)+z\bar{\Delta}'(\bar{z})+\bar{\delta}'(\bar{z})+\tfrac{1}{2}\Gamma(z,\bar{z})+$$
$$+\tfrac{1}{2}\int^z[\Omega'(z)]^2\,dz-\int^{\bar{z}}\bar{\Omega}'(\bar{z})\bar{\omega}''(\bar{z})\,d\bar{z}-\tfrac{1}{2}z[\bar{\Omega}'(\bar{z})]^2, \qquad (6.1.16)$$

where

$$\Gamma(z, \bar{z}) = [z\bar{\Omega}''(\bar{z})+\bar{\omega}''(\bar{z})][\bar{z}\Omega'(z)+\omega'(z)-\bar{\Omega}(\bar{z})]+$$
$$+[\Omega'(z)+\bar{\Omega}'(\bar{z})][z\bar{\Omega}'(\bar{z})+\bar{\omega}'(\bar{z})-\Omega(z)]$$
$$= -\left({}^0D\frac{\partial}{\partial z}+{}^0\bar{D}\frac{\partial}{\partial \bar{z}}\right)\frac{\partial{}^0\phi}{\partial \bar{z}}, \qquad (6.1.17)$$

and $\bar{\delta}(\bar{z})$ is a further arbitrary function of $\bar{z}$. From (6.1.15) and (6.1.16) we have

$$^1D(z, \bar{z}) = \Delta(z)-z\overline{\Delta}'(\bar{z})-\bar{\delta}'(\bar{z})-\tfrac{1}{2}\Lambda(z, \bar{z}), \qquad (6.1.18)$$

where

$$\Lambda(z, \bar{z}) = [z\bar{\Omega}''(\bar{z})+\bar{\omega}''(\bar{z})][\bar{z}\Omega'(z)+\omega'(z)-\bar{\Omega}(\bar{z})]-$$
$$-[\Omega'(z)-\bar{\Omega}'(\bar{z})][z\bar{\Omega}'(\bar{z})+\bar{\omega}'(\bar{z})-\Omega(z)]$$
$$= \left({}^0D\frac{\partial}{\partial z}+{}^0\bar{D}\frac{\partial}{\partial \bar{z}}\right){}^0D. \qquad (6.1.19)$$

For problems which are non-dislocational in character the complex potential functions $\Omega(z)$, $\omega(z)$, $\Delta(z)$, and $\delta(z)$ must be chosen so that the stress and displacement components are single-valued. This implies that ${}^0D(z, \bar{z})$, ${}^1D(z, \bar{z})$,..., and all their derivatives with respect to z and $\bar{z}$ and similarly the second- and higher-order derivatives of ${}^0\phi(z, \bar{z})$, ${}^1\phi(z, \bar{z})$,..., if they exist, must be single-valued at interior points of the body, so that from (6.1.12)

$$[\Omega'(z)]_C = 0, \quad [\omega''(z)]_C = 0, \quad [\Omega(z)]_C = [\bar{\omega}'(\bar{z})]_C, \qquad (6.1.20)$$

where $[\ ]_C$ denotes the change in value of the function inside the brackets during a complete circuit of a closed contour C lying entirely within the deformed body. From these conditions, with (6.1.16) to (6.1.19), it follows that

$$[\Delta'(z)]_C = 0, \quad [\delta''(z)]_C = 0, \quad [\Delta(z)]_C = [\bar{\delta}'(\bar{z})]_C. \qquad (6.1.21)$$

Alternative forms for $\partial^1\phi/\partial \bar{z}$ and ${}^1D$ may be obtained from (6.1.16) and (6.1.18) by replacing $\Delta(z)$ by $\Delta(z)+l\int^z [\Omega'(z)]^2\, dz$ and $\delta'(z)$ by $\delta'(z)+k\int^z \Omega'(z)\omega''(z)\, dz$ where k and l are constants. We thus obtain

$$\frac{\partial^1\phi(z, \bar{z})}{\partial \bar{z}} = \Delta(z)+z\overline{\Delta}'(\bar{z})+\bar{\delta}'(\bar{z})+\tfrac{1}{2}\Gamma(z, \bar{z})+$$
$$+k_1\int^{\bar{z}} \bar{\Omega}'(\bar{z})\bar{\omega}''(\bar{z})\, d\bar{z}+k_2\int^z [\Omega'(z)]^2\, dz+k_3 z[\bar{\Omega}'(\bar{z})]^2, \qquad (6.1.22)$$

$$^1D(z, \bar{z}) = \Delta(z)-z\overline{\Delta}'(\bar{z})-\bar{\delta}'(\bar{z})-\tfrac{1}{2}\Lambda(z, \bar{z})-$$
$$-k\int^{\bar{z}} \bar{\Omega}'(\bar{z})\bar{\omega}''(\bar{z})\, d\bar{z}+l\int^z [\Omega'(z)]^2\, dz-lz[\bar{\Omega}'(\bar{z})]^2, \qquad (6.1.23)$$

where $\qquad k_1 = k-1, \qquad k_2 = l+\tfrac{1}{2}, \qquad k_3 = l-\tfrac{1}{2}.$ \hfill (6.1.24)

The integral terms may then, if desired, be removed from the expression for $\partial^1\phi/\partial \bar{z}$ by choosing $k = 1$, $l = -\tfrac{1}{2}$, so that $k_1 = k_2 = 0$, $k_3 = -1$. The conditions (6.1.21) for single-valued stresses and displacements now, however, become

$$[\Delta'(z)]_C = 0, \qquad [\delta''(z)]_C = 0$$
$$[\Delta(z)-\bar{\delta}'(\bar{z})]_C = \left[k \int^{\bar{z}} \bar{\Omega}'(\bar{z})\bar{\omega}''(\bar{z})\, d\bar{z} - l \int^z \{\Omega'(z)\}^2\, dz \right]_C. \qquad (6.1.25)$$

6.2. Coordinates in the undeformed body

For some problems it is convenient to choose the complex coordinates ζ, $\bar{\zeta}$ in the undeformed body as independent variables. The theory may be developed directly from the results of §§ 3.1–3.3 by choosing the curvilinear coordinate system θ_α to coincide with the complex coordinates ζ, $\bar{\zeta}$. It is somewhat simpler, however, to derive the expressions for the stress and displacement functions by making a change of independent variable in the results of § 6.1.

From (3.4.4) and (3.4.8) we obtain, with $\lambda = 1$, $I_3 = 1$,

$$\frac{\partial}{\partial z} = \left(1+\frac{\partial \bar{D}}{\partial \bar{\zeta}}\right)\frac{\partial}{\partial \zeta} - \frac{\partial D}{\partial \zeta}\frac{\partial}{\partial \bar{\zeta}}, \qquad (6.2.1)$$

together with the complex conjugate of this equation. If now we express D in the form

$$D = \epsilon[^0D'(\zeta, \bar{\zeta})] + \epsilon^2[^1D'(\zeta, \bar{\zeta})] + \cdots, \qquad (6.2.2)$$

so that $\qquad z = \zeta + \epsilon[^0D'(\zeta, \bar{\zeta})] + \epsilon^2[^1D'(\zeta, \bar{\zeta})] + \cdots, \qquad (6.2.3)$

and introduce the latter expansion into (6.1.5), we obtain

$$D = \epsilon[^0D(\zeta, \bar{\zeta})] + \epsilon^2\left[{}^1D(\zeta, \bar{\zeta}) + {}^0D'(\zeta, \bar{\zeta})\frac{\partial\, {}^0D(\zeta, \bar{\zeta})}{\partial \zeta} + {}^0\bar{D}'(\bar{\zeta}, \zeta)\frac{\partial\, {}^0D(\zeta, \bar{\zeta})}{\partial \bar{\zeta}}\right] + O(\epsilon^3). \qquad (6.2.4)$$

Comparing (6.2.2) and (6.2.4) and making use of (6.1.19) we have

$$\left.\begin{array}{l} {}^0D'(\zeta, \bar{\zeta}) = {}^0D(\zeta, \bar{\zeta}) \\ {}^1D'(\zeta, \bar{\zeta}) = {}^1D(\zeta, \bar{\zeta}) + \Lambda(\zeta, \bar{\zeta}) \end{array}\right\}. \qquad (6.2.5)$$

Similarly, from (6.1.9), (6.2.3), (6.2.5), and (6.1.17) we obtain

$$\left.\begin{array}{l} \dfrac{\partial\, {}^0\phi(z, \bar{z})}{\partial \bar{z}} = \dfrac{\partial\, {}^0\phi(\zeta, \bar{\zeta})}{\partial \bar{\zeta}} \\[2mm] \dfrac{\partial\, {}^1\phi(z, \bar{z})}{\partial \bar{z}} = \dfrac{\partial\, {}^1\phi(\zeta, \bar{\zeta})}{\partial \bar{\zeta}} - \Gamma(\zeta, \bar{\zeta}) \end{array}\right\}. \qquad (6.2.6)$$

In (6.2.5) and (6.2.6), $\Lambda(\zeta, \bar\zeta)$ and $\Gamma(\zeta, \bar\zeta)$ are derived from $\Lambda(z, \bar z)$ and $\Gamma(z, \bar z)$ respectively merely by replacing $z, \bar z$ by $\zeta, \bar\zeta$ throughout.

Equations (6.2.5) and (6.2.6), when combined with (6.1.12), (6.1.22), and (6.1.23), yield

$$\left.\begin{aligned}{}^0\phi(\zeta, \bar\zeta) &= \zeta\bar\Omega(\bar\zeta)+\bar\zeta\Omega(\zeta)+\omega(\zeta)+\bar\omega(\bar\zeta)\\ {}^0D'(\zeta, \bar\zeta) &= \Omega(\zeta)-\zeta\bar\Omega'(\bar\zeta)-\bar\omega'(\bar\zeta)\end{aligned}\right\}, \quad (6.2.7)$$

$$\left.\begin{aligned}\frac{\partial {}^1\phi(z, \bar z)}{\partial \bar z} &= \Delta(\zeta)+\zeta\bar\Delta'(\bar\zeta)+\delta'(\bar\zeta)-\tfrac{1}{2}\Gamma(\zeta, \bar\zeta)+\\ &\quad +k_1\int^{\bar\zeta}\bar\Omega'(\bar\zeta)\bar\omega''(\bar\zeta)\,d\bar\zeta+k_2\int^{\zeta}[\Omega'(\zeta)]^2\,d\zeta+k_3\zeta[\bar\Omega'(\bar\zeta)]^2\\ {}^1D'(\zeta, \bar\zeta) &= \Delta(\zeta)-\zeta\bar\Delta'(\bar\zeta)-\delta'(\bar\zeta)+\tfrac{1}{2}\Lambda(\zeta, \bar\zeta)-\\ &\quad -k\int^{\bar\zeta}\bar\Omega'(\bar\zeta)\bar\omega''(\bar\zeta)\,d\bar\zeta+l\int^{\zeta}[\Omega'(\zeta)]^2\,d\zeta-l\zeta[\bar\Omega'(\bar\zeta)]^2\end{aligned}\right\}. \quad (6.2.8)$$

The conditions (6.1.20), (6.1.25) for single-valued stresses and displacements continue to apply with z replaced by ζ throughout. The integral terms may again be eliminated from the expression for $\partial {}^1\phi/\partial \bar z$ by choosing $k = 1$, $l = -\tfrac{1}{2}$ and remembering the relations (6.1.24). Similarly, by choosing $k = l = 0$ relations may be obtained with the integral terms absent from the expression for ${}^1D'$.

6.3. Force and couple resultants

Expressions for the force and couple resultants across a simple closed curve C in the deformed body in terms of the complex potentials follow from § 6.1. If we assume expansions

$$\left.\begin{aligned}P &= {}^0\mathscr{H}\epsilon({}^0P+\epsilon{}^1P+...)\\ M &= {}^0\mathscr{H}\epsilon({}^0M+\epsilon{}^1M+...)\end{aligned}\right\} \quad (6.3.1)$$

for the force P and couple M across a curve C in the deformed body, we obtain from (3.5.6), (3.5.7), and (6.1.9)

$$\left.\begin{aligned}{}^rP &= 2i\left[\frac{\partial({}^r\phi)}{\partial \bar z}\right]_C\\ {}^rM &= \left[z\frac{\partial({}^r\phi)}{\partial z}+\bar z\frac{\partial({}^r\phi)}{\partial \bar z}-{}^r\phi\right]_C \quad (r = 0, 1,...)\end{aligned}\right\}, \quad (6.3.2)$$

$[\ ]_C$, as before, denoting the change in value of the quantity inside the brackets along the curve C.

From (6.3.2) and (6.1.12), with the help of the single-valued conditions (6.1.20), we obtain for the first-order terms

$$\begin{aligned}{}^0P &= 2i[\Omega(z)+\bar{\omega}'(\bar{z})]_C = 4i[\Omega(z)]_C \\ {}^0M &= [z\omega'(z)+\bar{z}\bar{\omega}'(\bar{z})-\omega(z)-\bar{\omega}(\bar{z})]_C\end{aligned} \quad (6.3.3)$$

Also, using (6.1.12), (6.1.17), and (6.1.24), and remembering that $^1\phi$ is real, we obtain from (6.1.22) by integration

$$\begin{aligned}{}^1\phi(z,\bar{z}) =\; & \bar{z}\Delta(z)+z\bar{\Delta}(\bar{z})+\delta(z)+\bar{\delta}(\bar{z})+\tfrac{1}{2}{}^0D(z,\bar{z})\,{}^0\bar{D}(\bar{z},z)- \\ & -\Omega(z)\bar{\Omega}(\bar{z})+\int^z \Omega'(z)\omega'(z)\,dz+\int^{\bar{z}} \bar{\Omega}'(\bar{z})\bar{\omega}'(\bar{z})\,d\bar{z}+ \\ & +k_1\Big\{z\int^z \Omega'(z)\omega''(z)\,dz+\bar{z}\int^{\bar{z}} \bar{\Omega}'(\bar{z})\bar{\omega}''(\bar{z})\,d\bar{z}- \\ & -\int^z z\Omega'(z)\omega''(z)\,dz-\int^{\bar{z}} \bar{z}\bar{\Omega}'(\bar{z})\bar{\omega}''(\bar{z})\,d\bar{z}\Big\}+ \\ & +k_2\Big\{\bar{z}\int^z [\Omega'(z)]^2\,dz+z\int^{\bar{z}} [\bar{\Omega}'(\bar{z})]^2\,d\bar{z}\Big\}, \end{aligned} \quad (6.3.4)$$

and hence, from (6.3.2), (6.1.20), (6.1.22), and (6.1.25),

$$\begin{aligned}{}^1P &= 2i\Big[\Delta(z)+\bar{\delta}'(\bar{z})+k_1\int^{\bar{z}} \bar{\Omega}'(\bar{z})\bar{\omega}''(\bar{z})\,d\bar{z}+k_2\int^z \{\Omega'(z)\}^2\,dz\Big]_C \\ &= 2i\Big[2\Delta(z)-\int^{\bar{z}} \bar{\Omega}'(\bar{z})\bar{\omega}''(\bar{z})\,d\bar{z}+(2l+\tfrac{1}{2})\int^z \{\Omega'(z)\}^2\,dz\Big]_C, \quad (6.3.5)\end{aligned}$$

$$\begin{aligned}{}^1M =\; & [z\delta'(z)+\bar{z}\bar{\delta}'(\bar{z})-\delta(z)-\bar{\delta}(\bar{z})]_C+ \\ & +k_1\Big[\int^z z\Omega'(z)\omega''(z)\,dz+\int^{\bar{z}} \bar{z}\bar{\Omega}'(\bar{z})\bar{\omega}''(\bar{z})\,d\bar{z}\Big]_C+ \\ & +\Big[\Omega(z)\bar{\Omega}(\bar{z})-\int^z \Omega'(z)\omega'(z)\,dz-\int^{\bar{z}} \bar{\Omega}'(\bar{z})\bar{\omega}'(\bar{z})\,d\bar{z}\Big]_C. \quad (6.3.6)\end{aligned}$$

When boundary conditions are formulated in terms of the initial configuration of the body, it is convenient to employ expressions for the resultant forces and couples in terms of the coordinates ζ, $\bar{\zeta}$. From (6.2.6)–(6.2.8) it follows that the formulae for ^0P, ^0M, and ^1P are obtained from (6.3.3) and (6.3.5) merely by replacing z, $\bar{z}$ by ζ, $\bar{\zeta}$ throughout. Also, by writing $z = \zeta+D$ in (6.1.5) and (6.1.9) and expanding each term of the series for ϕ and D by Taylor's theorem, we may obtain

$$^0\phi(z, \bar{z}) = {}^0\phi(\zeta, \bar{\zeta}), \qquad {}^0D(z, \bar{z}) = {}^0D(\zeta, \bar{\zeta}),$$

$$^1\phi(z, \bar{z}) = {}^1\phi(\zeta, \bar{\zeta}) + {}^0D(\zeta, \bar{\zeta})\frac{\partial {}^0\phi(\zeta, \bar{\zeta})}{\partial \zeta} + {}^0\bar{D}(\zeta, \bar{\zeta})\frac{\partial {}^0\phi(\zeta, \bar{\zeta})}{\partial \bar{\zeta}},$$

and when these formulae are combined with (6.3.1), (6.3.2), and single-valued terms discarded, we have

$$^1M = \left[\zeta\frac{\partial {}^1\phi(\zeta, \bar{\zeta})}{\partial \zeta} + \bar{\zeta}\frac{\partial {}^1\phi(\zeta, \bar{\zeta})}{\partial \bar{\zeta}} - {}^1\phi(\zeta, \bar{\zeta})\right]_C, \qquad (6.3.7)$$

which implies that the formula for ^1M for coordinates in the undeformed body is also obtained from the corresponding expression (6.3.6) merely by replacing z, $\bar{z}$ by ζ, $\bar{\zeta}$ respectively.

If the resultant force and couple on a closed contour are zero, (6.3.3) and (6.1.20) yield

$$[\Omega(z)]_C = 0, \quad [\omega'(z)]_C = 0, \quad [\omega(z) + \bar{\omega}(\bar{z})]_C = 0, \qquad (6.3.8)$$

while, from these conditions with (6.1.25), (6.3.5), and (6.3.6), we have

$$\left.\begin{aligned}
&[\Delta'(z)]_C = 0, \qquad [\delta''(z)]_C = 0, \\
&2[\Delta(z)]_C = \left[\int^{\bar{z}} \bar{\Omega}'(\bar{z})\bar{\omega}''(\bar{z})\,d\bar{z} - (2l+\tfrac{1}{2})\int^z \{\Omega'(z)\}^2\,dz\right]_C \\
&2[\bar{\delta}'(\bar{z})]_C = -\left[(2k-1)\int^{\bar{z}} \bar{\Omega}'(\bar{z})\bar{\omega}''(\bar{z})\,d\bar{z} + \tfrac{1}{2}\int^z \{\Omega'(z)\}^2\,dz\right]_C \\
&[\delta(z) + \bar{\delta}(\bar{z})]_C = [z\delta'(z) + \bar{z}\bar{\delta}'(\bar{z})]_C + \\
&\qquad + k_1\left[\int^z z\Omega'(z)\omega''(z)\,dz + \int^{\bar{z}} \bar{z}\bar{\Omega}'(\bar{z})\bar{\omega}''(\bar{z})\,d\bar{z}\right]_C - \\
&\qquad - \left[\int^z \Omega'(z)\omega'(z)\,dz + \int^{\bar{z}} \bar{\Omega}'(\bar{z})\bar{\omega}'(\bar{z})\,d\bar{z}\right]_C
\end{aligned}\right\} \qquad (6.3.9)$$

If in the latter expressions the integral terms are also single-valued we obtain the forms

$$[\Delta(z)]_C = 0, \quad [\delta'(z)]_C = 0, \quad [\delta(z) + \bar{\delta}(\bar{z})]_C = 0, \qquad (6.3.10)$$

analogous to (6.3.8).

Formulae for ^1P, ^1M corresponding to the cases where the integral terms are absent from ^1D or $\partial ^1\phi/\partial \bar{z}$ are obtained by making an appropriate choice for the constants k, l.

6.4. Complex potentials for an infinite body

In classical elasticity theory, general forms of the complex potential functions have been examined for various systems of forces and couples applied to an infinite body containing various boundaries.† The corresponding results for the second-order theory developed in §§ 6.1–6.3 are of much greater complexity, and we here confine attention to the comparatively simple case where there is a uniform distribution of stress at infinity, and the body contains a single finite boundary consisting of a closed contour which, before and after deformation, encloses the origin and which is subject to a self-equilibrating system of forces so that $P = M = 0$. We again restrict attention to incompressible materials in plane strain.

Since we are considering the behaviour of the complex potentials at infinity, it is to be understood that the relations of the present section apply only for large values of x_α, y_α, $|\zeta|$, and $|z|$. Also, the problem will be made definite by assuming that at infinity there is zero rotation. The deformation, for large values of x_α or y_α, therefore reduces to a pure homogeneous strain defined by relations of the type

$$x_\alpha = c_{\alpha\beta} y_\beta \quad (\alpha, \beta = 1, 2; c_{\alpha\beta} = c_{\beta\alpha}),$$

where $c_{\alpha\beta}$ are constants, and these imply that

$$\frac{\partial x_2}{\partial y_1} = \frac{\partial x_1}{\partial y_2} \quad \text{or} \quad \frac{\partial y_2}{\partial x_1} = \frac{\partial y_1}{\partial x_2}. \tag{6.4.1}$$

In terms of the complex variables ζ, z, equations (6.4.1) become

$$\frac{\partial \zeta}{\partial z} = \frac{\partial \bar\zeta}{\partial \bar z} \quad \text{or} \quad \frac{\partial z}{\partial \zeta} = \frac{\partial \bar z}{\partial \bar \zeta},$$

which, with (3.4.8), (6.1.5), and (6.2.2), yield the conditions

$$\frac{\partial (^rD)}{\partial z} = \frac{\partial (^r\bar D)}{\partial \bar z}, \quad \frac{\partial (^rD')}{\partial \zeta} = \frac{\partial (^r\bar D')}{\partial \bar \zeta} \quad (r = 0, 1, \ldots). \tag{6.4.2}$$

The first-order terms in the expansions for ϕ and D in powers of ϵ correspond to the stress and displacement functions of the classical theory of elasticity, and to this degree of approximation it is immaterial whether we employ the coordinates ζ, $\bar\zeta$ or z, $\bar z$ as independent variables. For convenience we use the latter system initially. For a uniform state of stress at infinity, it follows from (3.5.2) and (3.5.4)

† See, for example, *T.E.*, chap. VIII.

that for large values of $|z|$ the second derivatives of ϕ must tend towards constant values. Equations (6.1.12) then imply that for a uniform state of stress and strain at infinity the expansions† of $\Omega(z)$, $\omega'(z)$ as doubly infinite power series in z cannot contain terms of higher order than the first. Also, the resultant force on any (large) circuit C_∞ surrounding the internal boundary is evidently zero. Hence, as $C \to C_\infty$, $^0P \to 0$, $^0M \to 0$ and the conditions (6.3.8) apply. Remembering (6.4.2) and (6.1.12), we may therefore write‡

$$\left.\begin{array}{l}\Omega(z) = A'z + a'/z + O(z^{-2}) \\ \omega(z) = \tfrac{1}{2} b' z^2 + B' \log z + O(z^{-1})\end{array}\right\}, \quad (6.4.3)$$

where A' and B' are real constants, a' and b' are constants which may be real or complex, and terms which do not contribute to the stresses have been omitted. From (6.1.12), (6.1.22)–(6.1.24), (6.1.17), and (6.1.19) we now have

$$\left.\begin{array}{l}\dfrac{\partial {}^0\phi(z,\bar z)}{\partial \bar z} = 2A'z + \bar b' \bar z + O(z^{-1}) \\ {}^0D(z,\bar z) = -\bar b' \bar z + O(z^{-1})\end{array}\right\}, \quad (6.4.4)$$

$$\left.\begin{array}{l}\dfrac{\partial {}^1\phi(z,\bar z)}{\partial \bar z} = \Delta(z) + z\bar\Delta'(\bar z) + \bar\delta'(\bar z) + L_1 z + L_2 \bar z + O(z^{-1}) \\ {}^1D(z,\bar z) = \Delta(z) - z\bar\Delta'(\bar z) - \bar\delta'(\bar z) + L_3 z + L_4 \bar z + O(z^{-1})\end{array}\right\}, \quad (6.4.5)$$

where
$$\left.\begin{array}{ll}L_1 = 2lA'^2 + \tfrac{1}{2} b'\bar b', & L_2 = kA'\bar b' \\ L_3 = -\tfrac{1}{2} b'\bar b', & L_4 = -kA'\bar b'\end{array}\right\}. \quad (6.4.6)$$

Using (6.4.3), the conditions (6.3.9) give

$$\left.\begin{array}{l}[\Delta(z)]_C = 0, \quad [\delta'(z)]_C = 0 \\ [\delta(z) + \bar\delta(\bar z)]_C = (2-k)[a'b' \log z + \bar a'\bar b' \log \bar z]_C \\ = 2\pi i(2-k)(a'b' - \bar a'\bar b') \quad (C \to C_\infty)\end{array}\right\}. \quad (6.4.7)$$

Since L_1 and L_3 are real, it follows from (6.4.5) and (6.4.7) that for large $|z|$ the complex potentials $\Delta(z)$ and $\delta(z)$ must take the forms

$$\left.\begin{array}{l}\Delta(z) = G'z + g'/z + O(z^{-2}) \\ \delta(z) = \tfrac{1}{2} h' z^2 + H' \log z + O(z^{-1})\end{array}\right\}, \quad (6.4.8)$$

where G' is a real constant in view of the condition (6.4.2) and g' and h' are constants which may be real or complex. Also, from (6.4.7) and

---

† We assume the deformation to be such that these expansions are valid.
‡ For a fuller discussion of these results see, for example, *T.E.*, chap. VI.

(6.4.8) we see that the imaginary part of H' is determined by the condition
$$H'-\overline{H}' = (2-k)(a'b'-\bar{a}'\bar{b}'), \qquad (6.4.9)$$
so that H' is real if $a'b' = \bar{a}'\bar{b}'$, that is, if $a' = \bar{b}'$ or if a' and b' are both real; H' is also real if the constant k in (6.1.24) is chosen to have the value 2 so that $k_1 = 1$. The formulae (6.4.5) now become

$$\left.\begin{array}{l}\dfrac{\partial^1\phi(z,\bar{z})}{\partial\bar{z}} = (2G'+L_1)z+(\bar{h}'+L_2)\bar{z}+O(z^{-1}) \\[2mm] ^1D(z,\bar{z}) = L_3 z-(\bar{h}'-L_4)\bar{z}+O(z^{-1})\end{array}\right\}. \qquad (6.4.10)$$

In (6.4.8) and (6.4.10) terms which do not affect the stresses have been omitted.

If N_1, N_2 are the principal stresses at infinity referred to coordinates in the deformed body, measured per unit area of the deformed body, and N_1 makes an angle ξ with the y_1-axis, we have
$$T^{11} = (N_1-N_2)e^{2i\xi}, \qquad T^{12} = N_1+N_2,$$
or with the help of (3.5.4)
$$4\frac{\partial\phi}{\partial\bar{z}} = (N_1+N_2)z - e^{2i\xi}(N_1-N_2)\bar{z}. \qquad (6.4.11)$$

Also, using (6.4.4) and (6.4.10), the expansion (6.1.9) for ϕ yields
$$\frac{\partial\phi}{\partial\bar{z}} = {}^0\mathscr{H}\epsilon\{(2A'z+\bar{b}'\bar{z})+\epsilon[(2G'+L_1)z+(\bar{h}'+L_2)\bar{z}]+\ldots\}, \qquad (6.4.12)$$
the terms neglected in the coefficient of each power of ϵ being $O(z^{-1})$. Choosing ϵ so that
$$4\,{}^0\mathscr{H}\epsilon = N_1+N_2, \qquad (6.4.13)$$
and comparing (6.4.11) and (6.4.12), we therefore have

$$\left.\begin{array}{l}A' = \tfrac{1}{2}, \quad b' = -Qe^{-2i\xi} \quad \left(Q = \dfrac{N_1-N_2}{N_1+N_2}\right) \\[2mm] G' = -\tfrac{1}{2}L_1, \quad h' = -\bar{L}_2\end{array}\right\}. \qquad (6.4.14)$$

The values of L_1 to L_4 are obtained from (6.4.6) by inserting the appropriate expressions for A' and b'.

A corresponding analysis, based on (6.2.7), (6.2.8), may be carried out in terms of the complex coordinates ζ, $\bar{\zeta}$ in the undeformed body. We again assume that there is a uniform stress at infinity, per unit area of the deformed body, represented by principal stresses N_1, N_2 with N_1 at an angle ξ to the y_1-axis. Then equation (6.4.11) still holds for large $|z|$ (or large $|\zeta|$) and we find that

$$\left.\begin{aligned}\Omega(\zeta) &= A'\zeta + O(\zeta^{-1}) \\ \omega(\zeta) &= \tfrac{1}{2}b'\zeta^2 + B'\log\zeta + O(\zeta^{-1}) \\ \Delta(\zeta) &= G'\zeta + O(\zeta^{-1}) \\ \delta(\zeta) &= \tfrac{1}{2}h'\zeta^2 + H'\log\zeta + O(\zeta^{-1})\end{aligned}\right\}, \quad (6.4.15)$$

where A', b', G', h', and H' are still given by (6.4.14), (6.4.6), and (6.4.9). It may also be inferred directly from (6.4.11) and (6.4.12) that the formulae (6.4.14) continue to apply. For if we write $z = \zeta + \epsilon\,{}^0\!D + \dots$ in the two expressions and expand the right-hand members by Taylor's theorem, coefficients of corresponding powers of ϵ in the resulting expressions are changed by equivalent amounts.

6.5. General theory for two-dimensional problems

Successive approximation formulae corresponding to those of §§ 6.1–6.4 may be developed for compressible materials in plane strain using (3.4.20) and (3.5.17) and for thin sheets of compressible or incompressible material in plane stress using the corresponding complex variable forms of the equations of §§ 4.1–4.4. The resulting formulae for ${}^0\!\phi$, ${}^0\!D$, $\partial^1\!\phi/\partial\bar{z}$, and ${}^1\!D$ in terms of the complex potential functions are similar in structure to those for incompressible materials in plane strain and are summarized here for convenience. For details of the calculations the reader is referred to the original papers.†

To examine the second-order theory for compressible materials we require the elastic constants

$$c_1 = 1 + \frac{A}{B}, \quad c_2 = \frac{C}{B}, \quad c_3 = \frac{F}{B}, \quad c_4 = \frac{H}{B}, \quad (6.5.1)$$

where A, B, F, and H are defined by (5.3.9), c_1, A, and B being related to the usual constants of classical elasticity by means of (5.3.13). The constant ${}^0\!\mathscr{H}$ occurring in (6.1.9) is chosen to have the values

$$\left.\begin{aligned}{}^0\!\mathscr{H} &= \mu \quad \text{(plane strain)} \\ {}^0\!\mathscr{H} &= 2h_0\mu \quad \text{(plane stress)}\end{aligned}\right\}. \quad (6.5.2)$$

In addition, it is convenient to define the constants

$$\left.\begin{aligned}\kappa &= (2c_1+3q)/(2c_1+q) \\ B_1 &= [6c_1+7q-4c_2-4(q+1)c_3]/(2c_1+q) \\ B_2 &= [(2c_1+q)\{4(c_1+q)^2-3q^2\}+12q(3q+4)c_2- \\ &\qquad -12q^2(q+1)c_3-8c_4]/(2c_1+q)^3\end{aligned}\right\}, \quad (6.5.3)$$

† Loc. cit., p. 200. Some minor modifications of notation have been made here.

where
$$q = -1 \quad \text{(plane strain)}$$
$$q = c_1 - 1 \quad \text{(plane stress)}$$
(6.5.4)

Also, we write
$$\begin{aligned} B_1' &= B_1-(\kappa+1), & B_1'' &= B_1+\tfrac{1}{2}B_1' \\ B_2' &= B_2-\tfrac{1}{2}(\kappa+1)^2, & B_3 &= B_1-2B_2/(\kappa+1) \\ B_4 &= \tfrac{1}{2}B_1'-B_2', & \gamma &= B_1/(\kappa+1) \end{aligned}$$
(6.5.5)

The corresponding values of these quantities for incompressible materials are obtained by introducing into (6.5.3) the limiting conditions

$$c_1 \to \infty, \quad c_2 \to \infty, \quad \frac{c_1}{c_2} \to 1, \quad \frac{c_4}{c_1^3} \to 0, \quad c_3 = \frac{C_1+2C_2}{C_1+C_2},$$
(6.5.6)

obtained from § 5.4, where C_1 and C_2 are the Mooney constants defined by (1.14.3). The constant c_4 then vanishes from the expressions for B_1, B_2.

In the series (6.1.5), (6.1.9) for D and ϕ the first- and second-order terms are given by

$$\begin{aligned} {}^0\phi(z,\bar{z}) &= \bar{z}\Omega(z)+z\bar{\Omega}(\bar{z})+\omega(z)+\bar{\omega}(\bar{z}) \\ {}^0D(z,\bar{z}) &= \kappa\Omega(z)-z\bar{\Omega}'(\bar{z})-\bar{\omega}'(\bar{z}) \end{aligned}$$
(6.5.7)

$$\begin{aligned} \frac{\partial {}^1\phi(z,\bar{z})}{\partial \bar{z}} &= \Delta(z)+z\bar{\Delta}'(\bar{z})+\delta'(\bar{z})+\gamma\Gamma(z,\bar{z})+ \\ &\quad +(B_3/\kappa)\bar{\Omega}'(\bar{z})\,{}^0D(z,\bar{z})+k_1\int^z \bar{\Omega}'(\bar{z})\bar{\omega}''(\bar{z})\,d\bar{z}+ \\ &\quad +k_2\int^z [\Omega'(z)]^2\,dz+k_3z[\bar{\Omega}'(\bar{z})]^2 \\ {}^1D(z,\bar{z}) &= \kappa\Delta(z)-z\bar{\Delta}'(\bar{z})-\delta'(\bar{z})-\gamma\Lambda(z,\bar{z})- \\ &\quad -(B_3/\kappa)\bar{\Omega}'(\bar{z})\,{}^0D(z,\bar{z})+k_1'\int^z \bar{\Omega}'(\bar{z})\bar{\omega}''(\bar{z})\,d\bar{z}+ \\ &\quad +k_2'\int^z [\Omega'(z)]^2\,dz+k_3'\,z[\bar{\Omega}'(\bar{z})]^2 \end{aligned}$$
(6.5.8)

where
$$\begin{aligned} \Gamma(z,\bar{z}) &= -\left[{}^0D\frac{\partial}{\partial z}+{}^0\bar{D}\frac{\partial}{\partial \bar{z}}\right]\frac{\partial {}^0\phi}{\partial z} \\ &= [z\bar{\Omega}''(\bar{z})+\bar{\omega}''(\bar{z})][\bar{z}\Omega'(z)+\omega'(z)-\kappa\bar{\Omega}(\bar{z})]+ \\ &\quad +[\Omega'(z)+\bar{\Omega}'(\bar{z})][z\Omega'(\bar{z})+\bar{\omega}'(\bar{z})-\kappa\Omega(z)] \\ \Lambda(z,\bar{z}) &= \left[{}^0D\frac{\partial}{\partial z}+{}^0\bar{D}\frac{\partial}{\partial \bar{z}}\right]{}^0D \\ &= [z\bar{\Omega}''(\bar{z})+\bar{\omega}''(\bar{z})][\bar{z}\Omega'(z)+\omega'(z)-\kappa\bar{\Omega}(\bar{z})]- \\ &\quad -[\kappa\Omega'(z)-\bar{\Omega}'(\bar{z})][z\bar{\Omega}'(\bar{z})+\bar{\omega}'(\bar{z})-\kappa\Omega(z)] \end{aligned}$$
(6.5.9)

and the constants k_r and k'_r may be chosen to have any real values which satisfy the equations

$$\left.\begin{array}{ll} k_1+k'_1 = B'_1, & \kappa k_2-k'_2 = B_4 \\ k_3-k_2 = B_3/\kappa-B_1, & \kappa k'_3+k'_2 = \kappa B''_1-B_3-B_4 \end{array}\right\}. \quad (6.5.10)$$

The integral terms may be excluded from the expression (6.5.8) for $\partial^1\phi/\partial\bar{z}$ by putting $k_1 = k_2 = 0$, the values of k'_1, k'_2, k_3, and k'_3 following from (6.5.10). Alternatively, by putting $k'_1 = k'_2 = 0$ and obtaining k_1, k_2, k_3, and k'_3 from (6.5.10) the integral terms may be excluded from the expression for ^1D.

For single-valued stress and displacement components it follows from (6.5.7) and (6.5.8) that

$$[\Omega'(z)]_C = 0, \quad [\omega''(z)]_C = 0, \quad [\kappa\Omega(z)-\bar{\omega}'(\bar{z})]_C = 0. \quad (6.5.11)$$

$$\left.\begin{array}{c} [\Delta'(z)]_C = 0, \quad [\delta''(z)]_C = 0 \\ [\kappa\Delta(z)-\delta'(\bar{z})]_C = -\left[k'_1\int^{\bar{z}}\bar{\Omega}'(\bar{z})\bar{\omega}''(\bar{z})\,d\bar{z}+k'_2\int^z\{\Omega'(z)\}^2\,dz\right]_C \end{array}\right\}, \quad (6.5.12)$$

C again being a closed contour lying entirely within the deformed body.

From (6.5.7), (6.5.8) we may, as before, deduce expressions for the force P and the couple M across a curve C in the deformed body. In the expansions (6.3.1) and (6.3.2) we have

$$\left.\begin{array}{l} ^0P = 2i[\Omega(z)+\bar{\omega}'(\bar{z})]_C = 2i(\kappa+1)[\Omega(z)]_C \\ ^0M = [z\omega'(z)+\bar{z}\bar{\omega}'(\bar{z})-\omega(z)-\bar{\omega}(\bar{z})]_C \end{array}\right\}, \quad (6.5.13)$$

$$^1P = 2i\left[\Delta(z)+\bar{\delta}'(\bar{z})+k_1\int^{\bar{z}}\bar{\Omega}'(\bar{z})\bar{\omega}''(\bar{z})\,d\bar{z}+k_2\int^z\{\Omega'(z)\}^2\,dz\right]_C$$

$$= 2i\left[(\kappa+1)\Delta(z)+B'_1\int^{\bar{z}}\bar{\Omega}'(\bar{z})\bar{\omega}''(\bar{z})\,d\bar{z}+(k_2+k'_2)\int^z\{\Omega'(z)\}^2\,dz\right]_C, \quad (6.5.14)$$

$$^1M = [z\delta'(z)+\bar{z}\bar{\delta}'(\bar{z})-\delta(z)-\bar{\delta}(\bar{z})]_C +$$

$$+k_1\left[\int^z z\Omega'(z)\omega''(z)\,dz+\int^{\bar{z}}\bar{z}\bar{\Omega}'(\bar{z})\bar{\omega}''(\bar{z})\,d\bar{z}\right]_C +$$

$$+\frac{B_3-\kappa B_1}{\kappa}\left[\int^z \Omega'(z)\omega'(z)\,dz+\int^{\bar{z}}\bar{\Omega}'(\bar{z})\bar{\omega}'(\bar{z})\,d\bar{z}-\kappa\Omega(z)\bar{\Omega}(\bar{z})\right]_C, \quad (6.5.15)$$

corresponding to (6.3.3), (6.3.5), and (6.3.6) respectively, if we make use of (6.5.11) and (6.5.12).

If C is a closed curve on which the resultant force and couple are zero we have, from (6.5.13),

$$[\Omega(z)]_C = 0, \quad [\omega'(z)]_C = 0, \quad [\omega(z)+\bar{\omega}(\bar{z})]_C = 0, \quad (6.5.16)$$

while these conditions, with (6.5.12), (6.5.14), and (6.5.15), yield

$$\begin{aligned}
[\Delta'(z)]_C &= 0, \quad [\delta''(z)]_C = 0 \\
(\kappa+1)[\Delta(z)]_C &= -\left[B_1' \int^z \bar{\Omega}'(\bar{z})\bar{\omega}''(\bar{z})\, d\bar{z} + (k_2+k_2') \int^z \{\Omega'(z)\}^2\, dz\right]_C \\
(\kappa+1)[\bar{\delta}'(\bar{z})]_C &= -\left[(\kappa k_1 - k_1') \int^{\bar{z}} \bar{\Omega}'(\bar{z})\bar{\omega}''(\bar{z})\, d\bar{z} + \right. \\
&\qquad \left. + (\kappa k_2 - k_2') \int^z \{\Omega'(z)\}^2\, dz\right]_C \\
[\delta(z)+\bar{\delta}(\bar{z})]_C &= [z\delta'(z) + \bar{z}\bar{\delta}'(\bar{z})]_C + \\
&\quad + k_1 \left[\int^z z\Omega'(z)\omega''(z)\, dz + \int^{\bar{z}} \bar{z}\bar{\Omega}'(\bar{z})\bar{\omega}''(\bar{z})\, d\bar{z}\right]_C + \\
&\quad + \frac{B_3 - \kappa B_1}{\kappa}\left[\int^z \Omega'(z)\omega'(z)\, dz + \int^{\bar{z}} \bar{\Omega}'(\bar{z})\bar{\omega}'(\bar{z})\, d\bar{z}\right]_C
\end{aligned} \quad .(6.5.17)$$

For complex coordinates ζ, $\bar{\zeta}$ in the undeformed body, the foregoing formulae are modified by the replacement of z, $\bar{z}$ by ζ, $\bar{\zeta}$ and γ by $\gamma' = \gamma - 1$ throughout. This result may be proved along the lines indicated for incompressible plane strain in §§ 6.2 and 6.3.

Forms of the complex potentials in an infinite body may be examined by using the procedure of § 6.4. We again assume that the elastic body contains a single closed contour surrounding the origin and that there is a uniform state of stress and strain, with zero rotation, at infinity. For complex coordinates in the deformed body, the complex potentials continue to take the forms

$$\left.\begin{aligned}\Omega(z) &= A'z + a'/z + O(z^{-2}) \\ \omega(z) &= \tfrac{1}{2}b'z^2 + B'\log z + O(z^{-1})\end{aligned}\right\}, \qquad (6.5.18)$$

$$\left.\begin{aligned}\Delta(z) &= G'z + g'/z + O(z^{-2}) \\ \delta(z) &= \tfrac{1}{2}h'z^2 + H'\log z + O(z^{-1})\end{aligned}\right\}, \qquad (6.5.19)$$

for large $|z|$, where A', B', and G' are real constants, a', b', g', and h' are constants which may be real or complex, and the imaginary part of the constant H' now satisfies the condition

$$H' - \bar{H}' = (\kappa B_1 - \kappa k_1 - B_3)(a'\bar{b}' - \bar{a}'b')/\kappa; \qquad (6.5.20)$$

H' is therefore real if a' and b' are both real, if $a' = \bar{b}'$, or if the constant k_1 is chosen to have the value $B_1 - B_3/\kappa$. From (6.5.7)–(6.5.9) we then obtain for the stress and displacement functions the expressions

$$\frac{\partial {}^0\phi(z,\bar{z})}{\partial \bar{z}} = 2A'z+\bar{b}'\bar{z}+O(z^{-1}) \Big\} \quad (6.5.21)$$
$${}^0D(z,\bar{z}) = (\kappa-1)A'z-\bar{b}'\bar{z}+O(z^{-1}) \Big\}$$

$$\frac{\partial {}^1\phi(z,\bar{z})}{\partial \bar{z}} = (2G'+L_1)z+(\bar{h}'+L_2)\bar{z}+O(z^{-1}) \Big\} \quad (6.5.22)$$
$${}^1D(z,\bar{z}) = [(\kappa-1)G'+L_3]z-(\bar{h}'-L_4)\bar{z}+O(z^{-1}) \Big\}$$

for large $|z|$, corresponding to (6.4.4) and (6.4.10) respectively. The constants L_1 to L_4 are now, however, given by

$$\begin{aligned} L_1 &= b'\bar{b}'\gamma+[k_2+k_3-2(\kappa-1)\gamma+(\kappa-1)B_3/\kappa]A'^2 \\ L_2 &= [k_1-(\kappa-3)\gamma-B_3/\kappa]A'\bar{b}' \\ L_3 &= -b'\bar{b}'\gamma+[k_2'+k_3'-(\kappa-1)^2\gamma-(\kappa-1)B_3/\kappa]A'^2 \\ L_4 &= [k_1'+2(\kappa-1)\gamma+B_3/\kappa]A'\bar{b}' \end{aligned} \quad (6.5.23)$$

When the stress at infinity, per unit area of the deformed body, is uniform and represented by principal stresses N_1, N_2 with N_1 at an angle ξ to the y_1-axis, the constants A', b', G', h' are given by (6.4.14) provided the appropriate values (6.5.23) are employed for L_1 to L_4. When the coordinates ζ, $\bar{\zeta}$ are used, equations (6.5.18) and (6.5.19) continue to apply with z, $\bar{z}$ replaced by ζ, $\bar{\zeta}$ respectively, and the constants A', b', G', h' are still given by (6.4.14) and (6.5.23), with γ unaltered.

6.6. Simultaneous solution of first and second boundary value problems

From (6.5.7) we observe that the expressions for $\partial {}^0\phi/\partial \bar{z}$ and ${}^0D$ in terms of the complex potential functions $\Omega(z)$ and $\omega(z)$ are similar in structure. When this feature is taken into account, it is seen that this similarity of structure extends to the second-order functions $\partial {}^1\phi/\partial \bar{z}$, ${}^1D$. This suggests that corresponding first and second boundary value problems may be solved simultaneously.

We confine attention to the case where boundary conditions are to be satisfied over a single closed curve C in the deformed body,† and

---

† This analysis also applies for an infinite plate in which stress or displacement conditions are specified along a simple internal boundary C and *stress and rotation* conditions are given over the outer boundary (i.e. at infinity). In this case the complex potentials are chosen to satisfy given conditions at infinity, such as those derived in §§ 6.4 and 6.5; these potentials must also be such that the functions ${}^0\chi$, ${}^1\chi$ satisfy the boundary conditions on C as discussed in the present section.

for the first-order problem define the function

$$^0\chi(z, \bar{z}, \alpha) = \alpha\Omega(z) + z\bar{\Omega}'(\bar{z}) + \bar{\omega}'(\bar{z}). \quad (6.6.1)$$

This function can be made to coincide with $\partial\,^0\phi/\partial\bar{z}$ and $-^0D$ by choosing the values $1, -\kappa$ respectively for the parameter α. Thus

$$^0\chi(z, \bar{z}, 1) = \frac{\partial\,^0\phi(z, \bar{z})}{\partial\bar{z}}, \qquad ^0\chi(z, \bar{z}, -\kappa) = -^0D(z, \bar{z}). \quad (6.6.2)$$

The problem in which $^0\chi(z, \bar{z}, \alpha)$ takes a prescribed set of values along the boundary curve C therefore includes as special cases, two problems in which stresses and displacements are prescribed along this curve.

To examine the second-order problem we construct the function

$$^1\chi(z, \bar{z}, \alpha, l_r) = \alpha\Delta(z) + z\bar{\Delta}'(\bar{z}) + \bar{\delta}'(\bar{z}) + F(z, \bar{z}, \alpha, \alpha, l_r), \quad (6.6.3)$$

where

$$\left.\begin{aligned}
F(z, \bar{z}, \alpha, \beta, l_r) &= \gamma Q(z, \bar{z}, \alpha, \beta) + l_1 \int^{\bar{z}} \bar{\Omega}'(\bar{z}, \alpha)\bar{\omega}''(\bar{z}, \alpha)\, d\bar{z} + \\
&\quad + l_2 \int^{z} [\Omega'(z, \alpha)]^2\, dz + l_3 z[\bar{\Omega}'(\bar{z}, \alpha)]^2 \\
Q(z, \bar{z}, \alpha, \beta) &= [z\bar{\Omega}''(\bar{z}, \alpha) + \bar{\omega}''(\bar{z}, \alpha)][\bar{z}\Omega'(z, \alpha) + \omega'(z, \alpha) - \kappa\bar{\Omega}(\bar{z}, \alpha)] + \\
&\quad + [\beta\Omega'(z, \alpha) + \nu\bar{\Omega}'(\bar{z}, \alpha)][z\bar{\Omega}'(\bar{z}, \alpha) + \bar{\omega}'(\bar{z}, \alpha) - \kappa\Omega(z, \alpha)]
\end{aligned}\right\}$$

$$(6.6.4)$$

$$\nu = 1 - B_3/(\gamma\kappa), \quad (6.6.5)$$

and the functions $\Omega(z, \alpha)$, $\omega(z, \alpha)$ represent the solution of the first-order problem, involving α, which is obtained by choosing the complex potential functions $\Omega(z)$, $\omega(z)$ so that $^0\chi(z, \bar{z}, \alpha)$ takes the prescribed set of values along the boundary curve C. From a comparison of (6.6.3) with (6.5.8) we see that the function $^1\chi(z, \bar{z}, \alpha, l_r)$ may be made to represent either $\partial^1\phi/\partial\bar{z}$ or $-^1D(z, \bar{z})$ by a suitable choice for the four parameters α, l_1, l_2, l_3. Thus

$$\left.\begin{aligned}
^1\chi(z, \bar{z}, 1, k_r) &= \frac{\partial\,^1\phi(z, \bar{z})}{\partial\bar{z}} \\
^1\chi(z, \bar{z}, -\kappa, -k'_r) &= -^1D(z, \bar{z})
\end{aligned}\right\}. \quad (6.6.6)$$

For a given set of boundary conditions prescribed along the curve C the function $^1\chi(z, \bar{z}, \alpha, l_r)$ gives rise to the four-parameter solution represented by the potential functions $\Delta(z, \alpha, l_r)$, $\delta(z, \alpha, l_r)$. This yields, as special cases, solutions to corresponding stress and displacement boundary value problems.

From these results it follows that the functions

$$\Omega(z, 1), \quad \omega(z, 1), \quad \Delta(z, 1, k_r), \quad \delta(z, 1, k_r), \qquad (6.6.7)$$

obtained by giving to the parameters α, l_r the values 1, k_r respectively, define a solution in which, as far as terms of the first and second orders are concerned, there is a given distribution of applied stress along C. Similarly, the functions

$$\Omega(z, -\kappa), \quad \omega(z, -\kappa), \quad \Delta(z, -\kappa, -k'_r), \quad \delta(z, -\kappa, -k'_r), \qquad (6.6.8)$$

derived by giving the parameters α, l_r the values $-\kappa$, $-k'_r$ respectively, define a solution in which a corresponding system of displacements is prescribed along this boundary. If along a single closed curve C, ${}^0\chi(z, \bar{z}, \alpha)$ and ${}^1\chi(z, \bar{z}, \alpha, l_r)$ are both zero, then it follows from (6.6.2) and (6.6.6) that (6.6.7) and (6.6.8) define solutions for a stress-free boundary and a fixed boundary respectively.

When the complex potential functions have been determined by the foregoing method, a similar procedure may be employed to evaluate simultaneously the stress and displacement components. For if we define, by analogy with (6.6.1), the two-parameter function

$$ {}^0\chi(z, \bar{z}, \alpha, \beta) = \beta\Omega(z, \alpha) + z\bar{\Omega}'(\bar{z}, \alpha) + \bar{\omega}'(\bar{z}, \alpha), \qquad (6.6.9)$$

and by analogy with (6.6.3) the eight-parameter function

$$ {}^1\chi(z, \bar{z}, \alpha, l_r, \beta, m_r) $$
$$ = \beta\Delta(z, \alpha, l_r) + z\bar{\Delta}'(\bar{z}, \alpha, l_r) + \bar{\delta}'(\bar{z}, \alpha, l_r) + F(z, \bar{z}, \alpha, \beta, m_r) \qquad (6.6.10)$$

where $F(z, \bar{z}, \alpha, \beta, m_r)$ is given by (6.6.4), we obtain by comparison with (6.5.7)–(6.5.9)

$$\left. \begin{array}{l} {}^0\chi(z, \bar{z}, \alpha, 1) = \dfrac{\partial\,{}^0\phi(z, \bar{z}, \alpha)}{\partial \bar{z}}, \qquad {}^1\chi(z, \bar{z}, \alpha, l_r, 1, k_r) \\[6pt] \qquad\qquad\qquad\qquad\qquad\qquad = \dfrac{\partial\,{}^1\phi(z, \bar{z}, \alpha, l_r)}{\partial \bar{z}} \\[6pt] {}^0\chi(z, \bar{z}, \alpha, -\kappa) = -{}^0D(z, \bar{z}, \alpha), \qquad {}^1\chi(z, \bar{z}, \alpha, l_r, -\kappa, -k'_r) \\[6pt] \qquad\qquad\qquad\qquad\qquad\qquad = -{}^1D(z, \bar{z}, \alpha, l_r) \end{array} \right\} . \qquad (6.6.11)$$

In principle it is possible to extend the method to problems involving several boundaries.† In this case a separate pair of functions

$$ {}^0\chi(z, \bar{z}, \alpha^{(i)}), \qquad {}^1\chi(z, \bar{z}, \alpha^{(i)}, l_r^{(i)}) $$

---

† For examples of this method in classical elasticity, reference may be made to J. E. Adkins, *J. Math. Phys. Solids* **4** (1956), 199.

is assigned to each of the n boundaries C_i, these functions being described by expressions of the type (6.6.1), (6.6.3). As before, by putting $\alpha^{(i)} = 1$, $l_r^{(i)} = k_r$ a boundary-stress condition may be satisfied along C_i; the choice $\alpha^{(i)} = -\kappa$, $l_r^{(i)} = -k_r'$ yields the solution for the corresponding boundary-displacement condition. The first-order complex potentials derived by this method are the n-parameter functions $\Omega(z, \alpha^{(i)})$, $\omega(z, \alpha^{(i)})$, while those of the second order are the $4n$-parameter functions $\Delta(z, \alpha^{(i)}, l_r^{(i)})$, $\delta(z, \alpha^{(i)}, l_r^{(i)})$. When these are introduced into formulae of the type (6.6.9), (6.6.10), the expression for ${}^0\chi$ involves the $(n+1)$ parameters β, $\alpha^{(i)}$ and the expression for ${}^1\chi$ contains the $(4n+4)$ parameters β, m_r, $\alpha^{(i)}$, $l_r^{(i)}$.

6.7. Infinite region containing a circular boundary

The effect of a prescribed system of forces or displacements along an internal boundary in an infinite elastic region may be examined by making use of the Hilbert problem.† In this section the general method is outlined for a circular boundary; applications to particular problems will be considered in §§ 6.8, 6.9. We use $z, \bar{z}$ as complex coordinates, but it is to be understood that the formulation of the present section applies equally well for complex coordinates in the deformed body or in the undeformed body.

Let S^+ be an infinite region, bounded internally by a smooth contour C which encloses the origin of coordinates in the z-plane and which does not intersect itself. The region enclosed by C is denoted by S^-. Let $\phi(t)$ be a complex function of position which satisfied the Hölder condition‡ for all points t on C and let $\Phi(z)$ be a function which is holomorphic in each of the regions S^+, S^- and which tends to the limits $\Phi^+(t)$, $\Phi^-(t)$ when z approaches t along any paths lying entirely in S^+ and S^- respectively. Then if $\Phi(z)$ satisfies the boundary condition

$$\Phi^+(t) - \Phi^-(t) = \phi(t) \tag{6.7.1}$$

for all points t on C and vanishes at infinity, we may write

$$\Phi(z) = \frac{1}{2\pi i} \int_C \frac{\phi(t)}{t-z} dt, \tag{6.7.2}$$

† For applications of this method to problems in classical elasticity see *T.E.* or N. I. Muskhelishvili, *Some Basic Problems of the Mathematical Theory of Elasticity* (edn. 2, Moscow, 1946), translated by J. R. M. Radok (Groningen-Holland: P. Noordhoff Ltd., 1953).
‡ See *T.E.*, § 1.14.

where the contour integral is taken in a *clockwise* direction round C.

We consider the functions
$$\begin{aligned}{}^{0}\chi(z,\bar{z},\alpha) &= \alpha\Omega(z)+z\bar{\Omega}'(\bar{z})+\bar{\omega}'(\bar{z}) \\ {}^{1}\chi(z,\bar{z},\alpha) &= \alpha\Delta(z)+z\bar{\Delta}'(\bar{z})+\bar{\delta}'(\bar{z})+F(z,\bar{z},\alpha)\end{aligned}. \tag{6.7.3}$$

By an appropriate choice for α as indicated in § 6.6, ${}^{0}\chi(z,\bar{z},\alpha)$ may be made to reduce to any of the expressions for $\partial{}^{0}\phi/\partial\bar{z}$, $-{}^{0}D$ given in the present chapter; for this value of α, ${}^{1}\chi(z,\bar{z},\alpha)$ may be made identical with the corresponding second-order function $\partial{}^{1}\phi/\partial\bar{z}$ or $-{}^{1}D$ when $F(z,\bar{z},\alpha)$ is suitably determined. For brevity $F(z,\bar{z},\alpha)$ denotes $F(z,\bar{z},\alpha,\alpha)$ and this latter function will, in general, contain further constants, such as the quantities k, l of § 6.1, k_r, k'_r of § 6.5, or the parameters l_r of § 6.6, but these do not affect the present analysis and are not exhibited explicitly.

The curve C is chosen to coincide with the internal boundary of the elastic body, which we here assume to be the circle $|z| = a$. On this boundary we suppose the functions ${}^{0}\chi$, ${}^{1}\chi$ to be known functions of position. Thus, putting $z = t$, $\bar{z} = a^2/t$, we have

$$\begin{aligned}{}^{0}\chi\!\left(t,\frac{a^2}{t},\alpha\right) &= {}^{0}\chi(t) \\ {}^{1}\chi\!\left(t,\frac{a^2}{t},\alpha\right) &= {}^{1}\chi(t)\end{aligned}, \tag{6.7.4}$$

where ${}^{0}\chi(t)$, ${}^{1}\chi(t)$ are known complex functions of t. It is often possible to formulate any given problem so that ${}^{1}\chi(t) = 0$, but this assumption is not essential for the development of the theory. The determination of the first- and second-order complex potential functions from the boundary conditions (6.7.4) follows exactly similar lines, the only difference arising from the fact that the additional function $F(z,\bar{z},\alpha)$ enters into the expression for ${}^{1}\chi(z,\bar{z},\alpha)$. We shall therefore suppose that the first-order functions $\Omega(z)$, $\omega(z)$ have been determined from the first of the conditions (6.7.4) and proceed to the determination of the second-order quantities.

We suppose that the complex potential functions $\Delta(z)$, $\delta(z)$ can be written in the form

$$\Delta(z) = \Delta_0(z)+\Delta_1(z), \qquad \delta(z) = \delta_0(z)+\delta_1(z). \tag{6.7.5}$$

In these expressions $\Delta_0(z)$, $\delta_0(z)$ are given everywhere and are such that when $|z|$ is small

$$\Delta'_0(z) = c+O(z), \qquad \delta'_0(z) = c'+O(z), \tag{6.7.6}$$

where c, c' are complex constants; the potentials $\Delta_0(z)$, $\delta_0(z)$ define second-order stress and displacement components which have no singularities in the region S^-, that is, inside the circle $|z| = a$, and have the prescribed forms for large $|z|$. For a uniform distribution of stress at infinity $\Delta(z)$, $\delta(z)$ must take the forms (6.5.19) and $\Delta_0'(z)$, $\delta_0''(z)$ then reduce to constants. The complex potentials $\Delta_1(z)$, $\delta_1(z)$ are chosen so that $\Delta(z)$, $\delta(z)$ give a system of stresses and displacements which satisfy the boundary conditions at $|z| = a$ and have the required forms for large $|z|$. At infinity, therefore, $\Delta_1(z)$ and $\delta_1'(z)$ give rise to zero rotation and stresses, and since the corresponding stresses over $|z| = a$ are self-equilibrating, they are both $O(z^{-1})$ for large $|z|$.

The functions $\Delta_1(z)$, $\delta_1(z)$, and hence also $\Delta(z)$, $\delta(z)$ are defined only for the region S^+ outside the circle $|z| = a$. To apply (6.7.2) we extend the definition of these quantities into the region S^- ($|z| < a$) by means of the relation

$$\alpha\Delta_1(z)+z\bar{\Delta}_1'\!\left(\frac{a^2}{z}\right)+\bar{\delta}_1'\!\left(\frac{a^2}{z}\right) = 0 \qquad (z \text{ in } S^-), \tag{6.7.7}$$

where
$$\bar{\Delta}_1'\!\left(\frac{a^2}{z}\right) = \overline{\Delta_1'\!\left(\frac{a^2}{\bar{z}}\right)}, \qquad \bar{\delta}_1'\!\left(\frac{a^2}{z}\right) = \overline{\delta_1'\!\left(\frac{a^2}{\bar{z}}\right)}.$$

We therefore have

$$\alpha\Delta_1\!\left(\frac{a^2}{\bar{z}}\right)+\frac{a^2}{\bar{z}}\bar{\Delta}_1'(\bar{z})+\bar{\delta}_1'(\bar{z}) = 0 \qquad (z \text{ in } S^+), \tag{6.7.8}$$

and with this relation and (6.7.5), the second of (6.7.3) yields

$$^1\chi(z,\bar{z},\alpha) = \alpha\!\left[\Delta_1(z)-\Delta_1\!\left(\frac{a^2}{\bar{z}}\right)\right]+\left(z-\frac{a^2}{\bar{z}}\right)\bar{\Delta}_1'(\bar{z})+$$
$$+\alpha\Delta_0(z)+z\bar{\Delta}_0'(\bar{z})+\bar{\delta}_0'(\bar{z})+F(z,\bar{z},\alpha). \tag{6.7.9}$$

The second of the boundary conditions (6.7.4) thus gives at $|z| = a$

$$\alpha[\Delta_1^+(t)-\Delta_1^-(t)] = -\alpha\Delta_0(t)-t\bar{\Delta}_0'\!\left(\frac{a^2}{t}\right)-\bar{\delta}_0'\!\left(\frac{a^2}{t}\right)-F\!\left(t,\frac{a^2}{t},\alpha\right)+{}^1\chi(t), \tag{6.7.10}$$

so that, by applying (6.7.2) we obtain

$$\Delta_1(z,\alpha) = -\frac{1}{2\pi i}\int_C \frac{\Delta_0(t)}{t-z}dt - \frac{1}{2\pi i\alpha}\int_C\!\left[t\bar{\Delta}_0'\!\left(\frac{a^2}{t}\right)+\bar{\delta}_0'\!\left(\frac{a^2}{t}\right)\right]\!\frac{dt}{t-z} -$$
$$-\frac{1}{2\pi i\alpha}\int_C\!\left[F\!\left(t,\frac{a^2}{t},\alpha\right)-{}^1\chi(t)\right]\!\frac{dt}{t-z}. \tag{6.7.11}$$

This relation, with (6.7.5)–(6.7.7), yields expressions for $\Delta(z)$ and $\delta(z)$ and when these have been found the stresses and displacements may be determined from the formulae of §§ 6.1 and 6.5.

With the assumptions of the present section, the first and second contour integrals in (6.7.11) can be evaluated explicitly.† If in addition we can decompose the functions F and $^1\chi$ so that

$$\left. \begin{aligned} F\!\left(t, \frac{a^2}{t}, \alpha\right) &= F_1(t, \alpha) + F_2(t, \alpha) \\ {}^1\chi(t) &= \chi_1(t) + \chi_2(t) \end{aligned} \right\}, \qquad (6.7.12)$$

where $F_1(z, \alpha)$ and $\chi_1(z)$ are holomorphic in the region S^+ and for large $|z|$ are $O(z^{-1})$, and $F_2(z, \alpha)$, $\chi_2(z)$ are holomorphic in S^-, an expression for the third integral can also be written down. Remembering (6.7.12), the general expression (6.7.11) for $\Delta_1(z, \alpha)$ then becomes

$$\left. \begin{aligned} \Delta_1(z, \alpha) &= -\frac{1}{\alpha}\left[z\bar{\Delta}_0'\!\left(\frac{a^2}{z}\right) + \bar{\delta}_0'\!\left(\frac{a^2}{z}\right) - \chi_1(z) + F_1(z, \alpha) - \bar{c}z\right] \quad (z \text{ in } S^+) \\ \Delta_1(z, \alpha) &= \Delta_0(z) - \frac{1}{\alpha}[\chi_2(z) - F_2(z, \alpha) - \bar{c}z] \quad (z \text{ in } S^-) \end{aligned} \right\}, \quad (6.7.13)$$

apart from non-essential constants, and hence, using (6.7.5) and (6.7.7), we have, for z in S^+,

$$\left. \begin{aligned} \Delta(z, \alpha) &= \Delta_0(z) - \frac{1}{\alpha}\!\left[z\bar{\Delta}_0'\!\left(\frac{a^2}{z}\right) + \bar{\delta}_0'\!\left(\frac{a^2}{z}\right) - \chi_1(z) + F_1(z, \alpha) - \bar{c}z\right] \\ \delta'(z, \alpha) &= \delta_0'(z) - \alpha\bar{\Delta}_0\!\left(\frac{a^2}{z}\right) + \frac{a^2}{\alpha z}\!\left[\bar{\Delta}_0'\!\left(\frac{a^2}{z}\right) - \frac{a^2}{z}\bar{\Delta}_0''\!\left(\frac{a^2}{z}\right) - \frac{a^2}{z^2}\bar{\delta}_0''\!\left(\frac{a^2}{z}\right)\right] - \\ &\quad -\frac{a^2}{\alpha z}[\chi_1'(z) - F_1'(z, \alpha)] + \bar{\chi}_2\!\left(\frac{a^2}{z}\right) - \bar{F}_2\!\left(\frac{a^2}{z}, \alpha\right) - (\bar{c} + \alpha c)\frac{a^2}{\alpha z} \end{aligned} \right\}. \quad (6.7.14)$$

6.8. Incompressible plane strain: infinite body containing a circular hole

To illustrate the foregoing theory we consider the deformation of an infinite body containing a single circular boundary. For simplicity we confine attention to problems in the plane strain of incompressible materials; the corresponding solutions for the compressible case and for plane stress have a similar structure but the constants entering into these solutions have more complicated forms.

† *T.E.*, § 8.19.

6.8 TWO-DIMENSIONAL PROBLEMS

We examine first the problem in which an infinite body containing a stress-free circular hole of radius a is subjected to a uniform distribution of stress and strain at infinity. The complex potentials giving rise to this distribution of stress may be denoted by

$$\left.\begin{array}{ll}\Omega_0(z) = A'z, & \omega_0'(z) = b'z \\ \Delta_0(z) = G'z, & \delta_0'(z) = h'z\end{array}\right\}\quad (6.8.1)$$

where A', b', G', and h' are given by (6.4.14) and (6.4.6).

It is convenient to use the formulae (6.1.22), (6.1.23) with the constants chosen so that the integral terms are absent from the expression for $\partial^1\phi(z,\bar{z})/\partial\bar{z}$. The constants in these expressions, together with the corresponding values of A', b', G', h', therefore have the values

$$\left.\begin{array}{l}k_1 = k_2 = 0,\quad k_3 = -1,\quad k = 1,\quad l = -\tfrac{1}{2} \\ A' = \tfrac{1}{2},\quad b' = -Qe^{-2i\xi},\quad G' = \tfrac{1}{8}-\tfrac{1}{4}b'\bar{b}',\quad h' = -\tfrac{1}{2}b'\end{array}\right\}.\quad (6.8.2)$$

The first and second of equations (6.7.3) then become identical with the value of $\partial^0\phi(z,\bar{z})/\partial\bar{z}$ derived from (6.1.12) and with (6.1.22) respectively if we take

$$\left.\begin{array}{l}\alpha = 1,\quad {}^0\chi(z,\bar{z},\alpha) = \partial^0\phi(z,\bar{z})/\partial\bar{z} \\ {}^1\chi(z,\bar{z},\alpha) = \partial^1\phi(z,\bar{z})/\partial\bar{z} \\ F(z,\bar{z},\alpha) = F(z,\bar{z}) = v\Gamma(z,\bar{z})-z[\bar{\Omega}'(\bar{z})]^2\end{array}\right\}.\quad (6.8.3)$$

By putting $v = \tfrac{1}{2}$ in the last of these relations we obtain the problem for coordinates in the deformed body in which the hole becomes the circular boundary $|z| = a$ after deformation. From (6.2.7), (6.2.8) we see that by putting $v = -\tfrac{1}{2}$ and replacing z, $\bar{z}$ by ζ, $\bar{\zeta}$ we obtain the corresponding problem for coordinates in the undeformed body. In this case the inner boundary is initially the circle $|\zeta| = a$.

Since the hole is free from applied stress we have, from (6.3.2), the boundary conditions

$$\partial^0\phi/\partial\bar{z} = \partial^1\phi/\partial\bar{z} = 0 \quad (|z| = a),$$

so that in the formulae of § 6.7 we may write

$$\left.\begin{array}{l}{}^0\chi\!\left(t,\dfrac{a^2}{t},\alpha\right) = 0,\quad {}^1\chi\!\left(t,\dfrac{a^2}{t},\alpha\right) = 0 \\ \chi_1(t) = \chi_2(t) = 0 \quad (|t| = a)\end{array}\right\}.\quad (6.8.4)$$

The first-order solution follows immediately from (6.7.14) by replacing Δ, δ by Ω, ω respectively and making use of (6.8.1) and the value $c = \bar{c} = A'$ derived from these equations. Remembering that $\alpha = 1$ and that the terms which arise from $F(z, \bar{z})$ are absent from the first-order solution, we obtain from (6.7.14)

$$\Omega(z) = A'z - \bar{b}'\frac{a^2}{z}, \qquad \omega'(z) = b'z - 2A'\frac{a^2}{z} - \bar{b}'\frac{a^4}{z^3}. \qquad (6.8.5)$$

From (6.8.3), (6.1.17), and (6.8.5) we obtain

$$F(z, \bar{z}) = v\left[b'\bar{b}'z + 2A'\bar{b}'\bar{z} + (2b'\bar{b}' - 4A'^2)\frac{a^2}{\bar{z}} + 2\bar{b}'^2\frac{a^2\bar{z}}{z^2} + \right.$$
$$+ 4A'b'\frac{a^2z}{\bar{z}^2} - 2b'^2\frac{a^2z^2}{\bar{z}^3} - 4A'^2\frac{a^4}{z\bar{z}^2} + 2A'b'\frac{a^4}{\bar{z}^3} +$$
$$+ 2\bar{b}'^2\frac{a^4z}{\bar{z}^4} - 2A'\bar{b}'\frac{a^6}{z^3\bar{z}^2} + 4b'\bar{b}'\frac{a^6}{z^2\bar{z}^3} - 6A'b'\frac{a^6}{z\bar{z}^4} +$$
$$\left. + 2b'^2\frac{a^6}{\bar{z}^5} - 3b'\bar{b}'\frac{a^8}{z^3\bar{z}^4}\right] - \left(A'^2z + 2A'b'\frac{a^2z}{\bar{z}^2} + b'^2\frac{a^4z}{\bar{z}^4}\right), \qquad (6.8.6)$$

and with the values $z = t$, $\bar{z} = a^2/t$ at the inner boundary this expression yields

$$\left. \begin{aligned} F_1(t) &= R_3\frac{a^4}{t^3} \\ F_2(t) &= S_1 t + S_3\frac{t^3}{a^2} + S_5\frac{t^5}{a^4} \\ \left[F_1(t) + F_2(t) \right. &= \left. F\left(t, \frac{a^2}{t}\right)\right] \end{aligned} \right\}, \qquad (6.8.7)$$

where
$$\left. \begin{aligned} R_3 &= 2v\bar{b}'^2, \qquad S_1 = \bar{S}_1 = 4v(b'\bar{b}' - 2A'^2) - A'^2 \\ S_3 &= -2A'b', \qquad S_5 = b'^2(2v - 1) \end{aligned} \right\}. \qquad (6.8.8)$$

By introducing (6.8.7) into (6.7.14), with $\alpha = 1$, and making use of (6.8.1), we obtain for the second-order potential functions

$$\left. \begin{aligned} \Delta(z) &= G'z - \bar{h}'\frac{a^2}{z} - R_3\frac{a^4}{z^3} \\ \delta'(z) &= h'z - (2G' + \bar{S}_1)\frac{a^2}{z} - (\bar{h}' + \bar{S}_3)\frac{a^4}{z^3} - (3R_3 + \bar{S}_5)\frac{a^6}{z^5} \end{aligned} \right\}. \qquad (6.8.9)$$

These solutions may be simplified by means of (6.4.6) and (6.4.14). Thus, remembering (6.8.2) and (6.8.8), we obtain from (6.8.5) and

(6.8.9) for coordinates z, $\bar{z}$ in the deformed body

$$\Omega(z) = \tfrac{1}{2}z - \bar{b}'\frac{a^2}{z}, \qquad \omega'(z) = b'z - \frac{a^2}{z} - \bar{b}'\frac{a^4}{z^3}, \tag{6.8.10}$$

$$\left.\begin{aligned}\Delta(z) &= \tfrac{1}{8}(1-2b'\bar{b}')z + \tfrac{1}{2}\bar{b}'\frac{a^2}{z} - \bar{b}'^2\frac{a^4}{z^3} \\ \delta'(z) &= -\tfrac{1}{2}b'z + (1-\tfrac{3}{2}b'\bar{b}')\frac{a^2}{z} + \tfrac{3}{2}\bar{b}'\frac{a^4}{z^3} - 3\bar{b}'^2\frac{a^6}{z^5} \\ (b' &= -Qe^{-2i\xi})\end{aligned}\right\}, \tag{6.8.11}$$

respectively. The first-order solution (6.8.10) for the coordinates ζ, $\bar{\zeta}$ in the undeformed body is unchanged apart from the replacement of z, $\bar{z}$ by ζ, $\bar{\zeta}$ respectively, but (6.8.11) is replaced by

$$\left.\begin{aligned}\Delta(\zeta) &= \tfrac{1}{8}(1-2b'\bar{b}')\zeta + \tfrac{1}{2}\bar{b}'\frac{a^2}{\zeta} + \bar{b}'^2\frac{a^4}{\zeta^3} \\ \delta'(\zeta) &= -\tfrac{1}{2}b'\zeta - (1-\tfrac{5}{2}b'\bar{b}')\frac{a^2}{\zeta} + \tfrac{3}{2}\bar{b}'\frac{a^4}{\zeta^3} + 5\bar{b}'^2\frac{a^6}{\zeta^5}\end{aligned}\right\}. \tag{6.8.12}$$

When the deformation is produced by means of a uniform tension T in the direction of the x_1-axis at infinity we have in (6.8.9)–(6.8.12)

$$N_1 = T, \quad N_2 = 0, \quad \xi = 0, \quad Q = 1, \quad b' = -1, \quad 4\,{}^0\!\mathscr{H}\epsilon = T.$$

Similarly, when the deformation is produced by means of a uniform all-round tension of magnitude T at infinity we have

$$N_1 = N_2 = T, \qquad Q = b' = 0, \qquad 2\,{}^0\!\mathscr{H}\epsilon = T,$$

and the complex potentials (6.8.10)–(6.8.12) reduce to the simple forms

$$\Omega(z) = \tfrac{1}{2}z, \qquad \omega'(z) = -a^2/z, \tag{6.8.10'}$$

$$\Delta(z) = \tfrac{1}{8}z, \qquad \delta'(z) = a^2/z, \tag{6.8.11'}$$

$$\Delta(\zeta) = \tfrac{1}{8}\zeta, \qquad \delta'(\zeta) = -a^2/\zeta, \tag{6.8.12'}$$

respectively.

The stress and displacement functions $\partial\,{}^0\!\phi/\partial\bar{z}$, ${}^0\!D$, $\partial\,{}^1\!\phi/\partial\bar{z}$, ${}^1\!D$ may be evaluated by introducing the foregoing expressions for the potential functions into the relevant formulae of § 6.1, and expressions for D and $\partial\phi/\partial\bar{z}$ follow from (6.1.5) or (6.2.2) and (6.1.9). The stress components may then be determined from (3.5.4) and (3.5.2).†

† These calculations have been carried out for the case when $N_1 = T$, $N_2 = 0$ by J. E. Adkins, A. E. Green, and R. T. Shield, loc. cit., p. 101.

6.9. Incompressible plane strain: infinite body containing a circular rigid inclusion

We consider now the problem in which the undeformed body contains a single circular boundary of radius a which is bonded to a circular rigid inclusion also of radius a. This body is subjected to a uniform distribution of stress and strain at infinity so that equations (6.8.1) again describe the complex potentials for large values of $|z|$. Since the displacement components are zero at the surface of the inclusion we may choose complex coordinates either in the undeformed body or in the deformed body. The latter system is somewhat more convenient for the evaluation of the stress components and will be employed in this section.

For the second-order stress and displacement functions we now use the formulae (6.1.16), (6.1.18) in which the integral terms are absent from the expression for ^1D. The first and second of equations (6.7.3) may then be identified with the second of (6.1.12) and (6.1.18) respectively if we choose

$$\left.\begin{aligned} \alpha = -1, \quad {}^0\chi(z, \bar{z}, \alpha) &= -{}^0D(z, \bar{z}) \\ {}^1\chi(z, \bar{z}, \alpha) &= -{}^1D(z, \bar{z}) \\ F(z, \bar{z}, \alpha) = F(z, \bar{z}) &= \tfrac{1}{2}\Lambda(z, \bar{z}) \end{aligned}\right\}. \tag{6.9.1}$$

The conditions at the internal boundary are

$$^0D(z, \bar{z}) = {}^1D(z, \bar{z}) = 0 \quad (|z| = a), \tag{6.9.2}$$

so that in the formulae of § 6.7 we again have

$$\left.\begin{aligned} {}^0\chi\!\left(t, \frac{a^2}{t}, \alpha\right) = 0, \quad {}^1\chi\!\left(t, \frac{a^2}{t}, \alpha\right) &= 0 \\ \chi_1(t) = \chi_2(t) = 0 \quad (|t| = a) \end{aligned}\right\}. \tag{6.9.3}$$

Furthermore, from (6.9.1), (6.9.2), and (6.1.19) it follows that

$$\left.\begin{aligned} F\!\left(t, \frac{a^2}{t}\right) &= 0 \\ F_1(t, \alpha) = F_1(t) = 0, \quad F_2(t, \alpha) &= F_2(t) = 0 \end{aligned}\right\}. \tag{6.9.4}$$

The complex potential functions follow from (6.8.1) and (6.7.14) with $\alpha = -1$. For the first-order solution we have $c = \bar{c} = A' = \tfrac{1}{2}$ and

$$\Omega(z) = \tfrac{1}{2}z + \bar{b}'\frac{a^2}{z}, \quad \omega'(z) = b'z + \bar{b}'\frac{a^4}{z^3}. \tag{6.9.5}$$

Similarly, for the second-order solution we take $c = \bar{c} = G'$ and obtain

$$\Delta(z) = G'z + h'\frac{a^2}{z} = -\tfrac{1}{4}b'\bar{b}'z, \qquad \delta'(z) = h'z + h'\frac{a^4}{z^3} = 0, \qquad (6.9.6)$$

remembering that $k = l = 0$ in (6.4.6) and (6.4.14).

Expressions for the stress and displacement functions may be found, as before, by combining (6.9.5) and (6.9.6) with (6.1.12), (6.1.16)–(6.1.19), (6.1.5), and (6.1.9). The stress components may then be derived by making use of (3.5.4) and (3.5.2).

Corrigenda. Paper by J. E. Adkins and A. E. Green, loc. cit., p. 200. Add to end of § 4, p. 566: In (4.8), p. 565, when z is replaced by ζ, the values of G, h are still given by (4.13) and (4.6), *without* changing γ to γ'.

After (7.14), p. 574, alter sentence to read: The results for the corresponding problem for coordinates ζ, $\bar{\zeta}$ in the undeformed body are obtained by replacing γ by γ', *except in the formulae for G and h*.

VII

REINFORCEMENT BY INEXTENSIBLE CORDS

In a number of applications of highly elastic materials, reinforcement is provided by the introduction of systems of cords which have a much higher modulus of elasticity than that of the material in which they are embedded. Such reinforcement forms a feature, for example, of the construction of such common articles as pneumatic tyres and fire hose, in which it is necessary to restrict the magnitude of the deformation in certain directions and to give added strength to the composite body. These materials are considered in the present chapter,† attention being confined to two-dimensional problems and to the cylindrically symmetrical and flexure problems of the type discussed in Chapter II. In § 7.18 the relationship is examined between such problems and the theory of materials subject to constraints given in §§ 1.17 and 1.18.

7.1. Physical assumptions

When elastic materials are reinforced with systems of cords a number of problems may be examined by regarding each cord as a means of rendering the material inextensible along the path in which it lies. If the paths of such a family of cords are such that they intersect at the most at a finite number of points and if the density of the cords at each point is such that they may be considered to fill the material continuously and completely, then directional properties are introduced into the body throughout its entire volume. If, on the other hand, the family of cords forms a virtually continuous system throughout a smooth surface in the body, then corresponding surface effects are produced.

These concepts give rise to a mathematical treatment of systems of cords based on the following assumptions:

---

† The theory of large elastic deformations for reinforced bodies has been developed by:

J. E. Adkins and R. S. Rivlin, *Phil. Trans. R. Soc.* A, **248** (1955), 201.

J. E. Adkins, *J. rat. Mech. Analysis* **5** (1956), 189; *Phil. Trans. R. Soc.* A **249** (1956), 125; *Q. Jl Mech. appl. Math.* **11** (1958), 88.

The related problem of deformation of networks of inextensible cords has been examined by:

R. S. Rivlin, *J. rat. Mech. Analysis* **4** (1955), 951; *Arch ration. Mech. Analysis* **2** (1959), 447.

(i) each cord is capable neither of extension nor of any form of irregular kinking which would have the effect of reducing the length of its path in the body during deformation—it merely restricts the strain to be zero along the path in which it lies;

(ii) each cord is ideally thin and perfectly flexible;

(iii) the cords occur in sets, those in each set being sufficiently close together for any irregularities in deformation of the material between adjacent cords to be negligible;

(iv) the cords adhere to the body in which they are inserted, so that there is no movement of any cord relative to the adjacent material. In certain cases, the cords may need to sustain a compressive force in order to maintain the assumed deformation. It is likely that some form of instability will then occur and such deformations will, in general, be regarded as unacceptable.

These assumptions imply that each family of cords introduces into the composite body a geometrical constraint of the type discussed in § 1.17 and the restrictions of that section continue to apply upon the number of constraints (i.e., sets of cords) which are admissible if the material is to be susceptible of continuous deformation. Thus, if the material is compressible and the entire volume is filled with cords, not more than five independent systems can be introduced; in the case of incompressible bodies this number is reduced to four. If the cords lie in a surface, not more than two independent families can be introduced if extension of the surface is to be possible; flexure of the surface is, of course, not prevented by the presence of cords.

The resultant stress system in the composite body is obtained by adding to the stresses which arise from the deformation of the elastic material, components due to the tensions in the cords. The relations thus derived frequently take forms similar to those which follow from § 1.18 for the corresponding problems for materials subject to constraints, and indeed may be regarded as limiting cases of these equations.† For reinforced bodies, however, the strain energy function of the elastic material is not subject to any of the restrictions described in § 1.18.

7.2. Uniform stretching of a plane reinforced sheet

Let the undeformed body be a thin, uniform, plane rectangular sheet of highly elastic material, bounded in the rectangular cartesian

† See § 7.18 and J. E. Adkins, loc. cit., p. 228.

reference frame x_i by the planes $x_1 = \pm A$, $x_2 = \pm B$, $x_3 = \pm h_0$, h_0 being small compared with A and B. This sheet is reinforced in the middle plane $x_3 = 0$ by two families of thin, straight, inextensible cords, the cords of each set being parallel to each other and intersecting those of the other set at a constant angle.

The sheet is subjected to the pure homogeneous strain defined by

$$y_i = \lambda_i x_i \quad (i \text{ not summed}), \tag{7.2.1}$$

where λ_i are constants, and y_i are the coordinates after deformation of the point of the body initially at x_i. This deformation is only possible if the cords are placed symmetrically with respect to the x_1-, x_2-axes so that the principal directions of strain bisect the angles between them. Thus, if an element of length ds_0 in the undeformed sheet with direction cosines (l, m, n) attains a length ds after deformation we have

$$\left(\frac{ds}{ds_0}\right)^2 = \lambda_1^2 l^2 + \lambda_2^2 m^2 + \lambda_3^2 n^2. \tag{7.2.2}$$

If the cords lie in the directions $(l_1, m_1, 0)$ and $(l_2, m_2, 0)$ so that $ds/ds_0 = 1$ in these directions, we obtain from (7.2.2)

$$\lambda_1^2 l_1^2 + \lambda_2^2 m_1^2 = 1, \qquad \lambda_1^2 l_2^2 + \lambda_2^2 m_2^2 = 1,$$

and since
$$l_1^2 + m_1^2 = l_2^2 + m_2^2 = 1,$$

these relations yield
$$l_1^2 = l_2^2 = \frac{1 - \lambda_2^2}{\lambda_1^2 - \lambda_2^2},$$

or
$$l_1 = \pm l_2, \qquad m_1 = \pm m_2,$$

which is the required condition of symmetry. This result also follows directly by symmetry from the consideration that diameters of the strain ellipse† defined by the directions of the cords are equal.

If the cords of each set are inclined at angles $\pm\alpha$, $\pm\beta$ to the x_1-axis before and after deformation respectively, we have

$$\cos\beta = \lambda_1 l, \qquad \sin\beta = \lambda_2 m, \tag{7.2.3}$$

and
$$l^2 \lambda_1^2 + m^2 \lambda_2^2 = 1, \tag{7.2.4}$$

where
$$l = \cos\alpha, \qquad m = \sin\alpha. \tag{7.2.5}$$

It is readily seen that (7.2.4) is identical with the constraint condition derived from (1.17.2). For if we choose the coordinate system θ'_i to

---

† See, for example, A. E. H. Love, *A Treatise on the Mathematical Theory of Elasticity* (edn. 4, Cambridge, 1927).

be an oblique system with the θ_1', θ_2' axes coincident with the paths followed by the two intersecting cords through the origin of the x_i reference frame, so that

$$x_1 = l(\theta_1' - \theta_2'), \quad x_2 = m(\theta_1' + \theta_2'), \quad x_3 = \theta_3',$$

then, remembering (7.2.1), we have

$$G_{11}' = G_{22}' = l^2\lambda_1^2 + m^2\lambda_2^2, \quad g_{11}' = g_{22}' = 1,$$

and the constraint conditions $\gamma_{11}' = 0$, $\gamma_{22}' = 0$† both reduce to (7.2.4).

The stress resultants in the reinforced sheet arise from the deformation of the elastic material and the tensions in the cords of the deformed layer. If we choose $\theta_i = y_i$, the physical components $n_e^{\alpha\beta}$ due to the elastic material are given by

$$n_e^{\alpha\beta} = 2\lambda_3 h_0 \sigma_{\alpha\beta}, \tag{7.2.6}$$

where $\sigma_{\alpha\beta}$ are the physical components of stress‡ in the cartesian system y_i and Greek indices take the values 1, 2.

The contribution due to the layer of cords can be obtained by a direct calculation. At any given point of the middle plane $x_3 = 0$ of the undeformed sheet let the cords making an angle α with the x_1-axis (i.e., following the lines $\theta_2' = $ constant), be spaced a distance δ_1 apart, this distance being measured along the cords of the other set ($\theta_1' = $ constant) at that point. We assume that these cords carry a tension τ_1 when the sheet is deformed and we write $\sigma_1 = \tau_1/\delta_1$. The corresponding quantities for the other set of cords are denoted by δ_2, τ_2, and $\sigma_2 = \tau_2/\delta_2$ respectively. Lines in the x_1-, x_2-directions are then intersected initially by $1/(2l\delta_1)$ and $1/(2m\delta_1)$ cords of the first set respectively, per unit length, these quantities being changed to $1/(2\lambda_1 l\delta_1)$, $1/(2\lambda_2 m\delta_1)$ in the final configuration. The corresponding quantities for the second set of cords are obtained by replacing δ_1 by δ_2 throughout. Denoting the (physical) stress resultants in the layer of cords by $n_c^{\alpha\beta}$ we obtain by a direct resolution of forces

$$\left.\begin{aligned}n_c^{11} &= \frac{1}{2\lambda_2 m\delta_1}\tau_1\cos\beta + \frac{1}{2\lambda_2 m\delta_2}\tau_2\cos\beta \\ n_c^{22} &= \frac{1}{2\lambda_1 l\delta_1}\tau_1\sin\beta + \frac{1}{2\lambda_1 l\delta_2}\tau_2\sin\beta \\ n_c^{12} &= \frac{1}{2\lambda_2 m\delta_1}\tau_1\sin\beta - \frac{1}{2\lambda_2 m\delta_2}\tau_2\sin\beta\end{aligned}\right\}. \tag{7.2.7}$$

† Cf. (7.4.10). ‡ See § 1.15.

The physical stress resultants $n_{\alpha\beta}$ ($= n^{\alpha\beta}$) in the composite sheet are given by
$$n_{\alpha\beta} = n_e^{\alpha\beta} + n_c^{\alpha\beta}, \tag{7.2.8}$$
and if we combine this result with (7.2.6), (7.2.7), and remember (7.2.3), we obtain
$$\left.\begin{aligned} n_{11} &= 2\lambda_3 h_0 \sigma_{11} + \frac{\lambda_1 l}{2\lambda_2 m}(\sigma_1 + \sigma_2) \\ n_{22} &= 2\lambda_3 h_0 \sigma_{22} + \frac{\lambda_2 m}{2\lambda_1 l}(\sigma_1 + \sigma_2) \\ n_{12} &= 2\lambda_3 h_0 \sigma_{12} + \tfrac{1}{2}(\sigma_1 - \sigma_2) \end{aligned}\right\}. \tag{7.2.9}$$

We observe that the preceding analysis does not exclude the possibility that τ_α and δ_α may be variable functions of position throughout the middle plane of the sheet. The spacing of adjacent cords of each family and the tensions in the individual cords would then vary from point to point. If, however, the elastic material is homogeneous and the stress resultants $n_{\alpha\beta}$ are constants, the quotients σ_1, σ_2 must also be constant throughout the sheet.

7.3. Simple extension of a reinforced incompressible isotropic sheet

If the elastic material of the sheet considered in the previous section is isotropic and incompressible, the deformation
$$y_i = \lambda_i x_i \quad (i \text{ not summed}) \tag{7.3.1}$$
gives rise to the stresses†
$$\left.\begin{aligned} \sigma_{ii} &= \lambda_i^2 \Phi - \frac{1}{\lambda_i^2}\Psi + p \quad (i \text{ not summed}) \\ \sigma_{ij} &= 0 \quad\quad\quad\quad (i \neq j) \end{aligned}\right\}, \tag{7.3.2}$$
where
$$I_1 = \lambda_1^2 + \lambda_2^2 + \lambda_3^2, \quad I_2 = \frac{1}{\lambda_1^2} + \frac{1}{\lambda_2^2} + \frac{1}{\lambda_3^2}, \tag{7.3.3}$$
and
$$\lambda_1 \lambda_2 \lambda_3 = 1. \tag{7.3.4}$$

We assume also that the major surfaces $x_3 = \pm h_0$ are free from applied tractions, so that $\sigma_{33} = 0$, and then from (7.3.2) and (7.3.4)
$$\left.\begin{aligned} \sigma_{11} &= (\lambda_1^2 - \lambda_3^2)(\Phi + \lambda_2^2 \Psi) \\ \sigma_{22} &= (\lambda_2^2 - \lambda_3^2)(\Phi + \lambda_1^2 \Psi) \end{aligned}\right\}. \tag{7.3.5}$$

† A derivation of these results may be found in *T.E.*, § 3.1. In (7.3.2) p is not identical with the corresponding quantity in Chapter I.

7.3 REINFORCEMENT BY INEXTENSIBLE CORDS

From (7.2.9) the stress resultants $n_{\alpha\beta}$ therefore become

$$\left.\begin{aligned} n_{11} &= 2\lambda_3 h_0(\lambda_1^2-\lambda_3^2)(\Phi+\lambda_2^2\Psi)+\frac{l\lambda_1}{2m\lambda_2}(\sigma_1+\sigma_2) \\ n_{22} &= 2\lambda_3 h_0(\lambda_2^2-\lambda_3^2)(\Phi+\lambda_1^2\Psi)+\frac{m\lambda_2}{2l\lambda_1}(\sigma_1+\sigma_2) \\ n_{12} &= \tfrac{1}{2}(\sigma_1-\sigma_2) \end{aligned}\right\}. \qquad (7.3.6)$$

When the deformation is a simple extension of ratio λ_1 produced by means of a tensile force in the x_1-direction, then $n_{12} = n_{22} = 0$ and from (7.3.6)

$$\sigma_1 = \sigma_2 = -\frac{2h_0 c(\lambda_2^2-\lambda_3^2)}{\lambda_2^2}(\Phi+\lambda_1^2\Psi) = \sigma \quad \text{(say)}, \qquad (7.3.7)$$

and

$$n_{11} = 2h_0\frac{\lambda_1}{\lambda_2}\{[1-c^2+\lambda_3^4(\lambda_1^2 c^2-\lambda_2^2)]\Phi+[\lambda_2^2-\lambda_1^2 c^2+\lambda_3^4(\lambda_1^4 c^2-\lambda_2^4)]\Psi\}, \quad (7.3.8)$$

where
$$c = l/m = \cot\alpha. \qquad (7.3.9)$$

From physical considerations we may assume $\Phi+\lambda_1^2\Psi$ to be positive† and it then follows from (7.3.7) that σ is positive if

$$\lambda_2^2-\lambda_3^2 < 0. \qquad (7.3.10)$$

The difference $\lambda_2^2-\lambda_3^2$ may be expressed in terms of λ_1 by means of (7.2.4) and (7.3.4). We obtain

$$\lambda_2^2-\lambda_3^2 = \frac{(\lambda_1^6-1)[l^2-1/(\lambda_1^2-\lambda_1+1)][l^2-1/(\lambda_1^2+\lambda_1+1)]}{\lambda_1^2 m^2(1-\lambda_1^2 l^2)}. \qquad (7.3.11)$$

In this expression, since we have assumed a state of simple extension, $\lambda_1 > 1$. Also, from (7.2.4), since $m\lambda_2 = \lambda_2 \sin\alpha$ is real, $1-\lambda_1^2 l^2$ must be positive, and therefore, with $\lambda_1 > 1$, $l^2-1/(\lambda_1^2-\lambda_1+1)$ must be negative. The sign of σ will therefore be that of $l^2-1/(\lambda_1^2+\lambda_1+1)$. Hence, remembering (7.2.4) it follows that provided α is taken as lying between 0 and $\tfrac{1}{2}\pi$, σ is positive if

$$\cos^{-1}(1/\lambda_1) < \alpha < \cos^{-1}(\lambda_1^2+\lambda_1+1)^{-\tfrac{1}{2}}$$

and negative if $\quad \cos^{-1}(\lambda_1^2+\lambda_1+1)^{-\tfrac{1}{2}} < \alpha < \tfrac{1}{2}\pi.$

† This follows from (7.3.2) if we assume that $\sigma_{22} > \sigma_{33}$ when $\lambda_2 > \lambda_3$. For rubber, Φ and Ψ are both positive.

These inequalities imply that σ is positive if λ_1 is restricted so that

$$\sec\alpha > \lambda_1 > \tfrac{1}{2}[(4\sec^2\alpha-3)^{\frac{1}{2}}-1]. \qquad (7.3.12)$$

If $\sec\alpha > \sqrt{3}$, then $\tfrac{1}{2}[(4\sec^2\alpha-3)^{\frac{1}{2}}-1] > 1$ and σ is negative if

$$\tfrac{1}{2}[(4\sec^2\alpha-3)^{\frac{1}{2}}-1] > \lambda_1 > 1. \qquad (7.3.13)$$

If $\sec\alpha \leqslant \sqrt{3}$, then $\tfrac{1}{2}[(4\sec^2\alpha-3)^{\frac{1}{2}}-1] \leqslant 1$ and σ is positive throughout the complete range of deformation defined by $\sec\alpha > \lambda_1 > 1$.

Since a negative value of σ would imply that the cords must sustain a compressive force, it is likely that some form of instability would then occur. Solutions within the ranges of λ_1 giving negative values of σ may therefore be regarded as unacceptable. We observe that if $\sec\alpha > \sqrt{3}$ there is a critical value of λ_1 below which $\sigma < 0$; this critical value increases as α is increased.

7.4. Plane deformation of a thin reinforced sheet

We now consider the problem of a thin reinforced sheet of elastic material subject to an arbitrary plane deformation. The undeformed body consists, as in the preceding sections, of a thin sheet of highly elastic material, bounded in the coordinate system x_i by the planes $x_3 = \pm h_0$ and containing in the middle plane $x_3 = 0$ a layer of thin, flexible, inextensible cords. This reinforcing layer consists of two independent sets of cords lying along the two families of coordinate curves of a curvilinear coordinate system θ'_α in the plane $x_3 = 0$. As indicated in § 7.1, the cords are assumed to lie sufficiently close together for any plane deformation to be restricted by the condition that there is no extension along any of the θ'_α coordinate curves in the undeformed body.

The sheet is assumed to undergo a finite deformation which carries the point of the material initially at x_i to the point y_i in the same coordinate system, this deformation being symmetrical about the plane $x_3 = 0$. This plane therefore becomes the middle plane $y_3 = 0$ of the deformed body. The major surfaces of the plate after deformation are given by $y_3 = \pm h$ where h is, in general, a function of y_1, y_2. We choose the curvilinear coordinate system θ_i so that

$$y_\alpha = y_\alpha(\theta_1, \theta_2), \qquad y_3 = \theta_3, \qquad (7.4.1)$$

Greek indices taking the values 1, 2. Apart from the existence of the reinforcing layer, the conditions of the problem are identical with those specified in §§ 4.1–4.4 for unreinforced plates and equations (4.1.2),

7.4 REINFORCEMENT BY INEXTENSIBLE CORDS

(4.1.3), (4.3.1), and (4.3.2) again apply. Thus

$$G_{ij} = \begin{bmatrix} A_{11} & A_{12} & 0 \\ A_{12} & A_{22} & 0 \\ 0 & 0 & 1 \end{bmatrix}, \quad G^{ij} = \begin{bmatrix} A^{11} & A^{12} & 0 \\ A^{12} & A^{22} & 0 \\ 0 & 0 & 1 \end{bmatrix}, \quad G = A, \tag{7.4.2}$$

with
$$A = |A_{\alpha\beta}|, \quad A^{\alpha\rho}A_{\rho\beta} = \delta^\alpha_\beta,$$

where $A_{\alpha\beta}$, $A^{\alpha\beta}$ are the covariant and contravariant metric tensors associated with coordinates θ_α in the middle plane $y_3 = 0$ of the deformed sheet. Again, since the deformation is assumed to be completely symmetrical about the middle plane $y_3 = 0$, we have on this plane

$$\left. \begin{array}{l} x_\alpha = x_\alpha(\theta_1, \theta_2), \quad \dfrac{\partial x^3}{\partial \theta^3} = \dfrac{1}{\lambda} \\ \dfrac{\partial x^\alpha}{\partial \theta^3} = 0, \quad \dfrac{\partial x^3}{\partial \theta^\alpha} = 0 \end{array} \right\} \quad (y_3 = 0), \tag{7.4.3}$$

and
$$\left. \begin{array}{l} g_{\alpha\beta} = a_{\alpha\beta}, \quad g_{33} = 1/\lambda^2, \quad g_{\alpha 3} = 0 \\ g^{\alpha\beta} = a^{\alpha\beta}, \quad g^{33} = \lambda^2, \quad g^{\alpha 3} = 0 \\ g = a/\lambda^2, \quad a = |a_{\alpha\beta}| \end{array} \right\} \quad (y_3 = 0), \tag{7.4.4}$$

$a_{\alpha\beta}$, $a^{\alpha\beta}$ being the covariant and contravariant metric tensors associated with curvilinear coordinates θ_α in the plane $x_3 = 0$ of the undeformed body, and λ a scalar function of θ_1, θ_2.

The curves in the undeformed sheet followed by the cords may be defined by relations of the form

$$\theta'_\alpha(x_1, x_2) = \text{constant}, \quad x_3 = 0, \tag{7.4.5}$$

and when the plate is deformed these, from (7.4.1) and (7.4.3), are carried into the curves

$$\theta'_\alpha[x_1(y_1, y_2), x_2(y_1, y_2)] = \text{constant}, \tag{7.4.6}$$

in the plane $y_3 = 0$ of the deformed body. We may therefore associate with the coordinate curves θ'_α before and after deformation metric tensors analogous to those defined for the moving coordinate system θ_α. Distinguishing quantities associated with the system θ'_α by primes we may write, in conformity with (7.4.2) and (7.4.4)

$$\left. \begin{array}{l} G'_{\alpha\beta} = A'_{\alpha\beta}, \quad G'^{\alpha\beta} = A'^{\alpha\beta} \\ G' = A', \quad A' = |A'_{\alpha\beta}| \end{array} \right\}, \tag{7.4.7}$$

$$\left. \begin{array}{l} g'_{\alpha\beta} = a'_{\alpha\beta}, \quad g'^{\alpha\beta} = a'^{\alpha\beta} \\ g' = a', \quad a' = |a'_{\alpha\beta}| \end{array} \right\}. \tag{7.4.8}$$

The restriction upon the deformation (7.4.3) imposed by the layer of cords follows from geometrical considerations. If corresponding elements of length in the middle plane of the plate before and after deformation are denoted by ds_0, ds respectively, we have

$$ds_0^2 = a_{\alpha\beta}\, d\theta^\alpha\, d\theta^\beta = a'_{\alpha\beta}\, d\theta'^\alpha\, d\theta'^\beta,$$
$$ds^2 = A_{\alpha\beta}\, d\theta^\alpha\, d\theta^\beta = A'_{\alpha\beta}\, d\theta'^\alpha\, d\theta'^\beta.$$

Since $ds_0 = ds$ along the curves $\theta'_\alpha =$ constant, we obtain from these expressions

$$A'_{11} = a'_{11}, \qquad A'_{22} = a'_{22}. \tag{7.4.9}$$

These relations may be written as

$$\gamma'_{11} = \gamma'_{22} = 0$$

or

$$\frac{\partial y^\alpha}{\partial \theta'^1}\frac{\partial y^\alpha}{\partial \theta'^1} = \frac{\partial x^\alpha}{\partial \theta'^1}\frac{\partial x^\alpha}{\partial \theta'^1}, \quad \frac{\partial y^\alpha}{\partial \theta'^2}\frac{\partial y^\alpha}{\partial \theta'^2} = \frac{\partial x^\alpha}{\partial \theta'^2}\frac{\partial x^\alpha}{\partial \theta'^2}, \tag{7.4.10}$$

the strain components $\gamma'_{\alpha\beta}$ being defined for the coordinate system θ'_α by analogy with the quantities $\gamma_{\alpha\beta}$ as

$$\gamma'_{\alpha\beta} = \tfrac{1}{2}(A'_{\alpha\beta} - a'_{\alpha\beta}). \tag{7.4.11}$$

If the paths of the cords in the undeformed body are expressed as functional relationships between θ'_α and x_β, equations (7.4.9) or (7.4.10) serve to determine the coordinates y_α and hence the displacement components in the middle plane of the deformed sheet, as functions of θ'_α. Conversely, if the configuration of the system of cords in the deformed body is given, so that y_1, y_2 may be regarded as known functions of θ'_1, θ'_2, equations (7.4.9) may be used to determine relations between x_α and θ'_β. Furthermore, it follows from the symmetry of the system that if the relations

$$y_\alpha = f_\alpha(x_1, x_2) \tag{7.4.12}$$

represent a possible solution of the constraint conditions (7.4.10), the equations

$$x_\alpha = f_\alpha(y_1, y_2) \tag{7.4.13}$$

also represent a possible solution of these equations. This duality property may be compared with the corresponding result obtained for the plane strain of unreinforced bodies in § 3.6.

We shall henceforth confine our attention to the case where the paths of the cords in the undeformed body are known. With the coordinates y_α as dependent variables equations (7.4.9) may be written

$$y_{2,\alpha} = [a'_{\alpha\alpha} - (y_{1,\alpha})^2]^{\frac{1}{2}} \quad (\alpha \text{ not summed}), \tag{7.4.14}$$

where a comma now denotes partial differentiation with respect to θ'_α and the sign of the square root in (7.4.14) is chosen so that the equation reduces to an identity when $y_\alpha = x_\alpha$. The significance of the alternative choice of signs for the square root requires further discussion; a different choice would not affect the subsequent argument. From (7.4.14) by differentiation we obtain

$$y_{2,12} = \frac{a'_{11,2}-2y_{1,1}y_{1,12}}{2[a'_{11}-(y_{1,1})^2]^{\frac{1}{2}}} = \frac{a'_{22,1}-2y_{1,2}y_{1,12}}{2[a'_{22}-(y_{1,2})^2]^{\frac{1}{2}}},$$

or

$$2\{y_{1,1}[a'_{22}-(y_{1,2})^2]^{\frac{1}{2}}-y_{1,2}[a'_{11}-(y_{1,1})^2]^{\frac{1}{2}}\}y_{1,12}$$
$$= a'_{11,2}[a'_{22}-(y_{1,2})^2]^{\frac{1}{2}}-a'_{22,1}[a'_{11}-(y_{1,1})^2]^{\frac{1}{2}},$$

and a similar equation holds for the derivative $y_{2,12}$ with the figures 1 and 2 interchanged throughout. It follows therefore that the equations for the determination of y_1, y_2 are hyperbolic, with the characteristic curves coinciding with the paths followed by the cords.

7.5. Plane deformation of a reinforced sheet: stress resultants

The stress resultants $n^{\alpha\beta}$ in the deformed sheet may be calculated, as for the problem of uniform strain described in § 7.2, by combining the contributions which arise from deformation of the elastic material and the tensions in the cords. Thus we write

$$n^{\alpha\beta} = n_e^{\alpha\beta}+n_c^{\alpha\beta}, \qquad (7.5.1)$$

where the components $n_e^{\alpha\beta}$ for the elastic material are calculated by the methods of Chapter IV and the quantities $n_c^{\alpha\beta}$ may be expressed in terms of the tensions in the individual cords and the geometrical configuration of the reinforcing layer.

At any given point x_α in the middle plane $x_3 = 0$ of the undeformed sheet let δ_1 be the distance between adjacent cords of the set lying along the curves $\theta'_2 = $ constant, this distance being measured along the curve $\theta'_1 = $ constant through that point. We may then regard $1/\delta_1$ as a measure of the density of the cords in the θ'_1-direction at the point x_1, x_2. After deformation each of the cords in the θ'_1-direction is assumed to carry a tension τ_1 and we write $\sigma_1 = \tau_1/\delta_1$. Corresponding quantities δ_2, τ_2, and $\sigma_2 = \tau_2/\delta_2$ may be defined for the other set of cords. The quantities τ_α, δ_α, and σ_α are, in general, functions of the coordinates θ'_α (or θ_α) in the middle plane of the sheet. It is not necessary to assume that these functions are continuous provided that the functions δ_α are at no point too large for the assumptions of § 7.1 to be justified. For a continuous distribution of stress $n_c^{\alpha\beta}$ in the reinforcing layer, however, the quotients σ_α must be continuous. We assume also that the applied force system is such that the tensions τ_α nowhere become negative. We observe that since there is assumed to be no relative movement between each cord and the adjacent elastic material, the distances δ_1, δ_2 are unchanged by the deformation.

From the definition of σ_1, σ_2 we see that these quantities are the physical components of the stresses $n_c^{\prime 11}$, $n_c^{\prime 22}$ referred to the coordinate system θ_α'. In view of (4.1.12), (7.4.7), and (7.4.9) we therefore have

$$\sigma_1 = n_c^{\prime 11}\sqrt{\left(\frac{A_{11}'}{A^{\prime 11}}\right)} = n_c^{\prime 11}\sqrt{\left(\frac{A_{11}'A'}{A_{22}'}\right)} = n_c^{\prime 11}\sqrt{\left(\frac{a_{11}'A'}{a_{22}'}\right)},$$

with similar results for σ_2. It follows that

$$\left.\begin{aligned} n_c^{\prime 11} &= \sigma_1\sqrt{\left(\frac{a_{22}'}{a_{11}'A'}\right)} = \sigma_1' \quad (\text{say}) \\ n_c^{\prime 22} &= \sigma_2\sqrt{\left(\frac{a_{11}'}{a_{22}'A'}\right)} = \sigma_2' \quad (\text{say}) \\ n_c^{\prime 12} &= 0 \end{aligned}\right\}, \qquad (7.5.2)$$

and by a tensor transformation

$$n_c^{\alpha\beta} = \left[\sigma_1 \frac{\partial\theta^\alpha}{\partial\theta^{\prime 1}}\frac{\partial\theta^\beta}{\partial\theta^{\prime 1}}\sqrt{\left(\frac{a_{22}'}{a_{11}'}\right)} + \sigma_2 \frac{\partial\theta^\alpha}{\partial\theta^{\prime 2}}\frac{\partial\theta^\beta}{\partial\theta^{\prime 2}}\sqrt{\left(\frac{a_{11}'}{a_{22}'}\right)}\right]/\sqrt{A'}. \qquad (7.5.3)$$

7.6. Plane deformation of a reinforced sheet: Airy's stress function

When body forces are zero and the major surfaces of the sheet are free from applied forces, the equations of equilibrium may be expressed in the forms given in § 4.2 for unreinforced plates, the analysis being unaffected by the layer of cords in the middle plane of the sheet. The equations of equilibrium (4.2.4) are again satisfied by introducing the Airy stress function ϕ such that

$$n^{\alpha\beta} = \epsilon^{\alpha\gamma}\epsilon^{\beta\rho}\phi\|_{\gamma\rho}, \qquad (7.6.1)$$

where $n^{\alpha\beta}$ are now the stress resultants given by (7.5.1) and (7.5.3). The quantities σ_1', σ_2', and hence also the tensions in the cords, are readily evaluated by choosing $\theta_\alpha = \theta_\alpha'$. We then have

$$\epsilon^{\prime\alpha\gamma}\epsilon^{\prime\beta\rho}\phi\|_{\gamma\rho}' = n^{\prime\alpha\beta} = n_e^{\prime\alpha\beta} + \sum_{\nu=1}^{2}\delta_\nu^\alpha \delta_\nu^\beta \sigma_\nu', \qquad (7.6.2)$$

where the primes are used to indicate quantities defined with respect to the curves θ'_α in the deformed body and the double bar with the additional prime now denotes covariant differentiation with respect to the coordinates θ'_α using the metric tensors $A'_{\alpha\beta}$, $A'^{\alpha\beta}$. From (7.6.2) it follows that

$$\sigma'_1 = \phi\|'_{22}/A' - n_e'^{11}, \qquad \sigma'_2 = \phi\|'_{11}/A' - n_e'^{22}, \tag{7.6.3}$$

and
$$n'^{12} = n_e'^{12} = -\phi\|'_{12}/A'. \tag{7.6.4}$$

If we regard the coordinates y_α as unknown variables which are determined when the deformation in the plane $y_3 = 0$ is known, it follows from (7.4.7) and (7.4.8) that the strain components γ_{ij} or γ'_{ij} may be regarded as functions of these quantities together with the additional parameter λ. For compressible bodies we may therefore infer from (1.1.33) that the stresses τ_e^{ij}, and hence also, from (4.1.7), the stress resultants $n_e^{\alpha\beta}$, are functions of y_α and λ.

The relations (7.4.9) and (7.6.4) may thus be regarded as three equations involving the four unknowns y_1, y_2, λ, and ϕ. An additional relation between y_1, y_2, and ϕ is provided by the condition that the major surfaces of the sheet are free from applied traction; for bodies transversely isotropic with respect to the x_3-direction this takes the form (4.4.8). When the material is incompressible, the condition for the major surfaces of the sheet to be stress free may be used to eliminate the arbitrary scalar function p occurring in the stress–strain relations (1.1.35) from the expressions for $n_e^{\alpha\beta}$. When the material is transversely isotropic, the resulting formulae for $n_e^{\alpha\beta}$ take the form (4.4.7). Equations (7.4.9) and (7.6.4) together with the incompressibility condition

$$I_3 = \lambda^2 A/a = \lambda^2 A'/a' = 1,$$

then yield four relations for the determination of y_1, y_2, λ, and ϕ. When these unknowns have been determined, the tensions τ_α in the cords may be evaluated using (7.6.3) and (7.5.2).

The equation (7.6.4) for ϕ is hyperbolic with the characteristic curves coinciding with the paths $\theta'_\alpha =$ constant followed by the cords, and resembles the constraint conditions (7.4.9) in this respect. A discussion of the solution of this system of equations for a general curvilinear system θ'_α has been given by Adkins.† We shall here confine attention to the case where the cords of each set lie initially in parallel straight lines so that the components $a'_{\alpha\beta}$, $a'^{\alpha\beta}$ of the metric tensors associated with the coordinates θ'_α in the undeformed body are constants.

† Loc. cit., p. 228.

7.7. Plane deformation of a sheet reinforced with systems of parallel straight cords

A simplification of the theory occurs when the coordinate frame θ'_α defining the configuration of the cords in the undeformed body is a rectilinear system, so that the cords of each set lie initially in the parallel straight lines $\theta'_\alpha = c_{\alpha\beta} x_\beta = $ constant, where $c_{\alpha\beta}$ are constants. The metric tensor components $a'_{\alpha\beta}$ are then constants and the constraint conditions (7.4.10) yield, by differentiation,

$$y_{1,1} y_{1,12} + y_{2,1} y_{2,12} = 0,$$
$$y_{1,2} y_{1,12} + y_{2,2} y_{2,12} = 0.$$

Since $\sqrt{A'} = y_{1,1} y_{2,2} - y_{1,2} y_{2,1} \neq 0,$

these imply that $y_{1,12} = 0, \quad y_{2,12} = 0.$

We therefore have the solution

$$y_1 = f_1(\theta'_1) + f_2(\theta'_2), \qquad y_2 = g_1(\theta'_1) + g_2(\theta'_2), \tag{7.7.1}$$

where f_1, f_2, g_1, and g_2 are functions of the arguments indicated and are arbitrary except for the conditions

$$f'^2_1 + g'^2_1 = a'_{11}, \qquad f'^2_2 + g'^2_2 = a'_{22}, \tag{7.7.2}$$

imposed upon their first derivatives by (7.4.10).

The metric tensors $A'_{\alpha\beta}$, $A'^{\alpha\beta}$ now become

$$\left. \begin{aligned}
A'_{11} &= a'_{11} = f'^2_1 + g'^2_1, \qquad A'_{22} = a'_{22} = f'^2_2 + g'^2_2 \\
A'_{12} &= y_{\alpha,1} y_{\alpha,2} = f'_1 f'_2 + g'_1 g'_2 \\
\sqrt{A'} &= \frac{\partial(y_1, y_2)}{\partial(\theta'_1, \theta'_2)} = f'_1 g'_2 - f'_2 g'_1 \\
A'^{11} &= A'_{22}/A', \qquad A'^{22} = A'_{11}/A', \qquad A'^{12} = -A'_{12}/A'
\end{aligned} \right\}, \tag{7.7.3}$$

and the Christoffel symbols $\Gamma'^\rho_{\alpha\beta}$ formed with these components reduce to

$$\left. \begin{aligned}
\Gamma'^1_{11} &= -A'_{12} A'_{12,1}/A' = (g'_2 f''_1 - f'_2 g''_1)/\sqrt{A'} \\
\Gamma'^2_{22} &= -A'_{12} A'_{12,2}/A' = (f'_1 g''_2 - g'_1 f''_2)/\sqrt{A'} \\
\Gamma'^2_{11} &= A'_{11} A'_{12,1}/A' = (f'_1 g''_1 - g'_1 f''_1)/\sqrt{A'} \\
\Gamma'^1_{22} &= A'_{22} A'_{12,2}/A' = (g'_2 f''_2 - f'_2 g''_2)/\sqrt{A'} \\
\Gamma'^1_{12} &= \Gamma'^2_{12} = 0
\end{aligned} \right\}, \tag{7.7.4}$$

where in deriving these results we have used the relations

$$f'_1 f''_1 + g'_1 g''_1 = 0, \qquad f'_2 f''_2 + g'_2 g''_2 = 0, \tag{7.7.5}$$

which follow from (7.7.2) by differentiation.

7.7 REINFORCEMENT BY INEXTENSIBLE CORDS

From (7.4.11) we see that the strain components γ'_{ij} and hence also γ_{ij} may be expressed in terms of f'_1, f'_2, g'_1, g'_2, and λ. If the form of the strain energy function is known, W and hence also the elastic stress resultant components $n'^{\alpha\beta}_e$ may be regarded as known functions of f'_1, f'_2, g'_1, g'_2, and λ. For incompressible bodies λ is determined as a function of f'_1, f'_2, g'_1, and g'_2 by the incompressibility condition

$$\lambda^2 = a'/A' = a'/(f'_1 g'_2 - f'_2 g'_1)^2. \tag{7.7.6}$$

When the material is compressible λ is determined from the condition for the surfaces $y_3 = \pm h$ to be free from applied traction. This takes the form (4.4.8) for transversely isotropic bodies.

With the help of (7.7.4), equation (7.6.4) becomes

$$\phi\|_{12} = \phi_{,12} = -A' n'^{12}_e. \tag{7.7.7}$$

The right-hand member of this equation may be regarded as a known function of f'_1, f'_2, g'_1, and g'_2 and hence, if a solution of the constraint conditions has been obtained, of θ'_1, θ'_2. A formal integration then yields

$$\phi = F(\theta'_1, \theta'_2) + h_1(\theta'_1) + h_2(\theta'_2), \tag{7.7.8}$$

where $F(\theta'_1, \theta'_2)$ is a function chosen so that

$$F_{,12} = -A' n'^{12}_e = -(f'_1 g'_2 - f'_2 g'_1)^2 n'^{12}, \tag{7.7.9}$$

and h_1, h_2 are arbitrary functions of their arguments. The remaining stress resultant components are given, from (7.6.2), (7.7.4), and (7.7.8) by

$$\begin{aligned}
n'^{11} &= \phi\|_{22}/A' \\
&= \{(F_{,22} + h''_2)(f'_1 g'_2 - f'_2 g'_1) - (F_{,2} + h'_2)(f'_1 g''_2 - g'_1 f''_2) + \\
&\qquad + (F_{,1} + h'_1)(f'_2 g''_2 - g'_2 f''_2)\}/(f'_1 g'_2 - f'_2 g'_1)^3 \\
n'^{22} &= \phi\|_{11}/A' \\
&= \{(F_{,11} + h''_1)(f'_1 g'_2 - f'_2 g'_1) - (F_{,1} + h'_1)(g'_2 f''_1 - f'_2 g''_1) + \\
&\qquad + (F_{,2} + h'_2)(g'_1 f''_1 - f'_1 g''_1)\}/(f'_1 g'_2 - f'_2 g'_1)^3
\end{aligned} \tag{7.7.10}$$

or

$$\begin{aligned}
n'^{11} &= \left\{ \frac{\partial}{\partial \theta'^2}\left(\frac{F_{,2} + h'_2}{\sqrt{A'}}\right) + \frac{f'_2 g''_2 - g'_2 f''_2}{A'}(F_{,1} + h'_1)\right\}\Big/\sqrt{A'} \\
n'^{22} &= \left\{ \frac{\partial}{\partial \theta'^1}\left(\frac{F_{,1} + h'_1}{\sqrt{A'}}\right) + \frac{g'_1 f''_1 - f'_1 g''_1}{A'}(F_{,2} + h'_2)\right\}\Big/\sqrt{A'}
\end{aligned} \tag{7.7.11}$$

The tensions τ_α may be evaluated from the relations

$$\left.\begin{aligned}\tau_1 &= \delta_1\sigma_1 = \delta_1(n'^{11}-n_e'^{11})\sqrt{\left(\frac{A'a_{11}'}{a_{22}'}\right)} \\ \tau_2 &= \delta_2\sigma_2 = \delta_2(n'^{22}-n_e'^{22})\sqrt{\left(\frac{A'a_{22}'}{a_{11}'}\right)}\end{aligned}\right\}, \quad (7.7.12)$$

obtained by combining (7.5.1) and (7.5.2).

For unreinforced plates, the formula (4.2.8) has been derived for the force across a surface in the deformed body. This applies without modification in the present instance. Thus, if AP is an arc of a curve AB in the plane $y_3 = 0$ (Fig. 3.1, p. 106) and $\mathbf{P}$ is the force exerted by the region 1 on the region 2 across the surface in the deformed body formed by the normals to $y_3 = 0$ along AP, then

$$\mathbf{P} = \epsilon^{\rho\beta}\frac{\partial\phi}{\partial\theta^\rho}\mathbf{G}_\beta. \qquad (7.7.13)$$

Furthermore, if $\mathbf{v}$ is the displacement vector in the plane $y_3 = 0$, we may write

$$\left.\begin{aligned}\mathbf{v} &= v^\alpha \mathbf{g}_\alpha \\ \mathbf{G}_\alpha &= \mathbf{g}_\alpha + \frac{\partial \mathbf{v}}{\partial \theta^\alpha} = \mathbf{g}_\alpha + v^\beta|_\alpha \mathbf{g}_\beta\end{aligned}\right\}, \qquad (7.7.14)$$

where $\mathbf{g}_\alpha$, $\mathbf{G}_\alpha$ are the covariant base vectors in the middle plane of the undeformed body and the deformed body respectively and the single line denotes covariant differentiation with respect to the undeformed body, that is, with respect to the coordinates θ_α and the metric tensor components $a_{\alpha\beta}$, $a^{\alpha\beta}$. The latter expression, with (7.7.13), yields

$$\begin{aligned}\mathbf{P} &= \epsilon^{\rho\beta}\frac{\partial\phi}{\partial\theta^\rho}(\mathbf{g}_\beta + v^\gamma|_\beta \mathbf{g}_\gamma) \\ &= \frac{1}{\sqrt{A}}\left\{-\left[(1+v^1|_1)\frac{\partial\phi}{\partial\theta^2} - v^1|_2\frac{\partial\phi}{\partial\theta^1}\right]\mathbf{g}_1 + \left[(1+v^2|_2)\frac{\partial\phi}{\partial\theta^1} - v^2|_1\frac{\partial\phi}{\partial\theta^2}\right]\mathbf{g}_2\right\}. \quad (7.7.15)\end{aligned}$$

If now we choose the coordinate system θ_α so that $\theta_\alpha = x_\alpha$, the base vectors $\mathbf{g}_\alpha$ coincide with the unit vectors $\mathbf{i}_\alpha$ parallel to the x_α-axes. Also,

$$v^\alpha = v_\alpha = y_\alpha - x_\alpha$$

are now the components of displacement in the coordinate system x_α, the covariant derivatives in (7.7.15) become ordinary partial derivatives with respect to x_α, and

$$\sqrt{A} = \frac{\partial(y_1, y_2)}{\partial(x_1, x_2)}.$$

Thus from (7.7.15) we obtain

$$\mathbf{P} = \left\{-\left(\frac{\partial y_1}{\partial x_1}\frac{\partial \phi}{\partial x_2} - \frac{\partial y_1}{\partial x_2}\frac{\partial \phi}{\partial x_1}\right)\mathbf{i}_1 + \left(\frac{\partial y_2}{\partial x_2}\frac{\partial \phi}{\partial x_1} - \frac{\partial y_2}{\partial x_1}\frac{\partial \phi}{\partial x_2}\right)\mathbf{i}_2\right\}\bigg/\frac{\partial(y_1, y_2)}{\partial(x_1, x_2)}$$

$$= -\left\{\frac{\partial(y_1, \phi)}{\partial(\theta_1', \theta_2')}\mathbf{i}_1 + \frac{\partial(y_2, \phi)}{\partial(\theta_1', \theta_2')}\mathbf{i}_2\right\}\bigg/\frac{\partial(y_1, y_2)}{\partial(\theta_1', \theta_2')}. \tag{7.7.16}$$

Writing
$$\mathbf{P} = P_1\mathbf{i}_1 + P_2\mathbf{i}_2, \tag{7.7.17}$$

and making use of (7.7.1) and (7.7.8), we therefore have

$$\left.\begin{aligned}P_1 &= (y_{1,2}\phi_{,1} - y_{1,1}\phi_{,2})/(y_{1,1}y_{2,2} - y_{1,2}y_{2,1}) \\ &= \{f_2'(F_{,1} + h_1') - f_1'(F_{,2} + h_2')\}/(f_1'g_2' - f_2'g_1') \\ P_2 &= \{g_2'(F_{,1} + h_1') - g_1'(F_{,2} + h_2')\}/(f_1'g_2' - f_2'g_1')\end{aligned}\right\}, \tag{7.7.18}$$

and these equations yield the solutions

$$\left.\begin{aligned}h_1' &= f_1'P_2 - g_1'P_1 - F_{,1} \\ h_2' &= f_2'P_2 - g_2'P_1 - F_{,2}\end{aligned}\right\}, \tag{7.7.19}$$

for h_1, h_2.

Furthermore, by differentiating (7.7.13) and remembering (7.7.14) we obtain

$$\frac{\partial \mathbf{P}}{\partial \theta^\alpha} = \epsilon^{\rho\beta}\phi\|_{\alpha\rho}\mathbf{G}_\beta = \epsilon_{\alpha\gamma}n^{\gamma\beta}\mathbf{G}_\beta = \epsilon_{\alpha\gamma}n^{\gamma\rho}(\delta_\rho^\beta + v^\beta|_\rho)\mathbf{g}_\beta,$$

and if
$$\mathbf{P} = P^\beta\mathbf{g}_\beta,$$
it follows that

$$P^\beta|_\alpha = \epsilon_{\alpha\gamma}n^{\gamma\rho}(\delta_\rho^\beta + v^\beta|_\rho) = \frac{\partial \theta'^\lambda}{\partial \theta^\alpha}\frac{\partial \theta^\rho}{\partial \theta'^\mu}\epsilon_{\lambda\gamma}'n'^{\gamma\mu}(\delta_\rho^\beta + v^\beta|_\rho). \tag{7.7.20}$$

Choosing $\theta_\alpha = x_\alpha$ so that $\delta_\rho^\beta + v^\beta|_\rho = \partial y_\beta/\partial x_\alpha$, we have

$$\frac{\partial P_\beta}{\partial x_\alpha} = \frac{\partial y_\beta}{\partial \theta'^\mu}\epsilon_{\lambda\gamma}'n'^{\gamma\mu}\frac{\partial \theta'^\lambda}{\partial x_\alpha},$$

or
$$P_{\beta,\alpha} = y_{\beta,\mu}\epsilon_{\alpha\gamma}'n'^{\gamma\mu}, \tag{7.7.21}$$

the comma again denoting differentiation with respect to θ_α'. Also, P_1, P_2 are now defined by (7.7.17).

7.8. Solutions of constraint equations for a sheet reinforced with initially straight cords

Boundary value problems for the system of equations (7.4.9) have been examined by Rivlin† by the following methods.

† Loc. cit., p. 228.

Consider first a parallelogram $ABCD$ bounded by two pairs of characteristic lines $\theta'_\lambda = \text{constant}$, the boundaries AB, CD, BC, DA being the lines $\theta'_1 = \alpha_1$, $\theta'_1 = \alpha_2$, $\theta'_2 = \beta_1$, $\theta'_2 = \beta_2$ respectively ($\alpha_1 < \alpha_2$, $\beta_1 < \beta_2$) (Fig. 7.1). If the displacement component in the x_1-direction, and hence also y_1, is specified along the sides AB and BC, we have along AB

$$y_1 = y_1(\alpha_1, \theta'_2) = f_1(\alpha_1) + f_2(\theta'_2) \qquad (\beta_1 \leqslant \theta'_2 \leqslant \beta_2), \qquad (7.8.1)$$

and along BC

$$y_1 = y_1(\theta'_1, \beta_1) = f_1(\theta'_1) + f_2(\beta_1) \qquad (\alpha_1 \leqslant \theta'_1 \leqslant \alpha_2), \qquad (7.8.2)$$

using the solution (7.7.1) for y_1. Also, at B we have

$$y_1 = y_1(\alpha_1, \beta_1) = f_1(\alpha_1) + f_2(\beta_1). \qquad (7.8.3)$$

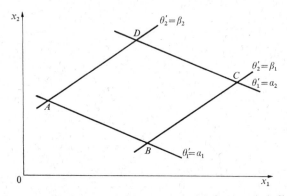

FIG. 7.1. Region bounded by two pairs of characteristic lines.

By adding (7.8.1) and (7.8.2) and subtracting (7.8.3) we thus obtain

$$y_1(\alpha_1, \theta'_2) + y_1(\theta'_1, \beta_1) - y_1(\alpha_1, \beta_1)$$
$$= f_1(\theta'_1) + f_2(\theta'_2) = y_1(\theta'_1, \theta'_2) \quad (\alpha_1 \leqslant \theta'_1 \leqslant \alpha_2, \beta_1 \leqslant \theta'_2 \leqslant \beta_2). \quad (7.8.4)$$

All terms in the left-hand member of this equation are specified within the ranges indicated and the displacement y_1 can therefore be determined throughout the parallelogram $ABCD$. Furthermore, by differentiation of (7.8.4), we may determine the quantities

$$f'_1(\theta'_1) = \frac{dy_1(\theta'_1, \beta_1)}{d\theta'^1}, \qquad f'_2(\theta'_2) = \frac{dy_1(\alpha_2, \theta'_2)}{d\theta'^2}, \qquad (7.8.5)$$

throughout the region $ABCD$. From (7.7.2) we then have

$$g'_1 = \pm(a'_{11} - f'^2_1)^{\frac{1}{2}}, \qquad g'_2 = \pm(a'_{22} - f'^2_2)^{\frac{1}{2}}. \qquad (7.8.6)$$

Two of the four possible combinations of signs in (7.8.6) may be excluded by using the condition

$$\frac{\partial(y_1, y_2)}{\partial(x_1, x_2)} > 0, \qquad (7.8.7)$$

and remembering that

$$\frac{\partial(y_1, y_2)}{\partial(x_1, x_2)} = \frac{\partial(y_1, y_2)}{\partial(\theta'_1, \theta'_2)} \frac{\partial(\theta'_1, \theta'_2)}{\partial(x_1, x_2)} = \frac{f'_1 g'_2 - f'_2 g'_1}{\sqrt{a'}}. \qquad (7.8.8)$$

In (7.8.8), $\sqrt{a'}$ is determined uniquely by the equations of the paths of the cords in the undeformed sheet. If, therefore, the inequality (7.8.7) can be satisfied by choosing both signs in (7.8.6) to be positive, it cannot be satisfied if both of the negative signs are taken. Similarly, one of the remaining pair of possible choices of signs in (7.8.6) may be excluded.

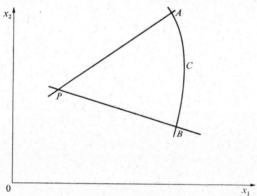

FIG. 7.2. Region enclosed by pair of characteristics and curve C not parallel to them.

Thus, if y_1 is specified on AB and BC it is completely determined throughout the parallelogram $ABCD$. Unless some other condition is given, however, there is an ambiguity of sign in the determination of g'_1 and g'_2 and this implies that we cannot determine y_2 unambiguously throughout the region $ABCD$. We observe from (7.7.2) that to ensure real values for the displacement components $y_1 - x_1$, $y_2 - x_2$ we must have

$$f'^2_1 \leqslant a'_{11}, \quad g'^2_1 \leqslant a'_{11}, \quad f'^2_2 \leqslant a'_{22}, \quad g'^2_2 \leqslant a'_{22}. \qquad (7.8.9)$$

Suppose now that both y_1 and y_2 are specified at all values of θ'_1, θ'_2 corresponding to points on a section AB of a curve C in the plane $x_3 = 0$ of the undeformed sheet, the curve C being at no point parallel to a characteristic line. The deformation in the plane $x_3 = 0$ may then be determined completely within the region bounded by AB and the characteristic lines PA, PB through its ends (Fig. 7.2). If we regard the coordinates of any point of C to be specified in terms of a parameter t,† so that $x_\alpha = x_\alpha(t)$ or $\theta'_\alpha = \theta'_\alpha(t)$, the quantities y_α and hence their derivatives dy_α/dt may be regarded as known functions of t along the arc AB. Writing for brevity

$$\frac{dy_\alpha}{dt} = \dot{y}_\alpha, \qquad \frac{d\theta'_\alpha}{dt} = p_\alpha(t) = p_\alpha, \qquad (7.8.10)$$

we then have, along AB,

$$\dot{y}_1 = f'_1 p_1 + f'_2 p_2, \qquad \dot{y}_2 = g'_1 p_1 + g'_2 p_2, \qquad (7.8.11)$$

from which

$$\left.\begin{aligned} p_1^2 f'^2_1 &= \dot{y}_1^2 - 2\dot{y}_1 p_2 f'_2 + p_2^2 f'^2_2 \\ p_1^2 g'^2_1 &= \dot{y}_2^2 - 2\dot{y}_2 p_2 g'_2 + p_2^2 g'^2_2 \end{aligned}\right\}. \qquad (7.8.12)$$

Adding and making use of (7.7.2) we have

$$2\dot{y}_2 p_2 g'_2 = \chi_1 - 2\dot{y}_1 p_2 f'_2, \qquad (7.8.13)$$

where

$$\chi_1 = \dot{y}_1^2 + \dot{y}_2^2 + p_2^2 a'_{22} - p_1^2 a'_{11}. \qquad (7.8.14)$$

By eliminating g'_2 between (7.8.13) and the second of the constraint conditions (7.7.2), we obtain

$$4p_2^2(\dot{y}_1^2 + \dot{y}_2^2)f'^2_2 - 4\dot{y}_1 \chi_1 p_2 f'_2 + \chi_1^2 - 4\dot{y}_2^2 p_2^2 a'_{22} = 0,$$

which may be solved to yield

$$f'_2 = \frac{\chi_1 \pm \gamma \chi_2}{2p_2(1+\gamma^2)\dot{y}_1}, \qquad (7.8.15)$$

where

$$\left.\begin{aligned} \chi_2 &= [4p_2^2(\dot{y}_1^2 + \dot{y}_2^2)a'_{22} - \chi_1^2]^{\frac{1}{2}} \\ \gamma &= \dot{y}_2/\dot{y}_1 \end{aligned}\right\} \qquad (7.8.16)$$

The values of the remaining first derivatives of f_α and g_α along AB may be determined from (7.8.11) and (7.8.13). If we choose the upper sign in (7.8.15) we obtain

$$g'_2 = \frac{\gamma\chi_1 - \chi_2}{2p_2(1+\gamma^2)\dot{y}_1}, \qquad (7.8.17)$$

while the lower sign yields

$$g'_2 = \frac{\gamma\chi_1 + \chi_2}{2p_2(1+\gamma^2)\dot{y}_1}, \qquad (7.8.18)$$

and in each case

$$f'_1 = (\dot{y}_1 - p_2 f'_2)/p_1, \qquad g'_1 = (\dot{y}_2 - p_2 g'_2)/p_1. \qquad (7.8.19)$$

† This notation is only used in the present section, so that no confusion need arise with the use of t to denote time elsewhere.

The correct choice of signs in (7.8.15) may be determined by making use of the condition (7.8.7). If the positive sign is chosen, we have from (7.8.15), (7.8.17), (7.8.19), and (7.8.8)

$$\frac{\partial(y_1, y_2)}{\partial(x_1, x_2)} = \frac{\partial(y_1, y_2)}{\partial(\theta'_1, \theta'_2)} \bigg/ \sqrt{a'} = -\frac{\chi_2}{2p_1 p_2 \sqrt{a'}},$$

while the choice of the negative sign in (7.8.15) leads to

$$\frac{\partial(y_1, y_2)}{\partial(x_1, x_2)} = \frac{\chi^2}{2p_1 p_2 \sqrt{a'}}.$$

Evidently only one set of solutions for f'_1, f'_2, g'_1 and g'_2 can satisfy (7.8.7). Thus, subject to this latter condition, f'_1, f'_2, g'_1, and g'_2 can be

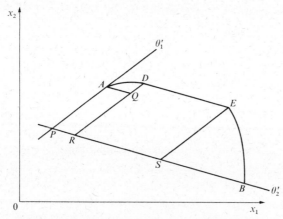

FIG. 7.3. Region bounded by pair of characteristics and curve $ADEB$ containing section DE parallel to a characteristic.

determined uniquely along the arc AB and, since the equations (7.4.10) are hyperbolic, y_1 and y_2 are also uniquely determined in the region bounded by the curve AB and the characteristic lines PA, PB.

If a section DE of the arc AB is parallel to one of the characteristic lines (Fig. 7.3), the deformation can still be determined uniquely within the region bounded by the composite line $ADEB$ and the characteristics PA, PB through its ends. For since y_1 and y_2 are specified on the arcs AD and EB they may again be uniquely determined throughout the two regions ADQ and BES bounded by the arcs AD, EB and the characteristic lines AQ, DQ, BS, and ES. Since both y_1 and y_2 are given on DE and can be determined along ES, it follows from the earlier discussion that they can be determined uniquely throughout the parallelogram bounded by the characteristics DE, ES, SR, and RD. Similarly, the values of y_1 and y_2 determined along AQ and QR are sufficient to determine y_1 and y_2 uniquely throughout the remaining region $AQRP$.

7.9. Plane deformation of a network of inextensible cords

The theory developed in the preceding sections may be employed to analyse the plane deformation of a network of thin, flexible, inextensible cords. It is assumed that in the undeformed state the cords lie in two families of parallel straight lines and that when the net is deformed, no two cords of a family are brought into contact. It is also assumed that intersecting cords of the two families cannot move relative to each other at the point of intersection during the deformation. Under these conditions the analysis of §§ 7.4–7.8 continues to apply provided that the terms due to the elastic material are omitted from the expressions for the stresses, and the metric tensors $a'_{\alpha\beta}$, $a'^{\alpha\beta}$ are assumed to have constant values appropriate to a system of initially straight cords. The solution (7.7.1), subject to the conditions (7.7.2), of the constraint conditions (7.4.9) again applies, and the treatment of the various types of boundary value problem given in § 7.8 is unaffected.

Remembering that $n'^{\alpha\beta}_e = 0$ and $n'^{\alpha\beta} = n'^{\alpha\beta}_c$, equation (7.7.7) now gives
$$\phi_{,12}/A' = -n'^{12} = 0, \qquad (7.9.1)$$

with the solution
$$\phi = h_1(\theta'_1) + h_2(\theta'_2) \qquad (7.9.2)$$

replacing (7.7.8). Also, from (7.7.11) and (7.5.2) we have

$$\left.\begin{aligned} n'^{11} &= \sigma_1 \sqrt{\left(\frac{a'_{22}}{a'_{11}A'}\right)} = \left\{\frac{\partial}{\partial \theta'^2}\left(\frac{h'_2}{\sqrt{A'}}\right) + \frac{h'_1}{A'}(f'_2 g''_2 - g'_2 f''_2)\right\}\Big/ \sqrt{A'} \\ n'^{22} &= \sigma_2 \sqrt{\left(\frac{a'_{11}}{a'_{22}A'}\right)} = \left\{\frac{\partial}{\partial \theta'^1}\left(\frac{h'_1}{\sqrt{A'}}\right) + \frac{h'_2}{A'}(g'_1 f''_1 - f'_1 g''_1)\right\}\Big/ \sqrt{A'} \end{aligned}\right\}, \quad (7.9.3)$$

and the formulae (7.7.18) and (7.7.19) reduce to

$$P_1 = \frac{f'_2 h'_1 - f'_1 h'_2}{f'_1 g'_2 - f'_2 g'_1}, \qquad P_2 = \frac{g'_2 h'_1 - g'_1 h'_2}{f'_1 g'_2 - f'_2 g'_1}, \qquad (7.9.4)$$

and
$$h'_1 = f'_1 P_2 - g'_1 P_1, \qquad h'_2 = f'_2 P_2 - g'_2 P_1, \qquad (7.9.5)$$

7.9 REINFORCEMENT BY INEXTENSIBLE CORDS

respectively. Again, by combining (7.5.2) with (7.7.21) we obtain

$$\left.\begin{array}{l}\sigma_1\sqrt{\left(\dfrac{a'_{22}}{a'_{11}}\right)} = -\dfrac{P_{1,2}}{f'_1} = -\dfrac{P_{2,2}}{g'_1} \\[2mm] \sigma_2\sqrt{\left(\dfrac{a'_{11}}{a'_{22}}\right)} = \dfrac{P_{1,1}}{f'_2} = \dfrac{P_{2,1}}{g'_2}\end{array}\right\}, \qquad (7.9.6)$$

since $n_e'^{12} = 0$.

From (7.9.6) we may infer that

$$a'_{22}\frac{\partial}{\partial \theta'^1}(f'_1\sigma_1) + a'_{11}\frac{\partial}{\partial \theta'^2}(f'_2\sigma_2) = 0,$$

$$a'_{22}\frac{\partial}{\partial \theta'^1}(g'_1\sigma_1) + a'_{11}\frac{\partial}{\partial \theta'^2}(g'_2\sigma_2) = 0,$$

and hence, remembering that $\sqrt{A'} = f'_1 g'_2 - f'_2 g'_1$,

$$\left.\begin{array}{l}\sigma_1 = \dfrac{a'_{11}}{a'_{22}(g'_1 f''_1 - f'_1 g''_1)}\dfrac{\partial}{\partial \theta'^2}(\sigma_2\sqrt{A'}) \\[2mm] \sigma_2 = \dfrac{a'_{22}}{a'_{11}(f'_2 g''_2 - g'_2 f''_2)}\dfrac{\partial}{\partial \theta'^1}(\sigma_1\sqrt{A'})\end{array}\right\}. \qquad (7.9.7)$$

These relations also follow from the equations of equilibrium

$$n'^{\alpha\beta}\|'_\alpha = 0, \qquad (7.9.8)$$

if we make use of (7.5.2) and (7.7.4). The elimination of σ_2 between equations (7.9.7) yields the hyperbolic differential equation

$$\sigma_1 = \frac{1}{g'_1 f''_1 - f'_1 g''_1}\frac{\partial}{\partial \theta'^2}\left[\frac{\sqrt{A'}}{f'_2 g''_2 - g'_2 f''_2}\frac{\partial}{\partial \theta'^1}(\sigma_1\sqrt{A'})\right], \qquad (7.9.9)$$

for σ_1; similarly by eliminating σ_1 we obtain a corresponding hyperbolic equation for σ_2.

By means of the formulae (7.9.5) and (7.9.6) the stresses and the deformation may be determined throughout a network if the applied tractions acting around its edges in its plane are specified. Suppose the network to be bounded by a closed curve C (Fig. 7.4) composed of simple smooth arcs. At the junctions of these arcs, the slope of the tangent may or may not change discontinuously. We assume that the sheet is deformed by forces in its plane acting around the curve C so that it remains plane after deformation. The vector $\mathbf{P}$ and hence also its components P_1, P_2 are therefore known functions of θ'_1, θ'_2 on the boundary curve C.

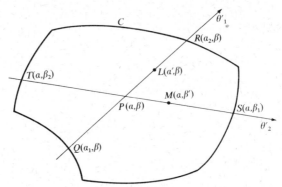

Fig. 7.4. Determination of deformation from stress boundary conditions.

Let P be any point within the region bounded by C with coordinates $(\theta_1', \theta_2') = (\alpha, \beta)$ and let the cords $\theta_2' = $ constant, $\theta_1' = $ constant through P intersect the curve C in the points Q, R, S, T, the coordinates of these points being (α_1, β), (α_2, β), (α, β_1), and (α, β_2), respectively. From (7.9.5) we then have at the points S, T

$$\left.\begin{aligned} h_1'(\alpha) &= f_1'(\alpha)P_2(\alpha, \beta_1) - g_1'(\alpha)P_1(\alpha, \beta_1) \\ h_1'(\alpha) &= f_1'(\alpha)P_2(\alpha, \beta_2) - g_1'(\alpha)P_1(\alpha, \beta_2) \end{aligned}\right\}, \quad (7.9.10)$$

and at the points Q, R

$$\left.\begin{aligned} h_2'(\beta) &= f_2'(\beta)P_2(\alpha_1, \beta) - g_2'(\beta)P_1(\alpha_1, \beta) \\ h_2'(\beta) &= f_2'(\beta)P_2(\alpha_2, \beta) - g_2'(\beta)P_1(\alpha_2, \beta) \end{aligned}\right\}. \quad (7.9.11)$$

From (7.9.10) we thus have

$$\left.\begin{aligned} \frac{g_1'(\alpha)}{f_1'(\alpha)} &= \frac{P_2(\alpha, \beta_1) - P_2(\alpha, \beta_2)}{P_1(\alpha, \beta_1) - P_1(\alpha, \beta_2)} \\ \frac{h_1'(\alpha)}{f_1'(\alpha)} &= \frac{P_1(\alpha, \beta_1)P_2(\alpha, \beta_2) - P_1(\alpha, \beta_2)P_2(\alpha, \beta_1)}{P_1(\alpha, \beta_1) - P_1(\alpha, \beta_2)} \end{aligned}\right\}, \quad (7.9.12)$$

and from (7.9.11)

$$\left.\begin{aligned} \frac{g_2'(\beta)}{f_2'(\beta)} &= \frac{P_2(\alpha_1, \beta) - P_2(\alpha_2, \beta)}{P_1(\alpha_1, \beta) - P_1(\alpha_2, \beta)} \\ \frac{h_2'(\beta)}{f_2'(\beta)} &= \frac{P_1(\alpha_1, \beta)P_2(\alpha_2, \beta) - P_1(\alpha_2, \beta)P_2(\alpha_1, \beta)}{P_1(\alpha_1, \beta) - P_1(\alpha_2, \beta)} \end{aligned}\right\}. \quad (7.9.13)$$

Since the values of P_1 and P_2 are known at all points Q, R, S, and T lying on the boundary C, the relations (7.9.12) and (7.9.13) in general uniquely determine the values of the ratios g_1'/f_1', g_2'/f_2', h_1'/f_1', h_2'/f_2' at all points of the sheet. Equations (7.9.4) then yield values for P_1 and P_2 at all points of the network and, remembering that a_{11}', a_{22}' are known constants, the quantities $f_1'\sigma_1$ and $f_2'\sigma_2$ (or $g_1'\sigma_1$ and $g_2'\sigma_2$) are uniquely determined from (7.9.6).

Since the ratio g_1'/f_1' is known throughout the sheet, f_1' and g_1' may be found from the first of (7.7.2) apart from an ambiguity which allows the sign of both of them to be changed simultaneously at any point; f_2' and g_2' may be found from the ratio g_2'/f_2' and the second of (7.7.2) apart from a similar ambiguity. Furthermore, since

$$\frac{\partial(y_1, y_2)}{\partial(x_1, x_2)} = \frac{f_1'g_2' - f_2'g_1'}{\sqrt{a'}} > 0$$

at each point, it follows that if the signs of f_1' and g_1' are changed at any point, the signs of f_2' and g_2' must also be reversed. This would also imply a reversal of the signs of σ_1, σ_2, h_1', and h_2', since the quantities $f_1'\sigma_1$, $f_2'\sigma_2$, h_1'/f_1', and h_2'/f_2' have already been uniquely determined. Thus f_1', f_2', g_1', g_2', h_1', h_2', σ_1, and σ_2 can be found uniquely at each point apart from an ambiguity which allows the signs of all of them to be changed simultaneously. Moreover, a reversal of sign at any one point is readily seen to involve a change of sign throughout the entire sheet. For if, in Fig. 7.4, L and M are any points (α', β), (α, β') respectively on the cords through $P(\alpha, \beta)$, then f_2', g_2', h_2' have the same values at L and P, and f_1', g_1', h_1' have the same values at M and P. A change of sign of these quantities at P therefore leads to a similar change of sign of all of the quantities at L and M. Since it is possible to reach any point of the sheet from any other by passing alternately along cords of the families θ_1' = constant and θ_2' = constant, a change of the signs of f_1', f_2', g_1', g_2', h_1', h_2', σ_1, and σ_2 at any one point implies a corresponding change at all points of the sheet.

Thus f_1', f_2', g_1', g_2', h_1', h_2', σ_1, and σ_2 are uniquely determined apart from the possibility that the signs of all of them may be simultaneously changed at all points of the sheet. Such a change of sign corresponds to a rigid-body rotation of the sheet through two right angles in its plane before the edge tractions are applied.† When f_1', f_2', g_1', g_2' are known, the displacements in the plane of the sheet may be found, except for an arbitrary rigid-body translation, by making use of (7.7.1).

† In the present case, a reversal of the signs of σ_1, σ_2 implies a change in the signs to be affixed to δ_1, δ_2 and does not affect the tensions τ_1, τ_2 in the individual cords.

7.10. Cylindrically symmetric deformations: constraint conditions

We examine here, for reinforced materials, the cylindrically symmetric deformation considered in § 2.1 for unreinforced bodies.

We again choose polar coordinate systems (ρ, ϑ, x_3), (r, θ, y_3) in the undeformed body and the deformed body respectively and consider the deformation defined by

$$r = r(\rho), \qquad \theta = a\vartheta + bx_3, \qquad y_3 = c\vartheta + dx_3, \qquad (7.10.1)$$

$a, b, c,$ and d being constants. We suppose the undeformed body to be bounded by the cylindrical surfaces $\rho = A_1$, $\rho = A_2$ and to contain layers of cords lying in the surfaces $\rho = \rho_i$ $(i = 1, 2, ..., n)$, where $A_1 \geqslant \rho_1 \geqslant \rho_2 \geqslant ... \geqslant \rho_n \geqslant A_2$. The boundary surfaces are carried by the deformation (7.10.1) into the surfaces $r = A_1'$, $r = A_2'$ respectively, and the layer of cords initially at $\rho = \rho_i$ is displaced to $r = r_i$. We assume that the cords in each layer follow systems of helical paths

$$x_3 = \rho_i \vartheta \cot \alpha_i + \text{constant}, \qquad \rho = \rho_i \quad (i = 1, 2, ..., n) \qquad (7.10.2)$$

in the undeformed body. Each cord in the layer at $\rho = \rho_i$ therefore cuts generators of this surface at an angle α_i, this angle being changed to β_i by the deformation. Two sets of cords i and j will be described as independent if typical paths of the sets i, j cannot coincide, that is, if $\rho_i \neq \rho_j$ or if $\rho_i = \rho_j$ and $\alpha_i \neq m\pi + \alpha_j$ where m is an integer.

As in previous sections, the cords of each family are assumed to be thin, inextensible, and perfectly flexible and to be sufficiently close together to have the effect of establishing a surface with directional properties in the undeformed body. If we assume that there is no relative motion between each cord and the adjacent elastic material during deformation, each family of cords introduces a functional relationship between the parameters a, b, c, d and the values which r may assume for a specified value of ρ. For if ds is an element of length lying along a path taken by one of the cords in the layer at $\rho = \rho_i$ we have before deformation

$$ds \cos \alpha_i = dx_3, \qquad ds \sin \alpha_i = \rho_i d\vartheta,$$

and after deformation

$$ds \cos \beta_i = dy_3 = c\, d\vartheta + d\, dx_3,$$
$$ds \sin \beta_i = r_i d\theta = r_i(a\, d\vartheta + b\, dx_3).$$

Combining these relations we obtain

$$\left. \begin{aligned} \cos \beta_i &= (c/\rho_i)\sin \alpha_i + d \cos \alpha_i \\ \sin \beta_i &= r_i[(a/\rho_i)\sin \alpha_i + b \cos \alpha_i] \end{aligned} \right\}, \qquad (7.10.3)$$

and hence

$$r_i^2(a \sin \alpha_i + b\rho_i \cos \alpha_i)^2 + (c \sin \alpha_i + d\rho_i \cos \alpha_i)^2 = \rho_i^2. \qquad (7.10.4)$$

It follows from § 1.17 that each surface $\rho = \rho_i$ must contain not more than two independent sets of cords if deformation within this surface is to be possible. This conclusion also follows from (7.10.4).†

When the elastic material between adjacent layers of cords is compressible, the geometrical constraints (7.10.4) do not necessarily impose a limitation upon the permissible number of surfaces containing a single family of cords; for given values of a, b, c, and d each equation of the type (7.10.4) merely determines a value which r must assume for certain specified values ρ_i of ρ. Henceforth we shall confine our attention to incompressible materials for which the conditions (2.1.12) apply. We thus have

$$\rho^2 = \lambda r^2 + K \qquad (\lambda = ad - bc > 0), \qquad (7.10.5)$$

and
$$\rho_i^2 = \lambda r_i^2 + K. \qquad (7.10.6)$$

If r_i is eliminated between (7.10.4) and (7.10.6) the resulting equation may be regarded as a relationship connecting the five constants a, b, c, d, and K which define the deformation. Five independent families of cords would give rise to five independent relationships of this kind which could, in principle, be solved to give a finite number of values for a, b, c, d, and K. It follows that a continuously varying deformation of the type (7.10.1) is usually only possible if there are not more than four independent sets of cords.

In general, each set of cords reduces by unity the number of degrees of freedom available to the deformation (7.10.1) and conversely each limitation upon this deformation imposed by restricting one of the constants a, b, c, d, K to have its initial value diminishes by unity the number of independent sets of cords permissible. An exception occurs when $b = c = 0$, so that (7.10.4) is unchanged by changing the sign of α_i. Each layer may then contain two independent families of cords making equal angles with generators of the surface in which they lie. For simple flexure, two such double layers are admissible; in the case of a cylindrical tube submitted to combined inflation and extension, the additional restriction $a = 1$ permits only one double layer.

† J. E. Adkins, loc. cit., p. 228.

7.11. Cylindrically symmetric problems: stress resultants in a layer of cords

As in the problems considered in §§ 7.4–7.9, the tensions in the cords occurring as a result of the deformation give rise to a system of stress resultants. Referred to the system of coordinates $(\theta_1, \theta_2, \theta_3) = (r, \theta, y_3)$ in the deformed body, we denote the stress resultants in the layer of cords at $r = r_i$ by n_i^{rs} so that n_i^{22}, n_i^{33} act in the transverse and longitudinal directions respectively and $n_i^{1s} = n_i^{s1} = 0$.† Initially, we assume each layer to contain only one set of cords, the set in the layer at $\rho = \rho_i$ following the paths (7.10.2). We denote by δ_i the distance between adjacent cords in the undeformed layer *measured along a curve orthogonal to them*. The tension in each cord is denoted by τ_i. The spacing of the cords need not be uniform throughout the layer, but in the class of problems considered the quotient $\sigma_i = \tau_i/\delta_i$ is constant.

In the deformed layer at $r = r_i$ let the lines $y_3 =$ constant, $\theta =$ constant respectively be intersected by l_{i2}, l_{i3} cords per unit length. These quantities may be expressed in terms of δ_i by observing that the helix

$$\rho = \rho_i, \qquad x_3 = \rho_i \vartheta \cot \alpha_i + k \qquad (7.11.1)$$

is carried by the deformation (7.10.1) into the curve

$$r = r_i, \qquad \theta = (a + b\rho_i \cot \alpha_i)\vartheta + kb,$$

$$y_3 = (c + d\rho_i \cot \alpha_i)\vartheta + kd,$$

or, when ϑ is eliminated,

$$r = r_i, \qquad y_3 = \frac{(c + d\rho_i \cot \alpha_i)\theta + k\lambda}{a + b\rho_i \cot \alpha_i}. \qquad (7.11.2)$$

In passing to a point on a neighbouring curve, the increments Δy_3, $\Delta \theta$, and Δk of y_3, θ, and k must therefore be connected by the relation

$$\Delta y_3 = \frac{(c + d\rho_i \cot \alpha_i) \Delta\theta + \lambda \Delta k}{a + b\rho_i \cot \alpha_i}. \qquad (7.11.3)$$

If neighbouring curves of the family (7.11.2) coincide with adjacent cords, $\Delta k = \delta_i \operatorname{cosec} \alpha_i$ and in this case, when $\Delta y_3 = 0, r_i \Delta\theta = -1/l_{i2}$, and when $\Delta\theta = 0$, $\Delta y_3 = 1/l_{i3}$. From (7.11.3) we thus obtain

$$\left. \begin{array}{l} l_{i2} = (c \sin \alpha_i + d\rho_i \cos \alpha_i)/(\lambda r_i \delta_i) \\ l_{i3} = (a \sin \alpha_i + b\rho_i \cos \alpha_i)/(\lambda \delta_i) \end{array} \right\}. \qquad (7.11.4)$$

Remembering (7.10.3), a simple resolution of forces in the layer

---

† Here and subsequently we use the notation $n_i^{\alpha\beta}$ ($\alpha, \beta = 2, 3$) to denote *physical* components of stress in the layer of cords at $r = r_i$.

$r = r_i$ now yields

$$\left.\begin{aligned}
n_i^{22} &= l_{i3}\tau_i \sin \beta_i = \sigma_i r_i (a \sin \alpha_i + b\rho_i \cos \alpha_i)^2/(\lambda \rho_i) \\
n_i^{33} &= l_{i2}\tau_i \cos \beta_i = \sigma_i (c \sin \alpha_i + d\rho_i \cos \alpha_i)^2/(\lambda \rho_i r_i) \\
n_i^{23} &= l_{i2}\tau_i \sin \beta_i = l_{i3}\tau_i \cos \beta_i \\
&= \sigma_i (a \sin \alpha_i + b\rho_i \cos \alpha_i)(c \sin \alpha_i + d\rho_i \cos \alpha_i)/(\lambda \rho_i)
\end{aligned}\right\}, \quad (7.11.5)$$

where i is not summed, and from these, with (7.10.4), we have

$$\left.\begin{aligned}
\lambda r_i(n_i^{22} + n_i^{33}) &= \rho_i \sigma_i \\
n_i^{22} n_i^{33} &= (n_i^{23})^2
\end{aligned}\right\}. \quad (7.11.6)$$

The stress resultants for a layer containing two families of cords may be obtained by adding expressions of the type (7.11.5) for each set of cords.

7.12. Cylindrical symmetry: forces on the composite body

To determine the resultant forces on the composite body, we consider separately the deformation of each section of the elastic material under the influence of surface tractions exerted by the adjacent layers of cords, and the equilibrium of each layer of cords under the influence of the forces exerted by the elastic material on either side. The normal components of the surface forces acting on the boundaries of the elastic material adjoining the layer of cords initially at $\rho = \rho_i$ are denoted by Π_{i1}, Π_{i2}. These components are regarded as positive when directed outwards from the material on which they act and are measured per unit area of deformed surface; the forces Π_{i1}, Π_{i2} act on the sections of the elastic material which lie in the undeformed body nearer to the surfaces $\rho = A_1$, $\rho = A_2$ respectively. The normal components of the surface traction on the latter surfaces are denoted by R_1, R_2 respectively.

For incompressible elastic materials, the stress components τ^{ij} are given by (2.1.20) and (2.1.18) or (2.10.6) respectively, according as the elastic body is transversely isotropic about the x_3-axis or exhibits a general type of cylindrical aeolotropy subject only to the restriction described in § 2.10. Alternatively, we may use the parametric forms (2.18.10)–(2.18.14). In each case we may write

$$\left.\begin{aligned}
\tau^{11} &= \int_{r_0}^{r} f_1(r)\, dr + \tau_0^{11} \\
r^2 \tau^{22} &= \tau^{11} + r^2 f_2(r) \\
\tau^{33} &= \tau^{11} + f_3(r) \\
\tau^{23} &= f_4(r)
\end{aligned}\right\}, \quad (7.12.1)$$

where $f_i(r)$ are continuous functions of r throughout each section of the elastic material; these functions depend also upon the parameters a, b, c, and d. If we use the forms of § 2.18 which are independent of symmetries in the elastic material we have, from (2.18.10)–(2.18.14),

$$\begin{aligned} f_1(r) = rf_2(r) &= \frac{2}{r}\left(a\frac{\partial W}{\partial a}+b\frac{\partial W}{\partial b}\right) \\ &= \frac{\lambda r}{K}\left(\rho\frac{\partial W}{\partial \rho}+c\frac{\partial W}{\partial c}-b\frac{\partial W}{\partial b}\right) \\ f_3(r) &= a\frac{\partial W}{\partial a}+b\frac{\partial W}{\partial b}+c\frac{\partial W}{\partial c}+d\frac{\partial W}{\partial d} \\ f_4(r) &= a\frac{\partial W}{\partial c}+b\frac{\partial W}{\partial d} = \frac{1}{r^2}\left(c\frac{\partial W}{\partial a}+d\frac{\partial W}{\partial b}\right) \end{aligned} \qquad (7.12.2)$$

The relationship between $f_1(r)$ and $f_2(r)$ follows from the first of the equations of equilibrium (2.1.19) which holds throughout each section of the elastic material and may be written

$$r^2\tau^{22} = \frac{d}{dr}(r\tau^{11}). \qquad (7.12.3)$$

We shall suppose that the elastic material possesses suitable symmetry properties, as discussed in § 2.10, so that for the deformation (7.10.1) $\tau^{12} = \tau^{13} = 0$. Materials transversely isotropic with respect to the x_3-direction satisfy the required conditions, and the appropriate forms of $f_i(r)$ for these may be deduced from a comparison of (7.12.1) and (2.1.18).

If there are n sets of cords, where for an incompressible body $n \leqslant 4$, by applying the first of equations (7.12.1) in turn to each section of the elastic material, we obtain

$$\begin{aligned} R_1 &= \int_{r_1}^{A_1'} f_1(r)\,dr + \Pi_{11} \\ \Pi_{i2} &= \int_{r_{i+1}}^{r_i} f_1(r)\,dr + \Pi_{i+1,1} \quad (i = 1, 2, \ldots, n-1) \\ \Pi_{n2} &= \int_{A_2}^{r_n} f_1(r)\,dr + R_2 \end{aligned} \qquad (7.12.4)$$

Also, we may regard the layer of cords at $r = r_i$ as a thin flexible shell in equilibrium under the influence of a uniform radial force $\Pi_{i1} - \Pi_{i2}$

7.12 REINFORCEMENT BY INEXTENSIBLE CORDS

and the tensions at its boundaries. Under these circumstances, the second of equations (4.11.10) applies with $\kappa_1 = 1/r_i$, $\kappa_2 = 0$, $T_1 = n_i^{22}$, and $P = \Pi_{i1} - \Pi_{i2}$, and we have

$$r_i(\Pi_{i1} - \Pi_{i2}) = n_i^{22} = \sigma_i r_i (a \sin \alpha_i + b\rho_i \cos \alpha_i)^2/(\lambda \rho_i) \qquad (i = 1, 2, \ldots, n).$$
(7.12.5)

Since the cords are thin, there is no discontinuity in ρ or r in crossing any of the surfaces in which they lie. Furthermore, for a given state of deformation, the parameters a, b, c, d have the same constant values throughout the entire reinforced body. It follows, therefore, from (7.10.5) that K takes the same value in each section of the elastic material and r may thus be regarded as a single continuous function of ρ which, with its derivatives, is continuous throughout the entire body. The strain components (2.10.5), the invariants (2.1.13), (2.1.14), the strain energy function W, and the functions $f_i(r)$ may also be regarded as functions of r or ρ, which with their derivatives are continuous throughout the body. The forms of these functions are, in fact, unaffected by the presence of the cords. The principal stresses τ^{ii}, however, are, in general, discontinuous at each layer of cords owing to the presence in (7.12.1) of the constant τ_0^{11}. If τ_0^{11} represents the normal stress at a layer of cords at $r = r_0$, it can assume different values on either side of this surface. Remembering the continuity property of $f_1(r)$ we may combine (7.12.4) and (7.12.5) to obtain

$$\left.\begin{aligned}
R_1 - R_2 &= \int_{A_2'}^{A_1'} f_1(r)\,dr + \sum_{s=1}^{n} \frac{n_s^{22}}{r_s} \\
&= \int_{A_2'}^{A_1'} f_1(r)\,dr + \frac{1}{\lambda}\sum_{s=1}^{n} \frac{\sigma_s}{\rho_s}(a \sin \alpha_s + b\rho_s \cos \alpha_s)^2 \\
R_1 - \Pi_{i1} &= \int_{r_i}^{A_1'} f_1(r)\,dr + \frac{1}{\lambda}\sum_{s=1}^{i-1} \frac{\sigma_s}{\rho_s}(a \sin \alpha_s + b\rho_s \cos \alpha_s)^2 \\
&\qquad\qquad (i = 2, n) \\
\Pi_{i2} - R_2 &= \int_{A_2'}^{r_i} f_1(r)\,dr + \frac{1}{\lambda}\sum_{s=i+1}^{n} \frac{\sigma_s}{\rho_s}(a \sin \alpha_s + b\rho_s \cos \alpha_s)^2 \\
&\qquad\qquad (i = 1, n-1)
\end{aligned}\right\}, \quad (7.12.6)$$

and Π_{11}, Π_{n2} are given by the first and last of equations (7.12.4).

The formulae for τ^{jk} and $n_i^{\alpha\beta}$ may be employed to write down expressions for the resultant forces and couples across planes in the deformed body analogous to those derived for an unreinforced material in § 2.18. On a portion of the plane $\theta = $ constant enclosed by the lines $y_3 = l_1$, $y_3 = l_2$ ($l_1 > l_2$), we have a resultant normal force N_2 given from (7.12.3) and (7.12.5) by

$$N_2 = l\left[\int_{r_1}^{A_1'} r^2\tau^{22}\,dr + \sum_{i=1}^{n-1}\int_{r_{i+1}}^{r_i} r^2\tau^{22}\,dr + \int_{A_2'}^{r_n} r^2\tau^{22}\,dr + \sum_{i=1}^{n} n_i^{22}\right]$$

$$= l\left[A_1'R_1 - r_1\Pi_{11} + \sum_{i=1}^{n-1}(r_i\Pi_{i2} - r_{i+1}\Pi_{i+1,1}) + r_n\Pi_{n2} - A_2'R_2 + \sum_{i=1}^{n} n_i^{22}\right]$$

$$= l(A_1'R_1 - A_2'R_2), \qquad (7.12.7)$$

where $l = l_1 - l_2$. Similarly, for the tangential force S_2 acting across this area in the y_3-direction we obtain from (7.12.1) and (7.11.5)

$$S_2 = l\left[\int_{r_1}^{A_1'} r\tau^{23}\,dr + \sum_{i=1}^{n-1}\int_{r_{i+1}}^{r_i} r\tau^{23}\,dr + \int_{A_2'}^{r_n} r\tau^{23}\,dr + \sum_{i=1}^{n} n_i^{23}\right]$$

$$= l\left[\int_{A_2'}^{A_1'} rf_4(r)\,dr + \frac{1}{\lambda}\sum_{i=1}^{n}\frac{\sigma_i}{\rho_i}(a\sin\alpha_i + b\rho_i\cos\alpha_i)(c\sin\alpha_i + d\rho_i\cos\alpha_i)\right], \quad (7.12.8)$$

while the resultant couple M_2 about the y_3-axis is given from (7.12.1), (7.12.3), and (7.12.5) by

$$M_2 = l\left[\int_{r_1}^{A_1'} r^3\tau^{22}\,dr + \sum_{i=1}^{n-1}\int_{r_{i+1}}^{r_i} r^3\tau^{22}\,dr + \int_{A_2'}^{r_n} r^3\tau^{22}\,dr + \sum_{i=1}^{n} r_i n_i^{22}\right]$$

$$= \tfrac{1}{2}l\left\{\int_{A_2'}^{A_1'} r^2 f_1(r)\,dr + [r^2\tau^{11}]_{r_1}^{A_1'} + \sum_{i=1}^{n-1}[r^2\tau^{11}]_{r_{i+1}}^{r_i} + [r^2\tau^{11}]_{A_2'}^{r_n} + 2\sum_{i=1}^{n} r_i n_i^{22}\right\}$$

$$= \tfrac{1}{2}l\left[\int_{A_2'}^{A_1'} r^2 f_1(r)\,dr + A_1'^2 R_1 - A_2'^2 R_2 + \frac{1}{\lambda}\sum_{i=1}^{n}\frac{\sigma_i r_i^2}{\rho_i}(a\sin\alpha_i + b\rho_i\cos\alpha_i)^2\right]. \qquad (7.12.9)$$

Again, on a section of a plane $y_3 = $ constant lying between the lines

$\theta = \theta_1$ and $\theta = \theta_2$ we have a normal force

$$N_3 = (\theta_1-\theta_2)\left[\int_{r_1}^{A_1'} r\tau^{33}\,dr + \sum_{i=1}^{n-1}\int_{r_{i+1}}^{r_i} r\tau^{33}\,dr + \int_{A_2'}^{r_n} r\tau^{33}\,dr + \sum_{i=1}^{n} r_i n_i^{33}\right]$$

$$= \frac{\theta_1-\theta_2}{2}\left\{\int_{A_2'}^{A_1'}[2f_3(r)-rf_1(r)]r\,dr + A_1'^2 R_1 - r_1^2\Pi_{11} + \right.$$

$$\left. + \sum_{i=1}^{n-1}(r_i^2\Pi_{i2} - r_{i+1}^2\Pi_{i+1,1}) + r_n^2\Pi_{n2} - A_2'^2 R_2 + 2\sum_{i=1}^{n} r_i n_i^{33}\right\}$$

$$= \frac{\theta_1-\theta_2}{2}\left\{\int_{A_2'}^{A_1'}[2f_3(r)-rf_1(r)]r\,dr + A_1'^2 R_1 - A_2'^2 R_2 + 2\sum_{i=1}^{n} r_i n_i^{33} - \sum_{i=1}^{n} r_i n_i^{22}\right\}$$

$$= \frac{\theta_1-\theta_2}{2}\left\{\int_{A_2'}^{A_1'}[2f_3(r)-rf_1(r)]r\,dr + A_1'^2 R_1 - A_2'^2 R_2 + \right.$$

$$\left. + \frac{1}{\lambda}\sum_{i=1}^{n}\frac{\sigma_i}{\rho_i}[3(c\sin\alpha_i + d\rho_i\cos\alpha_i)^2 - \rho_i^2]\right\}, \quad (7.12.10)$$

and a couple M_3 about the y_3-axis given by

$$M_3 = (\theta_1-\theta_2)\left[\int_{A_2'}^{A_1'} r^3 f_4(r)\,dr + \sum_{i=1}^{n} r_i^2 n_i^{23}\right]$$

$$= (\theta_1-\theta_2)\left[\int_{A_2'}^{A_1'} r^3 f_4(r)\,dr + \frac{1}{\lambda}\sum_{i=1}^{n}\frac{\sigma_i r_i^2}{\rho_i}(a\sin\alpha_i + b\rho_i\cos\alpha_i)(c\sin\alpha_i + d\rho_i\cos\alpha_i)\right].$$
$$(7.12.11)$$

Remembering (7.10.6), equations (7.10.4), (7.12.8)–(7.12.11), and the first of (7.12.6) may be regarded as $n+5$ equations for the determination of the $n+5$ unknown quantities σ_i, a, b, c, d, and K when the forces applied to the body are known.

7.13. Flexure of an initially curved reinforced block

By taking $n = 2$ and setting

$$\left.\begin{array}{ll} \rho_1 = \rho_2 = \rho_0 & r_1 = r_2 = r_0 \\ \alpha_1 = -\alpha_2 = \alpha, & \beta_1 = -\beta_2 = \beta \\ b = c = 0, & ad = \lambda \end{array}\right\} \quad (7.13.1)$$

in the results of §§ 7.10–7.12, we may examine the simple flexure of an initially curved block containing two independent families of cords symmetrically placed in the surface $\rho = \rho_0$. Before deformation the cords make angles $\pm\alpha$ with the x_3-direction, these angles being changed to $\pm\beta$ by the flexure. The constraint condition (7.10.4) becomes

$$r_0^2 a^2 \sin^2\alpha + \rho_0^2(d^2 \cos^2\alpha - 1) = 0, \tag{7.13.2}$$

and with (7.10.6) this yields the relation

$$\rho_0^2(a \sin^2\alpha + d^3 \cos^2\alpha - d) = Ka \sin^2\alpha \tag{7.13.3}$$

connecting a, d, and K.

We suppose the undeformed body to be bounded by the cylindrical surfaces $\rho = A_1$, $\rho = A_2$ and by the planes $\vartheta = \pm\vartheta_0$, $x_3 = \pm L$. We assume further that $\sigma_1 = \sigma_2 = \sigma$ and that the elastic material possesses suitable symmetry properties, as discussed at the end of § 2.10, so that for the deformation described by (7.10.1) and (7.13.1) $\tau^{12} = \tau^{23} = \tau^{31} = 0$. Then, from (7.11.5),

$$\left.\begin{aligned} n_1^{22} + n_2^{22} &= \frac{2ar_0}{d\rho_0}\sigma \sin^2\alpha \\ n_1^{33} + n_2^{33} &= \frac{2d\rho_0}{ar_0}\sigma \cos^2\alpha \\ n_1^{23} + n_2^{23} &= 0 \end{aligned}\right\}, \tag{7.13.4}$$

and the expressions (7.12.8) and (7.12.11) for S_2 and M_3 vanish. The deformation is thus maintained by normal tractions only acting upon the surfaces of the deformed block.

If the deformation can be maintained without the application of normal tractions to the curved surfaces of the block, so that

$$R_1 = R_2 = 0,$$

it follows from (7.12.7) that $N_2 = 0$, while from (7.13.1), the first of (7.12.6), (7.12.2), and (2.1.12)

$$\sigma = -\frac{d\rho_0}{2a \sin^2\alpha}\int_{A_2}^{A_1'} f_1(r)\, dr$$

$$= -\frac{d\rho_0}{2Ka \sin^2\alpha}\int_{A_2}^{A_1} \rho^2\frac{\partial W}{\partial \rho}\, d\rho. \tag{7.13.5}$$

An expression for the resultant flexural couple M_2 on each of the planes initially at $\vartheta = \pm \vartheta_0$ follows by combining (7.13.5) and (7.13.1) with (7.12.9). We obtain

$$M_2 = Ld \int_{A_2'}^{A_1'} (r^2 - r_0^2) f_1(r)\, dr$$

$$= \frac{L}{Ka} \int_{A_2}^{A_1} (\rho^2 - \rho_0^2) \rho^2 \frac{\partial W}{\partial \rho} d\rho. \tag{7.13.6}$$

Similarly, from (7.12.10) the resultant normal force on the ends initially at $x_3 = \pm L$ is given by

$$N_3 = a\vartheta_0 \left\{ \int_{A_2'}^{A_1'} [2f_3(r) - rf_1(r)] r\, dr - \frac{\rho_0^2 (3d^2 \cos^2\alpha - 1)}{a^2 \sin^2 \alpha} \int_{A_2'}^{A_1'} f_1(r)\, dr \right\}$$

$$= \frac{\vartheta_0}{Ka} \int_{A_2}^{A_1} \left[2Ka \frac{\partial W}{\partial d} - (2d^2 \rho_0^2 \cot^2\alpha - a^2 r_0^2)\rho \frac{\partial W}{\partial \rho} \right] \rho\, d\rho. \tag{7.13.7}$$

When the material is transversely isotropic with respect to the x_3-direction, we obtain from (7.12.1) and (2.5.3)

$$\left.\begin{aligned} f_1(r) &= \frac{1}{r}\left(a^2\mu^2 - \frac{1}{\lambda^2\mu^2}\right)(\Phi + d^2\Psi) \\ f_3(r) &= \left(d^2 - \frac{1}{\lambda^2\mu^2}\right)(\Phi + a^2\mu^2\Psi) + d^2\Theta \\ f_4(r) &= 0 \end{aligned}\right\}, \tag{7.13.8}$$

where $\mu = r/\rho$.

7.14. Inflation and extension of a cylindrical tube

As in the preceding section, we take $n = 2$ and write

$$\left.\begin{aligned} \rho_1 = \rho_2 = \rho_0, \quad & r_1 = r_2 = r_0 \\ \alpha_1 = -\alpha_2 = \alpha, \quad & \beta_1 = -\beta_2 = \beta \end{aligned}\right\}. \tag{7.14.1}$$

By putting $\vartheta_0 = \pi$ and giving the constants in (7.10.1) the values

$$a = 1, \quad b = c = 0, \quad d = \lambda, \tag{7.14.2}$$

we may then examine the simple extension and inflation of a cylindrical tube reinforced with two sets of cords symmetrically inclined at an angle α to the generators of the surface $\rho = \rho_0$ in which they lie. From

(7.10.4), (7.10.5), (7.14.1), and (7.14.2) we have

$$\left.\begin{array}{r}r_0^2\sin^2\alpha = \rho_0^2(1-\lambda^2\cos^2\alpha)\\ K\sin^2\alpha = \rho_0^2(\lambda-1)[(\lambda^2+\lambda+1)\cos^2\alpha-1]\end{array}\right\}. \quad (7.14.3)$$

or

We again assume that $\sigma_1 = \sigma_2 = \sigma$ and that the elastic material possesses symmetry properties, as discussed at the end of § 2.10, sufficient to ensure that τ^{12}, τ^{23}, and τ^{31} are zero when $b = c = 0$. The conditions (7.14.1), (7.14.2), and (7.11.5) now yield

$$\left.\begin{array}{c}n_1^{22}+n_2^{22} = \dfrac{2r_0}{\lambda\rho_0}\sigma\sin^2\alpha, \quad n_1^{33}+n_2^{33} = \dfrac{2\lambda\rho_0}{r_0}\sigma\cos^2\alpha\\ n_1^{23}+n_2^{23} = 0\end{array}\right\}, \quad (7.14.4)$$

and, from (2.1.12), (7.12.2), and (7.12.6),

$$\left.\begin{array}{r}R_1-R_2 = \displaystyle\int_{A_2'}^{A_1'} f_1(r)\,dr + \dfrac{2\sigma}{\lambda\rho_0}\sin^2\alpha\\ = \dfrac{1}{K}\displaystyle\int_{A_2}^{A_1}\rho^3\dfrac{\partial W}{\partial \rho}d\rho + \dfrac{2\sigma}{\lambda\rho_0}\sin^2\alpha\end{array}\right\}. \quad (7.14.5)$$

The resultant longitudinal force N_3 on each end of the tube is given from (7.12.10) by

$$\left.\begin{array}{r}N_3 = \pi\left\{\displaystyle\int_{A_2'}^{A_1'}[2f_3(r)-rf_1(r)]r\,dr + A_1'^2 R_1 - A_2'^2 R_2 + \dfrac{2\sigma\rho_0}{\lambda}(3\lambda^2\cos^2\alpha-1)\right\}\\ = 2\pi\left\{\displaystyle\int_{A_2}^{A_1}\rho\dfrac{\partial W}{\partial \lambda}d\rho + \tfrac{1}{2}(A_1'^2 R_1 - A_2'^2 R_2) + \dfrac{\sigma\rho_0}{\lambda}(3\lambda^2\cos^2\alpha-1)\right\}\end{array}\right\}. \quad (7.14.6)$$

If the deformation can be produced by a simple tensile force N_3 alone, we may put $R_1 = R_2 = 0$ in (7.14.5) and (7.14.6), and make use of (7.14.3) to obtain

$$\left.\begin{array}{r}\sigma = -\dfrac{\lambda\rho_0}{2\sin^2\alpha}\displaystyle\int_{A_2'}^{A_1'}f_1(r)\,dr = -\dfrac{\lambda\rho_0}{2K\sin^2\alpha}\displaystyle\int_{A_2}^{A_1}\rho^2\dfrac{\partial W}{\partial \rho}d\rho\\ N_3 = \dfrac{\pi}{\lambda}\left\{\lambda\displaystyle\int_{A_2'}^{A_1'}[2f_3(r)-rf_1(r)]r\,dr - (\chi\rho_0^2+K)\displaystyle\int_{A_2'}^{A_1'}f_1(r)\,dr\right\}\\ = \dfrac{\pi}{K\lambda}\displaystyle\int_{A_2}^{A_1}\left[2K\lambda\dfrac{\partial W}{\partial \lambda}-(\chi\rho_0^2+K)\rho\dfrac{\partial W}{\partial \rho}\right]\rho\,d\rho\end{array}\right\} \quad (7.14.7)$$

where
$$\chi = 2\lambda^3 \cot^2\alpha - 1. \tag{7.14.8}$$

Similarly, if the deformation is produced only by means of a uniform inflating pressure P acting on the inner surface, we have $R_1 = N_3 = 0$, $R_2 = -P$, and

$$\left.\begin{aligned}\sigma &= -\frac{\lambda\rho_0\int_{A_2}^{A_1}[(A_2^2-K)\rho(\partial W/\partial\rho)+2K\lambda(\partial W/\partial\lambda)]\rho\,d\rho}{2K(\chi\rho_0^2+A_2^2)\sin^2\alpha} \\ P &= -\frac{\int_{A_2}^{A_1}[2K\lambda(\partial W/\partial\lambda)-(\chi\rho_0^2+K)\rho(\partial W/\partial\rho)]\rho\,d\rho}{K(\chi\rho_0^2+A_2^2)}\end{aligned}\right\}. \tag{7.14.9}$$

When the elastic material is transversely isotropic with the axis of anisotropy coinciding with the x_3-axis, the appropriate forms of $f_1(r)$, $f_3(r)$ may be obtained from a comparison of (7.12.1) with (2.1.18). Writing $\mu = r/\rho$ and making use of the conditions (7.14.2), we obtain

$$\left.\begin{aligned}f_1(r) &= \frac{1}{r}\left(\mu^2 - \frac{1}{\lambda^2\mu^2}\right)(\Phi + \lambda^2\Psi) \\ f_3(r) &= \left(\lambda^2 - \frac{1}{\lambda^2\mu^2}\right)(\Phi + \mu^2\Psi) + \lambda^2\Theta\end{aligned}\right\}. \tag{7.14.10}$$

Also, from (7.10.5) and (7.10.6) we have

$$K = \rho^2(1-\lambda\mu^2) = \rho_0^2(1-\lambda\mu_0^2),$$

where $\mu_0 = r_0/\rho_0$, and the expression for $f_1(r)$ becomes

$$f_1(r) = (\lambda\mu^2+1)(\lambda\mu_0^2-1)(\Phi+\lambda^2\Psi)\frac{\rho_0^2}{\lambda^2 r^3}. \tag{7.14.11}$$

If λ is assumed to be positive, the sign of $f_1(r)$ depends upon that of the factor $\lambda\mu_0^2-1$, the quantity $\Phi+\lambda^2\Psi$ being assumed positive as discussed in § 7.3. For a given state of deformation, $f_1(r)$ therefore has a constant sign for all values of r. When the deformation is produced by a tensile force N_3, we see from the first of (7.14.7) that the sign of σ is dependent upon that of $f_1(r)$ and, from (7.14.11), is positive if $\lambda\mu_0^2-1 < 0$ or if $K > 0$. Also, from (7.14.3),

$$\lambda\mu_0^2-1 = (\lambda^3-1)[1/(\lambda^2+\lambda+1)-\cos^2\alpha]\csc^2\alpha.$$

If therefore α is taken to lie between 0 and $\tfrac{1}{2}\pi$, σ is positive if

$$\lambda > 1 \quad \text{and} \quad \cos^{-1}(1/\lambda) < \alpha < \cos^{-1}(\lambda^2+\lambda+1)^{-\frac{1}{2}},$$

or
$$\lambda < 1 \quad \text{and} \quad \cos^{-1}(\lambda^2+\lambda+1)^{-\frac{1}{2}} < \alpha < \tfrac{1}{2}\pi,$$

and negative if
$$\lambda > 1 \quad \text{and} \quad \cos^{-1}(\lambda^2+\lambda+1)^{-\frac{1}{2}} < \alpha < \tfrac{1}{2}\pi,$$
or $\quad \lambda < 1 \quad \text{and} \quad \cos^{-1}(1/\lambda) < \alpha < \cos^{-1}(\lambda^2+\lambda+1)^{-\frac{1}{2}},$

the lower limit for α being determined from (7.14.3) by the consideration that μ_0 and hence $(1-\lambda^2\cos^2\alpha)^{\frac{1}{2}}$ must be real. As in § 7.3 these inequalities imply restrictions on the ranges of λ in which σ is positive. We find that σ is positive if

$$\sec\alpha > \lambda > \tfrac{1}{2}[(4\sec^2\alpha-3)^{\frac{1}{2}}-1] > 1 \quad (\sec\alpha > \sqrt{3}),$$
$$\sec\alpha > \lambda > 1 \quad (\sec\alpha \leqslant \sqrt{3}),$$

or if
$$0 < \lambda < \tfrac{1}{2}[(4\sec^2\alpha-3)^{\frac{1}{2}}-1] < 1 \quad (\sec\alpha < \sqrt{3}),$$
$$0 < \lambda < 1 \quad (\sec\alpha \geqslant \sqrt{3}),$$

and negative if
$$\tfrac{1}{2}[(4\sec^2\alpha-3)^{\frac{1}{2}}-1] > \lambda > 1 \quad (\sec\alpha > \sqrt{3}),$$
or if $\quad \tfrac{1}{2}[(4\sec^2\alpha-3)^{\frac{1}{2}}-1] < \lambda < 1 \quad (\sec\alpha < \sqrt{3}).$

A positive value for σ indicates that each of the individual cords are in tension; a negative value would suggest that the deformation becomes unstable. The conditions for σ to be positive may be compared with the analogous results of § 7.3 for the simple extension of a sheet.

7.15. Combined flexure and homogeneous strain of a reinforced cuboid

We take as the undeformed body a cuboid bounded by the planes $x_1 = A_1$, $x_1 = A_2$, $x_2 = \pm B$, $x_3 = \pm C$, which is reinforced in planes normal to the x_1-axis by families of thin inextensible cords lying in the paths
$$x_3 = x_2 \cot\alpha_i + \text{constant}, \quad x_1 = X_i \quad (i = 1, 2,\ldots, n), \quad (7.15.1)$$
where, for the constants X_i we have
$$A_1 \geqslant X_1 \geqslant X_2 \geqslant \ldots \geqslant X_n \geqslant A_2 > 0.$$
This cuboid is subjected to the deformation
$$r = r(x_1), \quad \theta = ax_2+bx_3, \quad y_3 = cx_2+dx_3, \quad \lambda = ad-bc > 0, \quad (7.15.2)$$
examined for unreinforced bodies in § 2.11 in which the planes $x_1 = A_1$, $x_1 = A_2$, $x_1 = X_i$ become the cylindrical surfaces $r = A_1'$, $r = A_2'$, $r = r$ respectively. We assume that
$$A_1' \geqslant r_1 \geqslant r_2 \geqslant \ldots \geqslant r_n \geqslant A_2' > 0.$$

7.15 REINFORCEMENT BY INEXTENSIBLE CORDS

The analysis of the problem may follow the lines of §§ 7.10–7.12. Alternatively, we may employ the limiting procedure of § 2.11. The latter again leads to the replacement of a/ρ, c/ρ, and λ/ρ by a, c, and λ respectively and r_ρ by dr/dx_1, in the preceding sections. Furthermore, we substitute a, c, and λ respectively for a/ρ_i, c/ρ_i, and λ/ρ_i to obtain the relations for the layer of cords at $x_1 = X_i$. The constraint conditions (7.10.4) then become

$$r_i^2(a\sin\alpha_i+b\cos\alpha_i)^2+(c\sin\alpha_i+d\cos\alpha_i)^2 = 1 \quad (i=1,2,\ldots,n). \quad (7.15.3)$$

We confine our attention to the case where the material is incompressible, so that

$$\lambda r \frac{dr}{dx_1} = 1 \quad \text{or} \quad 2x_1 = \lambda r^2 + K, \quad (7.15.4)$$

and the restrictions of § 7.10 on the number of sets of cords continue to apply. Thus, for a continuously varying deformation of the type (7.15.2) to be possible, the cuboid must contain at the most four independent families of cords, and not more than two of these can lie in any one layer.

The physical stress resultant components n_i^{rs}, defined as in § 7.11, are given from (7.11.5) by

$$\left.\begin{aligned}n_i^{22} &= \sigma_i r_i(a\sin\alpha_i+b\cos\alpha_i)^2/\lambda \\ n_i^{33} &= \sigma_i(c\sin\alpha_i+d\cos\alpha_i)^2/(\lambda r_i) \\ n_i^{23} &= \sigma_i(a\sin\alpha_i+b\cos\alpha_i)(c\sin\alpha_i+d\cos\alpha_i)/\lambda\end{aligned}\right\}, \quad (7.15.5)$$

with
$$\lambda r_i(n_i^{22}+n_i^{33}) = \sigma_i, \quad n_i^{22}n_i^{33} = (n_i^{23})^2. \quad (7.15.6)$$

The stress components τ^{ij} may again be written in the forms (7.12.1), and the parametric expressions (7.12.2) for $f_3(r)$, $f_4(r)$ are unaltered. However, the identity (2.19.4), which now applies, yields the expressions

$$f_1(r) = \frac{2}{r}\left(a\frac{\partial W}{\partial a}+b\frac{\partial W}{\partial b}\right) = \frac{\partial W_1}{\partial r} = \lambda r\frac{\partial W}{\partial x_1} \quad (7.15.7)$$

for $f_1(r)$, where here and subsequently we write $W(x_1) = W_1(r)$.

Equations (7.12.3), (7.12.4) are unchanged, but in place of (7.12.5) we have

$$r_i(\Pi_{i1}-\Pi_{i2}) = n_i^{22} = \sigma_i r_i(a\sin\alpha_i+b\cos\alpha_i)^2/\lambda. \quad (7.15.8)$$

The first of equations (7.12.6) is replaced by

$$\begin{aligned}R_1-R_2 &= \int_{A_2'}^{A_1'} f_1(r)\,dr + \frac{1}{\lambda}\sum_{i=1}^{n}\sigma_i(a\sin\alpha_i+b\cos\alpha_i)^2 \\ &= W_1(A_1')-W_1(A_2')+\frac{1}{\lambda}\sum_{i=1}^{n}\sigma_i(a\sin\alpha_i+b\cos\alpha_i)^2, \quad (7.15.9)\end{aligned}$$

if we make use of (7.15.7). Corresponding expressions may be obtained for $R_1 - \Pi_{i1}$, $\Pi_{i2} - R_2$.

Expressions for the resultant forces and couples acting on planes in the deformed body may be deduced without difficulty from (7.12.7)–(7.12.11). The formula for N_2 is unchanged and by making use of (7.15.7) the expression for M_2 may be reduced to

$$M_2 = \tfrac{1}{2}l\bigg\{A_1'^2[R_1+W_1(A_1')]-A_2'^2[R_2+W_1(A_2')]-$$

$$-\frac{2}{\lambda}\int_{A_2}^{A_1} W(x_1)\,dx_1 + \frac{1}{\lambda}\sum_{i=1}^{n}\sigma_i r_i^2 (a\sin\alpha_i + b\cos\alpha_i)^2\bigg\}. \quad (7.15.10)$$

If there is a single layer of cords at $x_1 = X_0$ (or $r = r_0$) containing either one or two independent families of cords and $R_1 = R_2 = 0$, equations (7.15.7), (7.15.9), and (7.15.10) may be combined to yield the alternative formulae

$$\left.\begin{aligned}M_2 &= \frac{l}{\lambda}\bigg\{(A_1-X_0)W(A_1)-(A_2-X_0)W(A_2)-\int_{A_2}^{A_1}W(x_1)\,dx_1\bigg\}\\ M_2 &= \tfrac{1}{2}l\int_{A_2'}^{A_1'}(r^2-r_0^2)\frac{\partial W_1(r)}{\partial r}\,dr\end{aligned}\right\}, \quad (7.15.11)$$

for M_2.

7.16. Flexure of a reinforced cuboid

We consider now the case where the cuboid of the preceding section is reinforced with a double layer of cords in the plane $x_1 = X_0$. The cords of the two families are symmetrically placed, so that in the results of § 7.15 we may take $n = 2$ and write $\alpha_1 = -\alpha_2 = \alpha$; after deformation the layer of cords lies in the surface $r = r_0$. We consider the simple flexure obtained by putting $b = c = 0$ in (7.15.2) and write

$$r^2 = 2r_0\lambda_1 x_1 + K', \qquad \theta = \frac{\lambda_2 x_2}{r_0}, \qquad y_3 = \lambda_3 x_3, \qquad (7.16.1)$$

with
$$\lambda_1\lambda_2\lambda_3 = 1, \qquad (7.16.2)$$
so that

$$a = \frac{\lambda_2}{r_0}, \qquad b = c = 0, \qquad d = \lambda_3, \qquad \lambda = \frac{1}{\lambda_1 r_0}, \qquad K = -\frac{K'}{\lambda_1 r_0}. \quad (7.16.3)$$

The deformation may then be considered to consist of a pure homogeneous strain, of extension ratio λ_1, λ_2, λ_3 parallel to the x_i-axes, upon which is superposed a simple flexure in which elements of length in the layer of cords remain unaltered. From the first of (7.16.1) we may then obtain

$$\left.\begin{array}{l} K' = r_0(r_0 - 2\lambda_1 X_0) \\ r^2 = r_0[r_0 + 2\lambda_1(x_1 - X_0)] \\ A_i'^2 = r_0[r_0 + 2\lambda_1(A_i - X_0)] \end{array}\right\}. \tag{7.16.4}$$

The constraint condition (7.15.3) becomes

$$\lambda_2^2 \sin^2\alpha + \lambda_3^2 \cos^2\alpha = 1, \tag{7.16.5}$$

and from this relation and (7.16.2) it follows that the deformation can be specified completely by the radius of flexure r_0 of the layer of cords and any one of the principal extension ratios λ_1, λ_2, λ_3 in this layer.

When the surfaces initially at $x_1 = A_1$, $x_1 = A_2$ are free from applied traction and $\sigma_1 = \sigma_2 = \sigma$, we obtain from (7.15.9)

$$\sigma = -\frac{\lambda_3 r_0}{2\lambda_2}[W(A_1) - W(A_2)]\operatorname{cosec}^2\alpha. \tag{7.16.6}$$

From (7.12.8) and (7.12.11) we may infer that S_2 and M_3 are zero if the elastic material possesses suitable symmetry properties as discussed in § 2.15, and the surface tractions are then everywhere normal to the boundaries of the deformed cuboid. From (7.12.7) the resultant normal force on each of the planes initially at $x_2 = \pm B$ vanishes if $R_1 = R_2 = 0$, and the normal tractions on each of these faces then reduce to a couple M_2 given by (7.15.11). In these formulae we have $l = 2\lambda_3 C$ and $\lambda = 1/(\lambda_1 r_0)$.

7.17. Simple extension and flexure of a thin reinforced isotropic sheet

By allowing the x_1-dimension of the cuboid of § 7.16 to become small compared with the radius of flexure, we may examine, by a limiting process, the bending of a thin reinforced sheet containing cords symmetrically placed with respect to the axis of flexure. Thus, if we write

$$\eta_i = A_i - X_0, \quad \epsilon_i = A_i' - r_0 \quad (i = 1, 2), \tag{7.17.1}$$

where η_i, ϵ_i are small compared with r_0, the undeformed body becomes a sheet of thickness $\eta_1 - \eta_2$ bounded in the x_2, x_3-plane by the lines $x_2 = \pm B$, $x_3 = \pm C$. We suppose that the deformation (7.16.1) can be maintained by means of a tensile force N_3 on the sides initially at $x_3 = \pm C$ and a flexural couple M_2 on the edges at $x_2 = \pm B$, the major

surfaces of the sheet being free from applied traction. We therefore write
$$R_1 = R_2 = 0, \quad N_2 = 0 \\ \sigma_1 = \sigma_2 = \sigma, \quad S_2 = M_3 = 0 \Bigg\}, \qquad (7.17.2)$$

consistent with the discussion of § 7.16. The couple M_2 is given by (7.15.11) and σ by (7.16.6).

If the material is isotropic, the strain energy function W is a function of I_1, I_2. Expressions for I_1, I_2 may be deduced from (2.1.13) by introducing the conditions (7.16.3) and using the limiting procedure of § 7.15. We have
$$I_1 = \frac{r_0^2 \lambda_1^2}{r^2} + \frac{\lambda_2^2 r^2}{r_0^2} + \lambda_3^2 \\ I_2 = \frac{r^2}{r_0^2 \lambda_1^2} + \frac{r_0^2}{\lambda_2^2 r^2} + \frac{1}{\lambda_3^2} \Bigg\}. \qquad (7.17.3)$$

An approximate expression for the flexural couple M_2 may now be obtained by expanding the second of equations (7.15.11) as a power series in ϵ_1, ϵ_2 and observing that for any function $f(r)$, subject to suitable restrictions, if $r = r_0 + \epsilon$ we have the expansion
$$\int_{A_2'}^{A_1'} f(r)\, dr = \int_{\epsilon_2}^{\epsilon_1} f(r_0 + \epsilon)\, d\epsilon$$
$$= (\epsilon_1 - \epsilon_2) f(r_0) + \tfrac{1}{2}(\epsilon_1^2 - \epsilon_2^2) f'(r_0) + \tfrac{1}{6}(\epsilon_1^3 - \epsilon_2^3) f''(r_0) + \ldots . \qquad (7.17.4)$$

Thus we obtain†
$$M_2 = \lambda_3 C \bigg\{ r_0 (\epsilon_1^2 - \epsilon_2^2) \left[\frac{dW_1}{dr}\right]_{r=r_0} + \tfrac{1}{3}(\epsilon_1^3 - \epsilon_2^3) \left(\left[\frac{dW_1}{dr}\right]_{r=r_0} + 2 r_0 \left[\frac{d^2 W_1}{dr^2}\right]_{r=r_0} \right) + \ldots \bigg\}. \qquad (7.17.5)$$

Also, from (7.16.4) and (7.17.1) we have
$$\frac{\epsilon_i}{r_0} = \left(1 + \frac{2\lambda_1 \eta_i}{r_0}\right)^{\frac{1}{2}} - 1 = \frac{\lambda_1 \eta_i}{r_0}\left(1 - \frac{\lambda_1 \eta_i}{2 r_0} + \frac{\lambda_1^2 \eta_i^2}{2 r_0^2} - \ldots\right),$$
and from (7.17.3)
$$\frac{dI_1}{dr} = \frac{1}{\lambda_3^2} \frac{dI_2}{dr} = -2 \left(\frac{r_0^2 \lambda_1^2}{r^3} - \frac{r \lambda_2^2}{r_0^2} \right).$$

† We now regard W, I_1, I_2 as functions of r only, so that partial differentiation signs are no longer used.

Introducing these expressions into (7.17.5) we obtain

$$M_2 = \frac{2\lambda_1 r_0^2}{\lambda_2} C \Big\{ (\lambda_2^2 - \lambda_1^2) \frac{\eta_1^2 - \eta_2^2}{r_0^2} DW + $$
$$+ \frac{4\lambda_1}{3} \frac{\eta_1^3 - \eta_2^3}{r_0^3} [2\lambda_1^2 DW + (\lambda_2^2 - \lambda_1^2)^2 D^2 W] + O\left(\frac{\eta^4}{r_0^4}\right) \Big\} \qquad (\eta = x_1 - X_0),$$
(7.17.6)

where $D \equiv \partial/\partial I_1 + \lambda_3^2 \partial/\partial I_2$ and the derivatives of W with respect to I_1 and I_2 are evaluated at $r = r_0$ (or $\eta = 0$). By a similar procedure we may obtain from (7.16.6) and (7.13.7) expressions for σ and N_3 which agree, to a first approximation, with the corresponding results for simple extension obtained in § 7.3.

From (7.17.6) it follows that the couple required to maintain a given state of flexure depends both upon the form of the strain energy function for the material of the sheet and upon the disposition of the cords. If we assume that DW is positive and that $\partial W/\partial I_1$ and $\partial W/\partial I_2$ are large compared with higher-order derivatives of W with respect to the invariants,† it follows from the definition of D that quantities containing these higher-order derivatives of W can only become important in any term of the expansion (7.17.6) if λ_3 is large. Since λ_3 is restricted by (7.16.5) to be appreciably less than sec α we shall suppose that it is sufficiently small for these higher-order terms not to assume overriding importance.

When the cords lie in a plane midway between the two major surfaces of the undeformed sheet we may put $\eta_1 = -\eta_2 = \eta_0$ in (7.17.6). If r_0 is sufficiently large compared with η_0 the value of M_2 is approximately equal to that of the second term in the expansion and we have

$$M_2 = \frac{16\lambda_1^2 \eta_0^3}{3\lambda_2 r_0} C\{2\lambda_1^2 DW + (\lambda_2^2 - \lambda_1^2)^2 D^2 W\}. \qquad (7.17.7)$$

To this order of approximation M_2 is directly proportional to the curvature of the deformed sheet. If r_0 is not large compared with η_0, higher-order terms may become important in the expansion for M_2 and the exact expression (7.15.11) must then be employed.

If the layer of cords does not lie exactly midway between the major surfaces of the undeformed sheet, $\eta_1 + \eta_2 \neq 0$, and for sufficiently large values of r_0 an approximation to M_2 is given by the first term of (7.17.6). The fact that this term is independent of the curvature $1/r_0$ of

† These assumptions are consistent with the experimental work of Rivlin and Saunders on vulcanized rubber described in Chapter IX and with the discussion of § 7.3.

the deformed sheet suggests that the sheet will bend under the application of the simple tensile force N_3 unless a suitable restoring couple is applied with its axis parallel to the direction of the force. The magnitude and sense of this couple depends upon the magnitude of N_3 and the disposition of the cords. The sign of the first term in the expansion (7.17.6) indicates the sense in which this couple must be applied. This is dependent upon the signs of the factors $\eta_1+\eta_2$ and $\lambda_2^2-\lambda_1^2$ since DW and $\eta_1-\eta_2$ are both positive. Remembering (7.16.2) and (7.16.5), it follows from analysis similar to that of § 7.3 that, if $\lambda_3 > 1$, then $\lambda_2^2-\lambda_1^2$ is positive if $\tfrac{1}{2}\pi > \alpha > \cos^{-1}(\lambda^2+\lambda_3+1)^{-\frac{1}{2}}$ and negative if
$$\cos^{-1}(\lambda_3^2+\lambda_3+1)^{-\frac{1}{2}} > \alpha > \cos^{-1}(1/\lambda_3).$$

Furthermore, σ is given to a first approximation by the expression (7.3.7) with λ_1 and λ_3 interchanged, or by (7.16.6), and therefore has the same sign as the factor $\lambda_1^2-\lambda_2^2$. We thus have†

$$\left.\begin{aligned}
&M_2 > 0,\quad \sigma < 0 \quad \text{if}\quad \eta_1+\eta_2 > 0 \text{ and}\\
&\qquad\qquad\qquad \tfrac{1}{2}\pi > \alpha > \cos^{-1}(\lambda_3^2+\lambda_3+1)^{-\frac{1}{2}}\\
&M_2 > 0,\quad \sigma > 0 \quad \text{if}\quad \eta_1+\eta_2 < 0 \text{ and}\\
&\qquad\qquad \cos^{-1}(\lambda_3^2+\lambda_3+1)^{-\frac{1}{2}} > \alpha > \cos^{-1}(1/\lambda_3)\\
&M_2 < 0,\quad \sigma > 0 \quad \text{if}\quad \eta_1+\eta_2 > 0 \text{ and}\\
&\qquad\qquad \cos^{-1}(\lambda_3^2+\lambda_3+1)^{-\frac{1}{2}} > \alpha > \cos^{-1}(1/\lambda_3)\\
&M_2 < 0,\quad \sigma < 0 \quad \text{if}\quad \eta_1+\eta_2 < 0 \text{ and}\\
&\qquad\qquad\qquad \tfrac{1}{2}\pi > \alpha > \cos^{-1}(\lambda_3^2+\lambda_3+1)^{-\frac{1}{2}}
\end{aligned}\right\}.\quad (7.17.8)$$

Hitherto we have assumed that
$$A_1 \geqslant X_0 \geqslant A_2 > 0,\qquad A_1' \geqslant r_0 \geqslant A_2' > 0$$
so that, from (7.17.1), if $\eta_1+\eta_2 > 0$ the plane containing the cords in the undeformed sheet lies nearer to the major surface which becomes concave after flexure, while if $\eta_1+\eta_2 < 0$ the cords lie initially nearer to the opposite boundary surface. Moreover, from (7.17.7), a positive couple must be applied to produce a flexure in which $A_1' \geqslant r_0 \geqslant A_2' > 0$ when the cords lie midway between the major surfaces of the sheet. If the first term of (7.17.6) is positive, M_2 is positive when $r_0 \to \infty$, that is, when the sheet is extended without flexure. A positive couple

---

† We observe that to the order of approximation being considered M_2 and σ vanish together, so that the difficulty inherent in negative values of σ need not arise when the flexural couple is removed. Alternatively, we may postulate that each cord, when forming part of a system embedded in an elastic material, is capable of sustaining a small compressive force.

must then be applied to prevent bending, and on removal of this couple flexure would occur in a sense opposite to that previously assumed. Hence the application of the simple tensile force N_3 alone will produce flexure in such a sense that the major surface of the undeformed sheet which lies nearer to the plane of the cords will become concave after deformation if

$$\cos^{-1}(\lambda_3^2+\lambda_3+1)^{-\frac{1}{2}} > \alpha > \cos^{-1}(1/\lambda_3),$$

that is, if

$$\sec\alpha > \lambda_3 > \tfrac{1}{2}[(4\sec^2\alpha-3)^{\frac{1}{2}} - 1] > 1 \quad (\sec\alpha > \sqrt{3}),$$

or

$$\sec\alpha > \lambda_3 > 1 \quad (\sec\alpha \leqslant \sqrt{3}),$$

and convex if

$$\tfrac{1}{2}\pi > \alpha > \cos^{-1}(\lambda_3^2+\lambda_3+1)^{-\frac{1}{2}},$$

that is, if $\tfrac{1}{2}[(4\sec^2\alpha-3)^{\frac{1}{2}}-1] > \lambda_3 > 1 \quad (\sec\alpha > \sqrt{3}).$

These inequalities suggest that if $\sec\alpha > \sqrt{3}$ and the reinforced sheet is slowly extended, the direction of flexure will reverse during the deformation, the change occurring at $\lambda_3 = \tfrac{1}{2}[(4\sec^2\alpha-3)^{\frac{1}{2}}-1]$.

If M_2 becomes zero for a sufficiently large value of r_0, the radius of flexure of the layer of cords in the deformed sheet may be obtained from the first and second terms of (7.17.6). Thus we have, approximately, $M_2 = 0$ if

$$r_0 = \frac{4\lambda_1(\eta_1^2+\eta_1\eta_2+\eta_2^2)[2\lambda_1^2 DW + (\lambda_1^2-\lambda_2^2)^2 D^2 W]}{3(\eta_1+\eta_2)(\lambda_1^2-\lambda_2^2)DW}. \tag{7.17.9}$$

This yields a large value of r_0 if $\eta_1^2+\eta_1\eta_2+\eta_2^2$ is large compared with $\eta_1+\eta_2$, and this occurs if the plane of the cords lies near to the middle plane of the undeformed sheet.

If the elastic material has the Mooney form of strain energy function, DW is a constant, $D^2W = 0$, and (7.17.9) becomes

$$r_0 = \frac{8\lambda_1^3(\eta_1^2+\eta_1\eta_2+\eta_2^2)}{3(\lambda_1^2-\lambda_2^2)(\eta_1+\eta_2)}. \tag{7.17.10}$$

7.18. Reinforced bodies and materials subject to constraints

Mathematically there is a close connexion between corresponding problems for reinforced bodies and materials subject to constraints. If we postulate a body in which geometrical constraints arise as an intrinsic property of the material, and these constraints are identical with those which would be introduced into a material free from intrinsic constraints by continuous systems of thin flexible inextensible cords, then we may expect the stress–strain relations for the two types of body to assume identical forms. For example, if we consider the problem of §§ 7.4–7.6 in which the undeformed body is a uniform plane sheet containing no reinforcing layer but in which the constraint conditions (7.4.9) occur as an intrinsic property of the elastic material, then by combining (7.4.9) with (1.18.5) and making use of the limiting procedure of Chapter IV we may derive expressions for the stress resultants $n^{\alpha\beta}$ which are similar in form to those obtained from (7.5.1) and (7.5.3). In this case, apart from a scalar factor, the Lagrangian multipliers q_1, q_2 replace the quantities σ_1, σ_2 which arise from the tensions in the cords in a reinforced sheet. The problem for a reinforced sheet may, in fact, be regarded as a limiting case of the corresponding problem for a sheet containing a layer of constrained material.† A similar procedure could be employed in the problems of §§ 7.10–7.17, each layer of cords being replaced by a layer of elastic material containing intrinsic constraints, the results for a reinforced body emerging as a special case when these layers are allowed to become indefinitely thin and the parameters q_1, q_2 to tend to infinity in a suitable manner. When the stress–strain relations have been formulated, the subsequent analysis for the solution of the constraint conditions (e.g., in § 7.8) and for the determination of the parameters σ_i (or q_i) proceeds along similar lines for corresponding problems. The parameters q_i are not, however, necessarily subject to the restrictions of § 7.1 on the quantities σ_i, which arise from the necessity for the tensions in the cords to be positive. Furthermore, it must be remembered that in the case of a material subject to intrinsic constraints the form of the strain energy function is restricted by the constraint conditions and may also be regarded as arbitrary to the extent indicated in (1.18.6). This is not the case for a material reinforced with cords.

† See J. E. Adkins, loc. cit., p. 228.

VIII

THERMO-ELASTICITY

In earlier parts of this book we have assumed the existence of an elastic potential or strain energy function. In the present chapter we consider the thermodynamics of the deformation.† We take the energy equation in § 8.1 and the entropy production inequality in § 8.2 as our basic thermodynamical postulates. The theory of thermo-elasticity is considered in § 8.3. The basic equations in this section show that if the temperature T is constant then the strain energy function is identical with the Helmholtz free energy function A. Alternatively, if the entropy is constant the strain energy is identical with the internal energy density U. However, the stress equations in terms of A or U in § 8.3 are not restricted to these two cases.

In the last part of this chapter the form of the heat conduction vector‡ is discussed with special reference to isotropic bodies and basic equations are given for the classical theory of thermo-elasticity in § 8.7.

8.1. Energy equation

We consider an arbitrary surface A enclosing a material volume τ of the continuum and we postulate an energy balance at time t in the form

$$\frac{d}{dt}\int_\tau (\tfrac{1}{2}\dot{\mathbf{v}}^2 + U)\rho \, d\tau = \int_\tau \rho(r + \mathbf{F}\cdot\dot{\mathbf{v}})\, d\tau + \int_A (\mathbf{t}\cdot\dot{\mathbf{v}} - \mathbf{Q}\cdot\mathbf{n})\, dA, \qquad (8.1.1)$$

where the scalar U is internal energy per unit mass, the scalar r is a heat supply function per unit mass per unit time (due to heat sources and radiation from external sources), and $\mathbf{Q}$ is the heat flux vector per unit area of the deformed body at time t. A superposed dot denotes time differentiation with respect to t holding θ fixed. Equation (8.1.1)

---

† For references see C. Truesdell, loc. cit., p. 1. See also §§ 2.7–2.9 of *T.E.*; B. D. Coleman and W. Noll, *Arch ration. Mech. Analysis* **13** (1963), 167.

‡ The general form of the heat conduction vector has been considered by R. S. Rivlin and A. C. Pipkin, *Arch ration. Mech. Analysis* **4** (1959), 129.

For classical thermoelasticity see

J. M. C. Duhamel, *J. Éc. polytech.* **15** (1837), 1.
W. Voight, *Lehrbuch Kristallphysik* (Leipzig, 1910).
H. Jeffreys, *Proc. Camb. phil. Soc. math. phys. Sci.* **26** (1930), 101.
M. A. Biot, *J. appl. Phys.* **27** (1956), 240.
P. Chadwick, *Prog. Solid Mech.* **1** (1960), 265.

states that the rate of increase of kinetic and internal energy in a material volume τ is equal to the rate of work of body and surface forces and the rate of work contributions due to a volume distribution of heat supply and the flux of heat across A.

If $\mathbf{n}$, $_0\mathbf{n}$ are unit outward normals to corresponding elements of surface dA, dA_0 of the deformed and undeformed bodies then, remembering (1.1.25) and (1.1.38), we have

$$\rho n_i dA = \rho_{00} n_i dA_0. \tag{8.1.2}$$

The heat flux term in (8.1.1) may now be written in the alternative forms

$$\int_A \mathbf{Q}.\mathbf{n}\, dA = \int_A Q^i n_i\, dA = \int_{A_0} q^i {}_0 n_i\, dA_0 = \int_{A_0} \mathbf{q} \cdot {}_0\mathbf{n}\, dA_0, \tag{8.1.3}$$

where
$$\mathbf{Q} = Q^i \mathbf{G}_i, \quad \mathbf{q} = q^i \mathbf{g}_i, \quad q^i = \rho_0 Q^i/\rho = Q^i \sqrt{I_3}, \tag{8.1.4}$$

so that $\mathbf{q}$ represents the heat conduction vector per unit area of the undeformed body. From (8.1.3) we have

$$\int_A \mathbf{Q}.\mathbf{n}\, dA = \int_\tau Q^i\|_i\, d\tau = \int_{\tau_0} q^i|_i\, d\tau_0, \tag{8.1.5}$$

where τ_0 is the volume in the undeformed state, and we observe that

$$\rho q^i|_i = \rho_0 Q^i\|_i = \frac{\rho_0}{\sqrt{G}} \frac{\partial}{\partial \theta^i}(Q^i \sqrt{G}) = \frac{\rho}{\sqrt{g}} \frac{\partial}{\partial \theta^i}(q^i \sqrt{g}). \tag{8.1.6}$$

With the help of (1.1.24), (1.1.26), and (1.1.27) equation (8.1.1) reduces to

$$\int_\tau (\tau^{ij}\dot{\gamma}_{ij} + \rho r - \rho \dot{U} - Q^i\|_i)\, d\tau = 0$$

for an arbitrary volume τ, so that, provided the integrand is continuous,

$$\rho r - \rho \dot{U} + \tau^{ij}\dot{\gamma}_{ij} - Q^i\|_i = 0. \tag{8.1.7}$$

Alternatively, using (8.1.6),

$$\rho_0 r - \rho_0 \dot{U} + s^{ij}\dot{\gamma}_{ij} - q^i|_i = 0, \tag{8.1.8}$$

where
$$s^{ij} = \tau^{ij}\sqrt{I_3}. \tag{8.1.9}$$

8.2. Entropy production inequality

If the scalar S is entropy per unit mass and the scalar T (>0) is temperature, we postulate an entropy production inequality for an arbitrary material volume τ bounded by a surface A in the form

$$\frac{d}{dt}\int_\tau \rho S\, d\tau - \int_\tau \rho \frac{r}{T}\, d\tau + \int_A \frac{\mathbf{Q}\cdot\mathbf{n}}{T}\, dA \geq 0. \qquad (8.2.1)$$

The left-hand side of this inequality represents the rate of increase of entropy minus the rate at which entropy is supplied by the volume distribution of heat minus the rate of supply of entropy across the surface due to heat conduction. If we reduce the surface integral in (8.2.1) to a volume integral and remember that the volume is arbitrary, we have, with the usual smoothness assumptions,

$$\rho T\dot{S} - \rho r + Q^i\|_i - \frac{Q^i T\|_i}{T} \geq 0. \qquad (8.2.2)$$

If we eliminate r from (8.2.2) with the help of (8.1.7) we obtain

$$\rho(T\dot{S} - U) + \tau^{ij}\dot{\gamma}_{ij} - \frac{Q^i T\|_i}{T} \geq 0. \qquad (8.2.3)$$

Alternatively, introducing the Helmholtz free energy function A where†

$$A = U - TS, \qquad (8.2.4)$$

the inequality (8.2.3) becomes

$$-\rho(S\dot{T} + \dot{A}) + \tau^{ij}\dot{\gamma}_{ij} - \frac{Q^i T\|_i}{T} \geq 0. \qquad (8.2.5)$$

With the help of (8.1.4), (8.1.6), and (8.1.9) this inequality may be replaced by

$$-\rho_0(S\dot{T} + \dot{A}) + s^{ij}\dot{\gamma}_{ij} - \frac{q^i T|_i}{T} \geq 0. \qquad (8.2.6)$$

8.3. Equations for an elastic body

We recall that the coordinates of corresponding points in the undeformed and deformed bodies, referred to the same rectangular cartesian axes, have been denoted by x_i, y_i respectively (see § 1.1). Let $\mathbf{i}_s$ denote unit vectors along these axes and let R_s be the components of the heat conduction vector $\mathbf{Q}$ at the point y_i along these vectors.

† The symbol A need not be confused with that used for a surface since no further surface integrals are employed in this chapter.

Thus
$$\mathbf{Q} = Q^i \mathbf{G}_i = R_s \mathbf{i}_s, \qquad (8.3.1)$$

and
$$Q^i = \frac{\partial \theta^i}{\partial y^s} R_s. \qquad (8.3.2)$$

Also, if σ_{ij} are the cartesian components of stress at the point y_i, we recall from (1.15.14) that

$$\sigma_{ij} = \frac{\partial y^i}{\partial \theta^r} \frac{\partial y^j}{\partial \theta^s} \tau^{rs}. \qquad (8.3.3)$$

We define an elastic continuum to be one for which the following constitutive equations hold at each material point of the continuum and for all time t:

$$A = A\left(y_i, \frac{\partial y_i}{\partial x_j}, \frac{\partial T}{\partial x_m}, T\right), \qquad (8.3.4)$$

$$S = S\left(y_i, \frac{\partial y_i}{\partial x_j}, \frac{\partial T}{\partial x_m}, T\right), \qquad (8.3.5)$$

$$\sigma_{rs} = \sigma_{rs}\left(y_i, \frac{\partial y_i}{\partial x_j}, \frac{\partial T}{\partial x_m}, T\right), \qquad (8.3.6)$$

$$R_s = R_s\left(y_i, \frac{\partial y_i}{\partial x_j}, \frac{\partial T}{\partial x_m}, T\right). \qquad (8.3.7)$$

We observe[†] that since these constitutive equations involve $\partial y_p/\partial x_q$ we could replace $\partial T/\partial x_m$ by $\partial T/\partial y_m$, but the above is more convenient for our purpose.[‡] Since A, S, σ_{ij}, and R_s must all be unchanged when a rigid body translation is imposed on the given motion it follows that y_p cannot occur explicitly in (8.3.4)–(8.3.7).

We now suppose that the deformed body is subjected to a rigid-body rotation so that the temperature T is unaltered at every point of the body and so that the point at y_i in the deformed state moves to $\bar{y}_i$ where
$$\bar{y}_i = a_{ij} y_j, \qquad (8.3.8)$$
and
$$a_{ik} a_{jk} = a_{ki} a_{kj} = \delta_{ij}, \qquad |a_{ij}| = 1. \qquad (8.3.9)$$

Then, following the same discussion as that which led from (1.3.1) to (1.3.8), we see that A may be replaced by the different function

$$A = A\left(e_{ij}, \frac{\partial T}{\partial x_m}, T\right). \qquad (8.3.10)$$

[†] The functions (8.3.4)–(8.3.7) will also depend on the initial density ρ_0 and functions of the initial coordinates x_i, but this is understood.
[‡] The alternative form of assumption is made in § 8.6.

Alternatively, in view of (1.1.15), A may be replaced by another function

$$A = A\left(\gamma_{ij}, \frac{\partial T}{\partial \theta^m}, T\right), \tag{8.3.11}$$

dependence on $\partial \theta^i/\partial x_r$ being understood. Again, since R_s is unaltered by the rigid-body rotation except for orientation, it follows that

$$R_s\left(\frac{\partial \bar{y}_i}{\partial x_j}, \frac{\partial T}{\partial x_m}, T\right) = a_{sr} R_r\left(\frac{\partial y_i}{\partial x_j}, \frac{\partial T}{\partial x_m}, T\right) \tag{8.3.12}$$

for all a_{sr} satisfying (8.3.9). Alternatively

$$R_s\left(\frac{\partial y_i}{\partial x_j}, \frac{\partial T}{\partial x_m}, T\right) = a_{rs} R_r\left(a_{ik}\frac{\partial y_k}{\partial x_j}, \frac{\partial T}{\partial x_m}, T\right). \tag{8.3.13}$$

This relation holds for a particular value of x_i and we may choose the special value R_{rs} for a_{sr}, where R_{rs} is defined in § 1.3, so that

$$R_s = R_{sr} R_r\left(M_{ij}, \frac{\partial T}{\partial x_m}, T\right). \tag{8.3.14}$$

The tensor M_{ij} in (8.3.14) is defined in § 1.3 and is a positive definite symmetric tensor satisfying the equation (1.3.6). Hence (8.3.14) may be replaced by

$$R_i = \frac{\partial y_i}{\partial x_r} L_r\left(e_{pq}, \frac{\partial T}{\partial x_m}, T\right), \tag{8.3.15}$$

where L_r is a vector function of the arguments stated in (8.3.15). It may be verified that (8.3.15) satisfies the condition (8.3.12) for all a_{rs} subject to (8.3.9).

In a similar way we may discuss the effect of invariance conditions under superposed rigid-body rotations on the entropy S and stress σ_{rs}, but we omit this since the final forms of these functions will emerge from our thermodynamical discussion.

In view of (8.3.11), the inequality (8.2.5) becomes

$$-\rho\left(S+\frac{\partial A}{\partial T}\right)\dot{T}+\left[\tau^{ij}-\tfrac{1}{2}\rho\left(\frac{\partial A}{\partial \gamma_{ij}}+\frac{\partial A}{\partial \gamma_{ji}}\right)\right]\dot{\gamma}_{ij}-\rho\frac{\partial A}{\partial T_{,m}}\dot{T}_{,m}-\frac{Q^i T\|_i}{T} \geqslant 0. \tag{8.3.16}$$

For given temperature and displacement gradients (8.3.16) is valid for all arbitrary values of $\dot{T}$, $\dot{\gamma}_{ij}$, $\dot{T}_{,m}$, and $T\|_i$ (subject to $\dot{\gamma}_{ij} = \dot{\gamma}_{ji}$) provided the body forces $\mathbf{F}$ and heat supply function r are suitably chosen so that the equations of motion and the energy equation are satisfied. In view of the constitutive assumptions on A, S, σ_{rs}, and R_s, and hence on τ^{ij} and Q^i, it follows that

$$\frac{\partial A}{\partial T_{,m}} = 0$$

so that A reduces to the form

$$A = A(\gamma_{ij}, T). \tag{8.3.17}$$

Also
$$S = -\frac{\partial A}{\partial T}, \tag{8.3.18}$$

$$\tau^{ij} = \tfrac{1}{2}\rho\left(\frac{\partial A}{\partial \gamma_{ij}} + \frac{\partial A}{\partial \gamma_{ji}}\right) = \frac{1}{2\sqrt{I_3}}\left(\frac{\partial W}{\partial \gamma_{ij}} + \frac{\partial W}{\partial \gamma_{ji}}\right), \tag{8.3.19}$$

where
$$W = \rho_0 A. \tag{8.3.20}$$

The inequality (8.3.16) now reduces to

$$-Q^i T\|_i \geqslant 0 \quad \text{or} \quad -q^i T|_i \geqslant 0. \tag{8.3.21}$$

For some purposes it is more convenient to regard U as a function of γ_{ij} and S, in which case (8.2.4), (8.3.18), and (8.3.19) may be used to deduce that

$$\left. \begin{array}{l} \tau^{ij} = \tfrac{1}{2}\rho\left(\dfrac{\partial U}{\partial \gamma_{ij}} + \dfrac{\partial U}{\partial \gamma_{ji}}\right) \\[1em] T = \dfrac{\partial U}{\partial S} \end{array} \right\} . \tag{8.3.22}$$

With the help of (8.2.4), (8.3.18), and (8.3.19) the energy equation (8.1.7) reduces to

$$\rho r - \rho T\dot{S} - Q^i\|_i = 0, \tag{8.3.23}$$

or
$$\rho_0 r - \rho_0 T\dot{S} - q^i|_i = 0. \tag{8.3.24}$$

8.4. Incompressible bodies

When the body is incompressible

$$I_3 = 1 \quad \text{or} \quad G^{ij}\dot{\gamma}_{ij} = 0, \tag{8.4.1}$$

and this condition must be remembered when using (8.2.5) which is replaced by

$$-\rho(\dot{A} + S\dot{T}) + (\tau^{ij} - pG^{ij})\dot{\gamma}_{ij} - \frac{Q^i T\|_i}{T} \geqslant 0, \tag{8.4.2}$$

where p, a Lagrangian multiplier, is an arbitrary scalar function. The constraint (8.4.1) causes a loss of one degree of freedom in our choice of constitutive equations and the inequality suggests that we replace the equation (8.3.6) by the assumption that $\sigma_{rs}-p\delta_{rs}$ depends on the variables stated. In view of invariance conditions under a rigid-body translation

$$\sigma_{rs}-p\delta_{rs} = f_{rs}\left(\frac{\partial y_i}{\partial x_j}, \frac{\partial T}{\partial x_m}, T\right). \tag{8.4.3}$$

As before, A is given by (8.3.17), and now

$$-\rho\left(S+\frac{\partial A}{\partial T}\right)\dot{T}+\left[\tau^{ij}-pG^{ij}-\tfrac{1}{2}\rho_0\left(\frac{\partial A}{\partial \gamma_{ij}}+\frac{\partial A}{\partial \gamma_{ji}}\right)\right]\dot{\gamma}_{ij}-\frac{Q^i T\|_i}{T} \geqslant 0, \tag{8.4.4}$$

for all arbitrary values of $\dot{T}$, $\dot{\gamma}_{ij}$, $T\|_i$ subject to $\dot{\gamma}_{ij} = \dot{\gamma}_{ji}$ and (8.4.1). Since at least one of the stress components τ^{ij} is arbitrary we choose p so that

$$\tau^{33} = pG^{33}+\rho_0\frac{\partial A}{\partial \gamma_{33}}. \tag{8.4.5}$$

The coefficient of $\dot{\gamma}_{33}$ in (8.4.4) is then zero and the remaining quantities $\dot{T}$, $\dot{\gamma}_{ij}$, $T\|_i$ may be chosen arbitrarily, their coefficients being independent of $\dot{T}$, $\dot{\gamma}_{ij}$, and $T\|_i$. Hence

$$S = -\frac{\partial A}{\partial T}, \tag{8.4.6}$$

$$\tau^{ij} = pG^{ij}+\tfrac{1}{2}\rho_0\left(\frac{\partial A}{\partial \gamma_{ij}}+\frac{\partial A}{\partial \gamma_{ji}}\right), \tag{8.4.7}$$

$$-Q^i T\|_i \geqslant 0 \quad \text{or} \quad -q^i T|_i \geqslant 0. \tag{8.4.8}$$

In view of (8.4.5) the result (8.4.7) is valid for all values of $i, j = 1, 2, 3$. With the help of (8.3.20), equation (8.4.7) becomes

$$\tau^{ij} = pG^{ij}+\tfrac{1}{2}\left(\frac{\partial W}{\partial \gamma_{ij}}+\frac{\partial W}{\partial \gamma_{ji}}\right). \tag{8.4.9}$$

Similarly, the equation corresponding to (8.3.22) are

$$\left.\begin{aligned}\tau^{ij} &= pG^{ij}+\tfrac{1}{2}\rho_0\left(\frac{\partial U}{\partial \gamma_{ij}}+\frac{\partial U}{\partial \gamma_{ji}}\right), \\ T &= \frac{\partial U}{\partial S}.\end{aligned}\right\} \tag{8.4.10}$$

Also the energy equation again reduces to (8.3.23) or (8.3.24).

8.5. Heat conduction vector for isotropic bodies†

The heat conduction vector for any elastic body is given by (8.3.1) where, from (8.3.15),

$$R_i = \frac{\partial y_i}{\partial x_r} L_r\left(e_{pq}, \frac{\partial T}{\partial x_m}, T\right),$$

and we assume that L_r is a polynomial in e_{pq} and $\partial T/\partial x_m$.‡ Further restrictions on the form of R_i will be necessary if the undeformed body belongs to one of the crystal classes discussed in Chapter I, and these restrictions may be found without difficulty. We shall, however, direct our attention here to the isotropic case. We suppose that the undeformed body is in a position that differs from its original position only by a rigid-body rotation, so that the point that was at x_i now becomes $\bar{x}_i$, where

$$\bar{x}_i = a_{ij} x_j \tag{8.5.1}$$

and a_{ij} satisfy the relations (8.3.9). The final position and state of the deformed body and, in particular, the heat conduction vector are unaltered. If the new position of the undeformed body is taken as the reference position, and the body is isotropic initially, then

$$R_i = \frac{\partial y_i}{\partial \bar{x}_r} L_r\left(\bar{e}_{pq}, \frac{\partial T}{\partial \bar{x}_m}, T\right) = \frac{\partial y_i}{\partial x_r} L_r\left(e_{pq}, \frac{\partial T}{\partial x_m}, T\right), \tag{8.5.2}$$

where

$$\left.\begin{aligned}\bar{e}_{pq} &= \frac{1}{2}\left(\frac{\partial y_r}{\partial \bar{x}_p}\frac{\partial y_r}{\partial \bar{x}_q} - \delta_{pq}\right) = a_{pr} a_{qs} e_{rs}, \\ \frac{\partial T}{\partial \bar{x}_m} &= a_{mn} \frac{\partial T}{\partial x_n}.\end{aligned}\right\} \tag{8.5.3}$$

From (8.5.2) and (8.5.3) we see that

$$L_r\left(\bar{e}_{pq}, \frac{\partial T}{\partial \bar{x}_m}, T\right) = a_{rs} L_s\left(e_{pq}, \frac{\partial T}{\partial x_m}, T\right). \tag{8.5.4}$$

We now define a symmetric matrix **e** and two antisymmetric matrices **b**, **P** by the formulae

$$\mathbf{e} = [e_{ij}], \quad \mathbf{b} = [b_{ij}], \quad \mathbf{P} = [P_{ij}], \tag{8.5.5}$$

where

$$b_{ij} = e_{ijk}\frac{\partial T}{\partial x_k}, \quad P_{ij} = e_{ijk} L_k\left(e_{pq}, \frac{\partial T}{\partial x_m}, T\right). \tag{8.5.6}$$

---
† See R. S. Rivlin and A. C. Pipkin, loc. cit., p. 273 for an alternative discussion.
‡ Extension to the non-polynomial case may be made with the help of the work of A. C. Pipkin and A. S. Wineman, loc. cit., p. 10.

It follows that

$$\bar{b}_{ij} = e_{ijk}\frac{\partial T}{\partial \bar{x}_k} = a_{ir}a_{js}e_{rsk}\frac{\partial T}{\partial x_k} = a_{ir}a_{js}b_{rs}, \quad (8.5.7)$$

$$\bar{P}_{ij} = e_{ijk}L_k\left(\bar{e}_{pq}, \frac{\partial T}{\partial \bar{x}_m}, T\right)$$
$$= a_{ir}a_{js}e_{rsk}L_k\left(e_{pq}, \frac{\partial T}{\partial x_m}, T\right) = a_{ir}a_{js}P_{rs}, \quad (8.5.8)$$

if we use (8.5.4). Hence **P** is an anti-symmetric matrix polynomial in the symmetric matrix **e** and the anti-symmetric matrix **b**. It follows from (A20), (A21) in the Appendix that **P** is a linear combination of the matrices

$$\begin{matrix} \mathbf{b}, & \mathbf{be}+\mathbf{eb}, & \mathbf{be}^2+\mathbf{e}^2\mathbf{b} \\ \mathbf{b}^2\mathbf{e}-\mathbf{eb}^2, & \mathbf{b}^2\mathbf{e}^2-\mathbf{e}^2\mathbf{b}^2, & \mathbf{eb}^2\mathbf{e}^2-\mathbf{e}^2\mathbf{b}^2 \end{matrix} \Bigg\}, \quad (8.5.9)$$

with coefficients which are polynomials in the invariants

$$\begin{matrix} \mathrm{tr}\ \mathbf{b}^2, & \mathrm{tr}\ \mathbf{e}, & \mathrm{tr}\ \mathbf{e}^2, & \mathrm{tr}\ \mathbf{e}^3 \\ \mathrm{tr}\ \mathbf{b}^2\mathbf{e}, & \mathrm{tr}\ \mathbf{b}^2\mathbf{e}^2, & \mathrm{tr}\ \mathbf{ebe}^2\mathbf{b}^2 \end{matrix} \Bigg\}. \quad (8.5.10)$$

From (8.5.6) we have

$$L_k = \tfrac{1}{2}e_{ijk}P_{ij}, \qquad b_{ik}b_{kj} = \frac{\partial T}{\partial x_i}\frac{\partial T}{\partial x_j} - \delta_{ij}\frac{\partial T}{\partial x_m}\frac{\partial T}{\partial x_m}, \quad (8.5.11)$$

so that L_k is a linear function of the vectors **A**, **B**, **C**, **A**×**B**, **A**×**C**, **B**×**C**, where

$$A_k = \frac{\partial T}{\partial x_k}, \qquad B_k = \frac{\partial T}{\partial x_r}e_{rk}, \qquad C_k = \frac{\partial T}{\partial x_r}e_{rs}e_{sk}, \quad (8.5.12)$$

and

$$(\mathbf{A}\times\mathbf{B})_k = e_{ijk}A_iB_j, \qquad (\mathbf{A}\times\mathbf{C})_k = e_{ijk}A_iC_j, \qquad (\mathbf{B}\times\mathbf{C})_k = e_{ijk}B_iC_j, \quad (8.5.13)$$

with coefficients which are scalar polynomials in the invariants

$$e_{ii}, \quad e_{ij}e_{ij}, \quad e_{ij}e_{jk}e_{ki}, \quad (8.5.14)$$

$$\left.\begin{matrix} A_iA_i = \dfrac{\partial T}{\partial x_i}\dfrac{\partial T}{\partial x_i}, & A_iB_i = \dfrac{\partial T}{\partial x_i}\dfrac{\partial T}{\partial x_j}e_{ij} \\ A_iC_i = B_iB_i = \dfrac{\partial T}{\partial x_i}\dfrac{\partial T}{\partial x_j}e_{jk}e_{ki} \end{matrix}\right\}, \quad (8.5.15)$$

and

$$[\mathbf{ABC}] = e_{ijk}A_iB_jC_k. \quad (8.5.16)$$

The coefficients in the polynomials are continuous functions of T. Hence

$$-L_k = \mathscr{C}_1 A_k + \mathscr{C}_2 B_k + \mathscr{C}_3 C_k + \mathscr{C}_4 (\mathbf{A}\times\mathbf{B})_k + \mathscr{C}_5 (\mathbf{A}\times\mathbf{C})_k + \mathscr{C}_6 (\mathbf{B}\times\mathbf{C})_k, \tag{8.5.17}$$

where $\mathscr{C}_1,\ldots,\mathscr{C}_6$ are polynomials in the invariants (8.5.14)–(8.5.16). Now $[\mathbf{ABC}]^2$ can be expressed as a polynomial in the invariants in (8.5.14) and (8.5.15) if we recall that the matrix $\mathbf{e}=[e_{ij}]$ satisfies the Cayley–Hamilton equation

$$\mathbf{e}^3 - \mathbf{e}^2 \operatorname{tr}\mathbf{e} + \tfrac{1}{2}\mathbf{e}[(\operatorname{tr}\mathbf{e})^2 - \operatorname{tr}\mathbf{e}^2] - \mathbf{I}\det\mathbf{e} = \mathbf{0}. \tag{8.5.18}$$

We see therefore that the polynomials $\mathscr{C}_1,\ldots,\mathscr{C}_6$ are linear in $[\mathbf{ABC}]$. Further,

$$\begin{aligned}
[\mathbf{ABC}]\mathbf{A} &= (\mathbf{A}\cdot\mathbf{C})(\mathbf{A}\times\mathbf{B}) + \mathbf{A}^2(\mathbf{B}\times\mathbf{C}) + (\mathbf{A}\cdot\mathbf{B})(\mathbf{C}\times\mathbf{A}) \\
[\mathbf{ABC}](\mathbf{A}\times\mathbf{B}) &= (\mathbf{A}\times\mathbf{B})\cdot(\mathbf{B}\times\mathbf{C})\mathbf{A} + (\mathbf{A}\times\mathbf{B})\cdot(\mathbf{C}\times\mathbf{A})\mathbf{B} + (\mathbf{A}\times\mathbf{B})^2\mathbf{C}
\end{aligned} \tag{8.5.19}$$

with similar equations obtained by cyclic permutation of $\mathbf{A}, \mathbf{B}, \mathbf{C}$. The coefficients of the vectors on the right-hand side of each equation in (8.5.19) are polynomials in the invariants (8.5.14) and (8.5.15). Hence, without loss of generality, we may express L_k in the form (8.5.17) in which $\mathscr{C}_1,\ldots,\mathscr{C}_6$ are scalar polynomials in the invariants (8.5.14) and (8.5.15) only, with coefficients which are continuous functions of T.

If $[\mathbf{ABC}] \neq 0$ we see that the second equation in (8.5.19) may be used to express $\mathbf{A}\times\mathbf{B}$, $\mathbf{B}\times\mathbf{C}$, $\mathbf{C}\times\mathbf{A}$ as $[\mathbf{ABC}]^{-1}$ times a linear function of $\mathbf{A}, \mathbf{B}, \mathbf{C}$ with coefficients which are polynomials in the invariants (8.5.14) and (8.5.15). This representation fails if $[\mathbf{ABC}] = 0$ and to avoid any degeneracies in the representation of L_k we retain the form (8.5.17).

If we recall (8.3.15) and (8.5.12) we see that (8.5.17) gives

$$-R_i = \frac{\partial y_i}{\partial x_k}\left(\mathscr{C}_1 \frac{\partial T}{\partial x_k} + \mathscr{C}_2 \frac{\partial T}{\partial x_r} e_{rk} + \mathscr{C}_3 \frac{\partial T}{\partial x_r} e_{rs} e_{sk}\right) + $$
$$+ e_{rsk}\frac{\partial y_i}{\partial x_k}\frac{\partial T}{\partial x_m}\frac{\partial T}{\partial x_n} e_{mt}(\mathscr{C}_4\,\delta_{rn}\,\delta_{st} + \mathscr{C}_5\,\delta_{rn}e_{st} + \mathscr{C}_6 e_{rn}e_{st}). \tag{8.5.20}$$

The value of Q^i referred to general coordinates θ may now be obtained from the formulae (8.3.2) and (8.5.20), using also (1.1.15) and (1.1.19). Thus

$$-Q^i = (\mathscr{C}_1 \delta^i_j + \mathscr{C}_2 \gamma^i_j + \mathscr{C}_3 \gamma^i_m \gamma^m_j) T|^j + $$
$$+ {}_0\epsilon^{ijk} T|_r T|_s \gamma^r_t (\mathscr{C}_4 \delta^s_j \delta^t_k + \mathscr{C}_5 \delta^s_j \gamma^t_k + \mathscr{C}_6 \gamma^s_j \gamma^t_k), \tag{8.5.21}$$

where

$$T|^j = g^{jr}T|_r = g^{jr}T\|_r, \qquad T\|_r = T|_r = \frac{\partial T}{\partial \theta^r}, \qquad {}_0\epsilon^{ijk} = e_{ijk}/\sqrt{g}. \qquad (8.5.22)$$

The invariants in (8.5.14) and (8.5.15) may be expressed in general coordinates θ in the form

$$\left.\begin{array}{ccc} \gamma_i^i, & \gamma_j^i\gamma_i^j, & \gamma_j^i\gamma_k^j\gamma_i^k \\ T|^iT|_i, & T|^iT|_j\gamma_i^j, & T|^iT|_j\gamma_k^j\gamma_i^k \end{array}\right\}. \qquad (8.5.23)$$

Alternatively, the first three invariants in (8.5.23) may be replaced by the basic invariants I_1, I_2, I_3 (or J_1, J_2, J_3) of Chapter I. Thus $\mathscr{C}_1,\ldots,\mathscr{C}_6$ are polynomials in I_1, I_2, I_3 and the three invariants

$$\left.\begin{array}{l} I_4 = T|^iT|_i \\ I_5 = T|^iT|_j\gamma_i^j \\ I_6 = T|^iT|_j\gamma_k^j\gamma_i^k \end{array}\right\}, \qquad (8.5.24)$$

with coefficients which are continuous functions of T.

From (8.3.21) and (8.5.21) we have

$$\mathscr{C}_1 I_4 + \mathscr{C}_2 I_5 + \mathscr{C}_3 I_6 + \mathscr{C}_6 I_7 \geqslant 0 \qquad (8.5.25)$$

where I_7 is the invariant

$$I_7 = {}_0\epsilon^{ijk}T|_kT|_rT|_s\gamma_i^r\gamma_j^m\gamma_m^s = [\mathbf{ABC}]. \qquad (8.5.26)$$

The definition of isotropy used here is for a body which initially does not necessarily have a centre of symmetry. If we add the condition that the undeformed body has a centre of symmetry then the formula (8.5.20) must change R_i into $-R_i$ when both x_i and y_i are replaced by $-x_i$ and $-y_i$ respectively. Hence for an isotropic body with a centre of symmetry the formula (8.5.20) is replaced by

$$-R_i = \frac{\partial y_i}{\partial x_k}\left(\mathscr{C}_1 \frac{\partial T}{\partial x_k} + \mathscr{C}_2 \frac{\partial T}{\partial x_r}e_{rk} + \mathscr{C}_3 \frac{\partial T}{\partial x_r}e_{rs}e_{sk}\right), \qquad (8.5.27)$$

with a corresponding change in the formula (8.5.21). Also (8.5.25) becomes

$$\mathscr{C}_1 I_4 + \mathscr{C}_2 I_5 + \mathscr{C}_3 I_6 \geqslant 0. \qquad (8.5.28)$$

8.6. Heat conduction vector for isotropic bodies: alternative form

When the body is isotropic initially, an alternative form for the heat conduction law may be deduced from (8.5.20) in terms of $\partial T/\partial y_i$ and the quantities C_{pq} defined in (1.15.15). Here we obtain the alternative form *ab initio*.

Omitting y_i because of invariance conditions we replace the basic assumption (8.3.7) by

$$R_i = R_i\left(\frac{\partial y_p}{\partial x_q}, \frac{\partial T}{\partial y_m}, T\right). \tag{8.6.1}$$

We now suppose that the undeformed body is in a position that differs from its original position only by a rigid-body rotation, the final position and state of the deformed body being unaltered. If the undeformed body is isotropic then, by an argument similar to that used in § 8.5, we have

$$R_i\left(\frac{\partial y_p}{\partial \bar{x}_q}, \frac{\partial T}{\partial y_m}, T\right) = R_i\left(\frac{\partial y_p}{\partial x_q}, \frac{\partial T}{\partial y_m}, T\right), \tag{8.6.2}$$

so that R_i is a function of $\partial y_p/\partial x_q$ which is invariant under all proper orthogonal transformations (8.5.1). By a reduction similar to that used in § 8.3 we see that R_i must be restricted to the form

$$R_i = R_i\left(C_{pq}, \frac{\partial T}{\partial y_m}, T\right), \tag{8.6.3}$$

where C_{pq} is defined in (1.15.15). We now assume that R_i is a polynomial in C_{pq} and $\partial T/\partial y_m$ with coefficients that are continuous functions of T.

Next we assume that the deformed body is subjected to a rigid-body rotation given by (8.3.8) and (8.3.9), the temperature T being unaltered at each point of the body. Also, the heat conduction vector **Q** is rotated with the body. Then by a discussion similar to that given in § 8.3 we obtain

$$R_i\left(\frac{\partial T}{\partial \bar{y}_m}, \bar{C}_{pq}\right) = a_{ij} R_j\left(\frac{\partial T}{\partial y_m}, C_{pq}\right), \tag{8.6.4}$$

where
$$\left.\begin{aligned} \bar{C}_{pq} &= \frac{\partial \bar{y}_p}{\partial x_u}\frac{\partial \bar{y}_q}{\partial x_u} = a_{pr}a_{qs}C_{rs} \\ \frac{\partial T}{\partial \bar{y}_m} &= a_{mn}\frac{\partial T}{\partial y_n} \end{aligned}\right\}. \tag{8.6.5}$$

Equation (8.6.4) is similar in form to (8.5.4) and by a reduction similar to that used in § 8.5 we see that

$$-R_i = \mathscr{C}'_1\frac{\partial T}{\partial y_i} + \mathscr{C}'_2\frac{\partial T}{\partial y_r}C_{ri} + \mathscr{C}'_3\frac{\partial T}{\partial y_r}C_{rs}C_{si} +$$
$$+ e_{rsi}\frac{\partial T}{\partial y_m}\frac{\partial T}{\partial y_n}C_{mt}(\mathscr{C}'_4\delta_{rn}\delta_{st} + \mathscr{C}'_5\delta_{rn}C_{st} + \mathscr{C}'_6C_{rn}C_{st}), \tag{8.6.6}$$

where $\mathscr{C}'_1,\ldots,\mathscr{C}'_6$ are polynomials in the invariants

$$C_{ii}, \qquad C_{ij}C_{ij}, \qquad C_{ij}C_{jk}C_{ki}, \qquad (8.6.7)$$

$$\frac{\partial T}{\partial y_i}\frac{\partial T}{\partial y_i}, \qquad \frac{\partial T}{\partial y_i}\frac{\partial T}{\partial y_j}C_{ij}, \qquad \frac{\partial T}{\partial y_i}\frac{\partial T}{\partial y_j}C_{jk}C_{ki}, \qquad (8.6.8)$$

with coefficients which are continuous functions of T.

Although the two expressions (8.5.20) and (8.6.6) are formally different they may be shown to be equivalent. The invariants (8.5.14) are identical with the invariants (8.6.7). Also, a polynomial in the invariants (8.5.15) may be expressed as a polynomial in the invariants (8.6.8) if we use the Cayley–Hamilton equation for C_{ij}. Further, the expression (8.5.20) for R_i may be reduced to the form (8.6.6) apart from a factor $\sqrt{I_3}$ in the second group of terms. This factor, however, may be replaced by a polynomial approximation in the invariants (8.6.7) so we recover the form (8.6.6), without loss of generality. Conversely, we may pass from (8.6.6) to (8.5.20) with the help of the Cayley–Hamilton theorem, and a polynomial approximation for $1/\sqrt{I_3}$.

Equations (8.3.2) and (8.6.6) now yield

$$-Q^i = (\mathscr{C}''_1\delta^i_j + \mathscr{C}''_2\gamma^i_j + \mathscr{C}''_3\gamma^i_m\gamma^m_j)T\|^j + \\ + \epsilon^{ijk}T\|_r T\|_s \gamma^r_i(\mathscr{C}''_4\delta^s_j\delta^t_k + \mathscr{C}''_5\delta^s_j\gamma^t_k + \mathscr{C}''_6\gamma^s_j\gamma^t_k), \quad (8.6.9)$$

where
$$\epsilon^{ijk} = e_{ijk}/\sqrt{G}, \qquad T\|^j = G^{jr}T\|_r, \qquad (8.6.10)$$

and $\mathscr{C}''_1,\ldots,\mathscr{C}''_6$ are polynomials in the invariants

$$\left. \begin{array}{ccc} I_1, & I_2, & I_3 \\ T\|_i T\|^i, & T\|_i T\|^j \gamma^i_j, & T\|_i T\|^j \gamma^i_k \gamma^k_j \end{array} \right\}, \qquad (8.6.11)$$

with coefficients which are continuous functions of T.

If the body is initially isotropic with a centre of symmetry then only the first three terms in (8.6.6) and (8.6.9) are retained, as in § 8.5.

8.7. Classical theory for isotropic bodies

When the displacements, displacement gradients, and changes of temperature from a standard temperature are all infinitesimals of the first order the above theory is simplified considerably. If T_0 is the standard temperature we replace T by T_0+T, where T is now an infinitesimal of the same order as the displacements and displacement gradients. The strain γ_{ij} becomes the classical strain tensor† and

† See *T.E.*

(8.3.18) and (8.3.19) reduce to

$$\left.\begin{array}{l}\tau^{ij} = \dfrac{\rho_0}{2}\left(\dfrac{\partial A}{\partial \gamma_{ij}} + \dfrac{\partial A}{\partial \gamma_{ji}}\right)\\[2mm] S = -\dfrac{\partial A}{\partial T}\end{array}\right\}, \qquad (8.7.1)$$

provided the initial stress is zero. From (8.1.4) we see that

$$q^i = Q^i, \qquad (8.7.2)$$

and (8.3.24) becomes the single equation

$$\rho_0 r - \rho_0 T_0 \frac{\partial S}{\partial t} - q^i|_i = 0. \qquad (8.7.3)$$

The formulae (8.7.1)–(8.7.3) are valid for an aeolotropic body. For an homogeneous isotropic body, equation (8.5.21) reduces to

$$q^i = -KT|^i, \qquad (8.7.4)$$

where K is a constant and, in view of (8.3.21), $K \geqslant 0$. Consistent with these approximations we may take A in the form

$$\rho_0 A = \rho_0 A_0 - S_0 T + \frac{1}{2}\left(2\mu\gamma^i_j\gamma^j_i + \lambda\gamma^r_r\gamma^s_s - 2\beta T\gamma^r_r - \frac{sT^2}{T_0}\right), \qquad (8.7.5)$$

where A_0, S_0 are the mean free energy and entropy, λ, μ are the usual Lamé constants, and β, s are constants. From (8.7.1) and (8.7.5) we deduce that

$$\tau^{ij} = \lambda g^{ij}\gamma^r_r + 2\mu\gamma^{ij} - \beta T g^{ij}, \qquad (8.7.6)$$

$$\rho_0 S - S_0 = \beta\gamma^r_r + \frac{sT}{T_0}, \qquad (8.7.7)$$

where we recall that all tensors are referred to the undeformed body and their associated forms are related by the metric tensor g_{ij}. Alternatively, (8.7.6) may be written in the form

$$\tau^i_j = \lambda\delta^i_j\gamma^r_r + 2\mu\gamma^i_j - \beta T\delta^i_j, \qquad (8.7.8)$$

or

$$\gamma^i_j = \frac{1+\eta}{E}\tau^i_j - \frac{\eta}{E}\tau^r_r\delta^i_j + \alpha T\delta^i_j, \qquad (8.7.9)$$

where

$$\beta = \frac{2\mu(1+\eta)\alpha}{1-2\eta} = \frac{E\alpha}{1-2\eta}, \qquad (8.7.10)$$

and E, η are Young's modulus and Poisson's ratio respectively. Also, α is the coefficient of linear expansion. Finally, equations (8.7.3), (8.7.4), and (8.7.7) yield

$$\rho_0 r - \beta T_0 \frac{\partial \gamma_r^r}{\partial t} - s\frac{\partial T}{\partial t} + KT|_i^i = 0. \qquad (8.7.11)$$

IX

EXPERIMENTAL APPLICATIONS

In this chapter we consider the application of the preceding theory to the determination of the elastic properties of real materials. Attention is mainly concentrated on the investigation of the form of the strain energy function for vulcanized rubber,† but this is preceded by a general discussion of the underlying principles involved.

9.1. Nature of the problem

Experiments on real materials may be linked with the theory of the preceding chapters for two main purposes. These are:

(i) the investigation of elastic properties;
(ii) the experimental verification of the theoretical predictions.

The first of these applications is concerned with the fundamental problem of finding a precise mathematical description of the elastic properties of any given material. This problem is solved if the strain energy function can be expressed as a known function of the quantities employed to describe the deformation, that is, as a function of the strain components defined in Chapter I or the combinations of them which occur for any given crystal class. A characterization of this type may find an immediate application in testing the adequacy of any description of the mechanical properties which has arisen from structural or molecular considerations. The more adequate description derived with the help of elasticity theory may, in fact, suggest modifications to, or an amplification of, any picture of the internal structure

---

† The reader is referred to the following papers for a fuller account and for details of the experimental techniques:

R. S. Rivlin, *J. appl. Phys.* **18** (1947), 444.
R. S. Rivlin and D. W. Saunders, *Phil. Trans. R. Soc.* A, **243** (1951), 251.
A. N. Gent and R. S. Rivlin, *Proc. phys. Soc.* B, **65** (1952), 118, 487.
For more recent developments see
L. J. Hart-Smith, *Z. angew Mech. Phys.* **17** (1966), 608.

A further discussion of the form of the strain energy function and its relation to the molecular structure of rubber is given by L. R. G. Treloar, *The Physics of Rubber Elasticity* (edn. 2, Clarendon Press, Oxford, 1958), and in the following papers:

R. S. Rivlin and D. W. Saunders, *Trans. Faraday Soc.* **48** (1952), 200.
A. G. Thomas, *Trans. Faraday Soc.* **51** (1955), 569.
A. N. Gent and A. G. Thomas, *J. Polymer Sci.* **28** (1958), 625.

of the material which has been built up on the basis of other evidence, such as the results of physical and chemical experiments and analysis. For example, molecular theory for rubber-like materials† yields as a first approximation a strain energy function of the form

$$W = C(I_1-3). \tag{9.1.1}$$

The experiments of Rivlin and Saunders described in § 9.5 suggest a form
$$W = C(I_1-3)+f(I_2-3), \tag{9.1.2}$$

where I_1 and I_2 are the usual strain invariants for isotropic incompressible materials, C is a constant, and f is a function whose general behaviour can be determined. Any extension of the statistical theory based on a more detailed molecular picture should yield a form for W which is consistent with (9.1.2).

The second application of experimental techniques becomes possible when a precise description of the elastic properties has been obtained. The system of forces required to maintain a given state of deformation may then, in principle, be determined from the stress–strain relations, the equations of equilibrium, and any relevant constraint conditions; the same system of equations forms a basis for the treatment of the converse problem of determining the deformation from a given system of applied loads. When, as is frequently the case owing to the non-linearity of the field equations, an exact solution to the elastic problem cannot be obtained, suitable experiments may serve a useful purpose in testing the adequacy and the range of validity of the theoretical analysis employed or the assumptions made in obtaining any approximate solution.

The analysis of Chapter I implies that in any given case a series of preliminary experiments may be necessary to determine the general characteristics of the material, such as homogeneity, the possible existence of curvilinear aeolotropy, symmetry properties, and the presence or otherwise of internal geometrical constraints of the type considered in Chapter VII. The nature of such experiments will, to a large extent, be governed by practical considerations such as the type of specimen which can be manufactured, ease of handling, and the accuracy with which it is possible to produce and measure certain types of deformation and the associated force systems. Again, the appearance or method

† This is a statistical theory based upon the molecular picture of vulcanized rubber as a network of long-chain molecules. An account of the theory, with references to original papers, is given by L. R. G. Treloar, loc. cit., p. 288.

of manufacture of the material may suggest that a particular type of structure is present and thus serve as a guide to the type of experiments which need to be performed. For example, one of the more common highly elastic materials, rubber, is isotropic, effectively incompressible† and, as usually manufactured, is substantially homogeneous. All of these properties are readily verified by experiment. We shall here discuss in general terms the physical concepts underlying the theory of Chapter I, and an experimental approach to the determination of the basic elastic properties of materials suggested by these ideas. We shall then examine in greater detail the more important experiments which have been carried out on rubber.

9.2. Preliminary investigations

The basic mechanical properties which any preliminary examination of the material would seek to investigate are as follows:

(i) *Elasticity*

For an elastic body, the mechanical properties are specified by means of a strain energy function W‡ which depends only upon the state of deformations§ and is independent of previous states of strain. The stresses thereby form a conservative system and are independent of previous strain history and of the rate at which the deformation has been produced or is varying at the instant under consideration. These considerations exclude the possibility of static or dynamic hysteresis and stress relaxation and creep effects, and are realized approximately in many types of vulcanized rubber.

(ii) *Homogeneity*

If equal cubes A and B are cut from the material at any two points as shown in Fig. 9.1 a, the edges of these cubes being parallel to a given

---

† Small volume changes in rubber during stretching have been measured by G. Gee, J. Stern, and L. R. G. Treloar, *Trans. Faraday Soc.* **46** (1950), 1101.

‡ More general types of material, for which a strain energy function does not exist, have been considered by W. Noll, *J. rat. Mech. Analysis* **4** (1955), 3, and by G. F. Smith and R. S. Rivlin, *Arch ration. Mech. Analysis* **1** (1957), 107. These materials are excluded from the present discussion.

§ We suppose the deformation to take place either isothermally or adiabatically so that thermodynamic variables may be omitted from the discussion. For a fuller discussion of the thermodynamic aspects, reference may be made to Chapter VIII. A more detailed examination of energy and entropy changes in rubber during deformation, with further references to original papers, is given by L. R. G. Treloar, chap. II, loc. cit., p. 288.

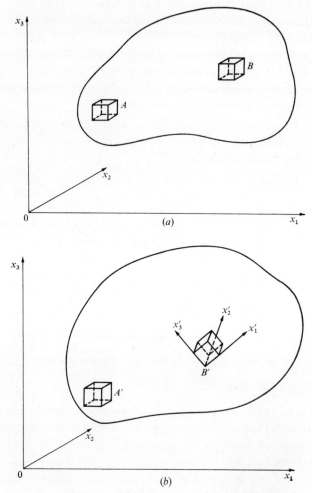

FIG. 9.1. To illustrate homogeneity (a) for rectilinearly aeolotropic bodies, (b) for curvilinearly aeolotropic bodies.

set of rectangular cartesian axes $Ox_1x_2x_3$, and if any arbitrarily given set of forces produces precisely the same state of strain when applied to cube A as when it is applied in exactly the same manner to cube B, we may regard the material as homogeneous. For a practical test of this hypothesis, it may be convenient to use test pieces which are not cubical in shape. If the material is aeolotropic the samples should have the same orientation relative to the system of axes x_i; if the body is isotropic the relative orientation of the two test pieces is unimportant.

When the foregoing condition is not satisfied it may be possible to describe the material as homogeneous in the following sense. Suppose that identical cubes A', B' are cut from the material at any two points, their edges being parallel to two different sets of rectangular cartesian axes x_i, x_i' respectively, the frame x_i' being obtained from x_i by a suitable rotation (Fig. 9.1 b). If the cubes thus formed possess identical elastic properties, as described above, the material may still be regarded as elastically homogeneous. When the axes are the tangent lines to an orthogonal curvilinear reference frame θ_i in the undeformed body at A', B' respectively, the material exhibits curvilinear aeolotropy with respect to the reference frame θ_i.

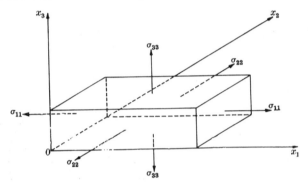

FIG. 9.2. Cube subjected to pure homogeneous strain.

(iii) *Symmetry properties*

The kind of experiment which is necessary to verify the existence of the more usual types of symmetry such as rhombic symmetry, transverse isotropy, or isotropy is suggested fairly readily by the stress–strain relations. It follows, for example, from the stress–strain relations (1.15.3), that if any material exhibits rhombic symmetry relative to a given set of directions, represented by the rectangular axes $Ox_1x_2x_3$ (Fig. 9.2), then a cube of this material with its edges parallel to these axes will undergo a pure homogeneous strain with principal directions parallel to the x_i-axes, under the influence of uniform normal tractions σ_{11}, σ_{22}, σ_{33} applied to its plane faces. This system of tractions applied to a cube of triclinic material, on the other hand, would produce a distortion of the cube in which shearing deformations are present. In practice it might be more convenient to confirm the existence of rhombic symmetry by the examination of thin sheets of material in pure homogeneous strain or tensile test pieces in simple elongation. If in Fig. 9.2 equal stresses ($\sigma_{11} = \sigma_{22}$) produce equal extensions in the

x_1-, x_2-directions, we may suspect the existence of transverse isotropy relative to the x_3-direction. Similarly, the production of a uniform dilatation (or compression) under the influence of equal and uniform normal tractions applied to all three faces of the cube is consistent with an assumption of isotropy.

(iv) *Constraints*

The type of geometrical constraint which occurs most frequently is that of incompressibility. This may be tested by the examination of volume changes during deformation. In particular a material is incompressible if a uniform applied pressure fails to produce any change in volume. The presence of other types of constraint may be suggested by structural considerations. For example, a material reinforced with thin extensible cords lying close together in parallel straight lines throughout its volume would effectively be inextensible in the direction of the cords and this assumption would be capable of easy experimental verification.

9.3. The functional form of W

Experiments based on the foregoing considerations serve to determine the variables which enter into the strain energy function W, each of these variables being functions of suitably defined components of strain; it remains to find the functional form of W in terms of the variables thus introduced. One method of approach to this problem is suggested by the form of the stress–strain relations. We consider first a compressible body in which, owing to symmetry considerations, the strain components e_{ij} may be grouped together in the known combinations

$$\Psi_r = \Psi_r(e_{ij}) \quad (r = 1, 2, ..., n). \tag{9.3.1}$$

These functions are defined so that $\Psi_r = 0$ in the unstrained state. In the case of transversely isotropic bodies, for example, when W has the form (1.13.12), we may define five quantities Ψ_r by the relations

$$\Psi_1 = I_1 - 3, \quad \Psi_2 = I_2 - 3, \quad \Psi_3 = I_3 - 1,$$

$$\Psi_4 = K_1, \quad \Psi_5 = K_2.$$

Initially we impose the restriction $n \leqslant 6$. Writing

$$W = W(\Psi_r) \tag{9.3.2}$$

in the stress–strain relations (1.15.1), we then have

$$\tau^{ij} = \frac{1}{2\sqrt{I_3}} \sum_{k=1}^{n} \left[\left(\frac{\partial \theta^i}{\partial x^r} \frac{\partial \theta^j}{\partial x^s} + \frac{\partial \theta^i}{\partial x^s} \frac{\partial \theta^j}{\partial x^r} \right) \frac{\partial \Psi_k}{\partial e_{rs}} \right] \frac{\partial W}{\partial \Psi_k}. \tag{9.3.3}$$

Consider now the homogeneous deformation

$$y_i = c_{ik} x_k, \tag{9.3.4}$$

for which

$$e_{rs} = \tfrac{1}{2}(c_{ir} c_{is} - \delta_{rs}), \qquad \sqrt{I_3} = |c_{ik}|. \tag{9.3.5}$$

Denoting the physical stress components referred to the rectangular cartesian system y_i in the deformed body by σ_{ij}, we have

$$\sigma_{ij} = \frac{1}{2\sqrt{I_3}} \sum_{k=1}^{n} \left[(c_{ir} c_{js} + c_{is} c_{jr}) \frac{\partial \Psi_k}{\partial e_{rs}} \right] \frac{\partial W}{\partial \Psi_k}. \tag{9.3.6}$$

The stresses σ_{ij} and the quantities c_{rs} defining the deformation, can, in principle, be measured. The derivatives $\partial \Psi_k / \partial e_{rs}$ are known functions of e_{rs} and can therefore be calculated from the measured values of c_{ik}. Equations (9.3.6) thus furnish a set of linear, simultaneous equations for the determination of the derivatives $\partial W / \partial \Psi_k$ for a given set of values of the arguments Ψ_k. Explicit values of these derivatives can therefore be calculated, provided the deformation (9.3.4) can be chosen so that n of the equations (9.3.6) are linearly independent. Consider now the expansion of $\partial W/\partial \Psi_k$ as a multiple Taylor series

$$\frac{\partial W}{\partial \Psi_k} = \left[\frac{\partial W}{\partial \Psi_k} \right]_0 + \Psi_r \left[\frac{\partial^2 W}{\partial \Psi_k \partial \Psi_r} \right]_0 + \tfrac{1}{2} \Psi_r \Psi_s \left[\frac{\partial^3 W}{\partial \Psi_k \partial \Psi_r \partial \Psi_s} \right]_0 + \ldots, \tag{9.3.7}$$

the symbol $[\ ]_0$ indicating that the derivative inside the brackets is to be evaluated at $\Psi_1 = \Psi_2 = \ldots = \Psi_n = 0$. For a given set of values of the constants c_{ik} the values of the functions Ψ_r may be calculated, while the left-hand member $\partial W/\partial \Psi_k$ of (9.3.7) is a quantity which may, in principle, be determined from (9.3.6). Moreover, if W is a polynomial in its arguments, the expansion (9.3.7) contains a finite number of terms and, for any given set of values c_{ik}, may be regarded as a linear equation relating a finite number of the derivatives of W evaluated at $\Psi_1 = \Psi_2 = \ldots = \Psi_n = 0$. A sufficient number of independent experiments would therefore furnish a sufficient number of independent linear equations of the type (9.3.7) to evaluate all of the functions $\partial W/\partial \Psi_k$ and their derivatives with respect to the arguments Ψ_r at $\Psi_r = 0$, and thus to determine explicitly the form of W as a function of the quantities Ψ_r.

If $n > 6$, equations (9.3.6) yield at the most six equations for the n unknowns $\partial W/\partial \Psi_k$ and the procedure breaks down. Since, however, there are only six independent components of strain there must then exist at least $n-6$ functional relationships between the functions Ψ_r. If these can be employed to express W as a function of six only of the quantities Ψ_r, and this function can be approximated by a polynomial in its arguments, the foregoing procedure can again be applied.

If the material is incompressible, equations (9.3.6) are replaced by

$$\sigma_{ij} = \frac{1}{2} \sum_{k=1}^{n} \left[(c_{ir}c_{js}+c_{is}c_{jr})\frac{\partial \Psi_k}{\partial e_{rs}} \right] \frac{\partial W}{\partial \Psi_k} + p\delta_{ij}, \qquad (9.3.8)$$

and when p is eliminated these yield only five equations for the determination of the derivatives $\partial W/\partial \Psi_k$. The incompressibility condition implies, however, that only five of the strain components e_{ij} can be chosen independently, and it follows that there are only five functionally independent strain invariants Ψ_r. In a similar manner each additional geometrical constraint reduces by one both the number of functionally independent invariants Ψ_r and the number of equations available for the determination of the derivatives $\partial W/\partial \Psi_k$.

Theoretically, it is thus possible to determine, by an experimental procedure, an approximate polynomial form for the strain energy function W for a given material in terms of a limited number of invariants Ψ_r. This representation may be adequate only for a limited range of deformation if the degree of the assumed polynomial form for W must be restricted owing to the limited number of independent experiments which can be performed. For example, a linear form for W may be adequate for reasonably small deformations but may be seriously in error if the strains are such as to produce large values for one or more of the arguments Ψ_r. Again, the form of W thus determined cannot safely be assumed to hold for deformations outside the range of measurement.

In practice, a procedure of the type outlined above might be expected to involve considerable experimental difficulties unless the number of functions Ψ_r and the number of significant derivatives of $\partial W/\partial \Psi_r$ are both quite small, or unless some other simplifying feature emerges. The measurement of tangential surface tractions may, for example, be difficult to perform without the introduction of considerable experimental error. Moreover, in the solution of any extensive system of simultaneous equations, even quite small errors in the values of some of the numerical coefficients may give rise to serious errors in the values of the unknowns determined from them. Nevertheless, the method outlined above is the basis of the procedure employed by Rivlin and Saunders to determine the form of W for rubber, but in this case only two functions Ψ_r identical with the invariants I_1-3, I_2-3 are involved, and the derivatives $\partial W/\partial I_1$, $\partial W/\partial I_2$ depend upon these in a particularly simple manner.

9.4. Experiments on rubber

Owing to its widespread use and convenience of handling, many of the experiments in the field of large elastic deformations have been carried out on vulcanized rubber. In the following sections we shall confine attention to experiments which have been concerned with the determination of the functional form of W for this material. In these it is assumed that the body is incompressible and isotropic, results which follow from earlier work; W is then the general function (1.14.2) of the two strain invariants I_1, I_2.

A simple experiment to perform is that of simple elongation, but this alone yields a limited amount of information about the form of the strain energy function and in some respects this information is misleading.† A more fruitful approach is provided by the experiment on the pure homogeneous strain of a thin sheet described in § 9.5. By means of this, the functional form of W may be determined for a limited range of deformation. The experiment is unsuitable for very large deformations, however, which are difficult to obtain in the manner described, and for small deformations in which I_1 and I_2 have values less then about 5·0. In the latter range the results obtained become very sensitive to small experimental errors.

Additional information about the form of W for low values of I_1, I_2 is provided by the experiment in pure shear described in § 9.6. This deformation is defined as a pure homogeneous deformation in which one of the principal extension ratios is unity and the volume is unchanged.‡ A large pure shear of a cuboid is easier to produce and more convenient to work with than the simple shear defined by (2.8.12); the latter requires a more complicated system of surface tractions to maintain it§ and is not easy to represent in terms of the principal extension ratios. Supplementary experiments which are convenient to carry out

---

† See § 9.7.
‡ For a discussion of this deformation see A. E. H. Love, chap. I, loc. cit., p. 230; L. R. G. Treloar, chap. V, loc. cit., p. 288.
§ See, for example, *T.E.*, chap. III.

and to analyse for low values of I_1 and I_2 include the torsion of a cylindrical rod and the torsion, inflation, and extension of a cylindrical tube. These are examined briefly in §§ 9.9–9.11.

The behaviour of the material for large strains is conveniently investigated by experiments in the uniform two-dimensional extension of a thin sheet. This type of strain occurs at the centre of a circular rubber sheet inflated as described in § 4.12. In this region the deformation is approximately uniform, and may be regarded as equivalent to a unidirectional compression in a direction normal to the middle surface of the sheet. By working with two-dimensional extension, however, difficulties due to instability are avoided and much larger strains are possible than would be obtained by compression.

9.5. Pure homogeneous strain of a thin rubber sheet

An approximate form of the strain energy function for rubber was determined by Rivlin and Saunders† from experiments on the pure homogeneous deformation of a thin sheet. We suppose the sheet to be of thickness $2h_0$ and to be bounded in the rectangular cartesian reference frame x_i by the planes

$$x_1 = \pm A, \quad x_2 = \pm B, \quad x_3 = \pm h_0 \quad (A, B \gg h_0).$$

This sheet is subjected to a pure homogeneous deformation with principal extension ratios λ_i in the direction of the x_i-axes, so that the coordinates y_i in the deformed state are given by

$$y_i = \lambda_i x_i \quad (i \text{ not summed}). \tag{9.5.1}$$

If the deformation is produced by means of edge tractions only, so that the major surfaces $x_3 = \pm h_0$ are free from applied stress, the formulae (7.3.3)–(7.3.5) apply. From these we have

$$I_1 = \lambda_1^2 + \lambda_2^2 + \frac{1}{\lambda_1^2 \lambda_2^2}, \quad I_2 = \frac{1}{\lambda_1^2} + \frac{1}{\lambda_2^2} + \lambda_1^2 \lambda_2^2 \quad (\lambda_1 \lambda_2 \lambda_3 = 1), \tag{9.5.2}$$

and

$$\left. \begin{aligned} \sigma_{11} &= 2\left(\lambda_1^2 - \frac{1}{\lambda_1^2 \lambda_2^2}\right)\left(\frac{\partial W}{\partial I_1} + \lambda_2^2 \frac{\partial W}{\partial I_2}\right) \\ \sigma_{22} &= 2\left(\lambda_2^2 - \frac{1}{\lambda_1^2 \lambda_2^2}\right)\left(\frac{\partial W}{\partial I_1} + \lambda_1^2 \frac{\partial W}{\partial I_2}\right) \\ \sigma_{33} &= 0, \quad \sigma_{ij} = 0 \quad (i \neq j) \end{aligned} \right\} , \tag{9.5.3}$$

† Loc. cit., p. 288.

σ_{ij} being the physical components of stress referred to the coordinates y_i.

The stresses σ_{11}, σ_{22} may be expressed in terms of the forces f_1, f_2 acting on the edges of the deformed sheet by means of the formulae

$$f_1 = 2h_0 \sigma_{11}/\lambda_1, \qquad f_2 = 2h_0 \sigma_{22}/\lambda_2, \tag{9.5.4}$$

where f_1, f_2 are measured per unit length of edge of sheet in the undeformed state. From measurements of these forces for any given values of λ_1 and λ_2, σ_{11} and σ_{22} can be calculated and the corresponding values of $\partial W/\partial I_1$ and $\partial W/\partial I_2$ found from the solution

$$\begin{aligned} \frac{\partial W}{\partial I_1} &= \left\{ \frac{\lambda_1^2 \sigma_{11}}{\lambda_1^2 - 1/(\lambda_1^2 \lambda_2^2)} - \frac{\lambda_2^2 \sigma_{22}}{\lambda_2^2 - 1/(\lambda_1^2 \lambda_2^2)} \right\} \Big/ 2(\lambda_1^2 - \lambda_2^2) \\ \frac{\partial W}{\partial I_2} &= -\left\{ \frac{\sigma_{11}}{\lambda_1^2 - 1/(\lambda_1^2 \lambda_2^2)} - \frac{\sigma_{22}}{\lambda_2^2 - 1/(\lambda_1^2 \lambda_2^2)} \right\} \Big/ 2(\lambda_1^2 - \lambda_2^2) \end{aligned} \tag{9.5.5}$$

of equations (9.5.3). If λ_1, λ_2 are varied in such a manner that I_2 remains constant, we obtain from (9.5.5) values of $\partial W/\partial I_1$ and $\partial W/\partial I_2$ for various values of I_1 at constant I_2. Similarly, variations of λ_1, λ_2 such that I_1 remains constant while I_2 varies give the dependence of $\partial W/\partial I_1$, $\partial W/\partial I_2$ upon I_2 for constant values of I_1.

Corresponding values of λ_1 and λ_2 for which I_1 is constant are given from (9.5.2) by

$$\lambda_2^2 = \tfrac{1}{2}\{(I_1 - \lambda_1^2) \pm [(I_1 - \lambda_1^2)^2 - 4/\lambda_1^2]^{\frac{1}{2}}\}; \tag{9.5.6}$$

the similar relation

$$\lambda_2^2 = \frac{1}{2\lambda_1^2}\left\{\left(I_2 - \frac{1}{\lambda_1^2}\right) \pm \left[\left(I_2 - \frac{1}{\lambda_1^2}\right)^2 - 4\lambda_1^2\right]^{\frac{1}{2}}\right\} \tag{9.5.7}$$

gives corresponding values of λ_1 and λ_2 for which I_2 is constant. The manner in which λ_1, λ_2 vary for certain constant values of I_1 and I_2 is shown in Fig. 9.3, the broken curves representing the lines

$$\lambda_2 = \lambda_1^{-\frac{1}{2}} (f_2 = 0), \qquad \lambda_1 = \lambda_2^{-\frac{1}{2}} (f_1 = 0), \tag{9.5.8}$$

which correspond to simple extensions parallel to the x_1-, x_2-axes respectively.

The experimental results obtained by this method by Rivlin and Saunders suggest a form of strain energy function

$$W = C_1(I_1 - 3) + f(I_2 - 3), \tag{9.5.9}$$

in which $\partial W/\partial I_1$ is a constant and $\partial W/\partial I_2 = df/dI_2$ is a function only of I_2. The ratio $(\partial W/\partial I_2)/(\partial W/\partial I_1)$ has an approximate value of 1/8 for

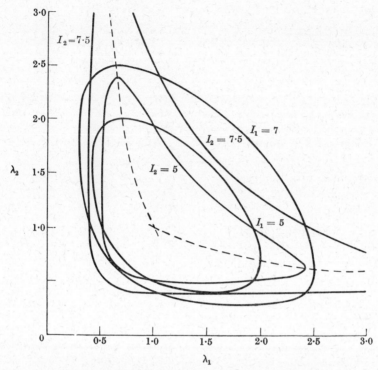

Fig. 9.3. The relation between λ_1 and λ_2 at constant I_1 or I_2.

low values of I_2 and decreases steadily with increasing I_2. There is, however, some scatter of the experimental readings and since the calculations based on (9.5.5) are sensitive to quite small experimental errors it is possible that the experimental results could equally well be represented by some slightly different dependence of $\partial W/\partial I_1$ and $\partial W/\partial I_2$ on I_1 and I_2. Nevertheless, the form (9.5.9) for W enables the theory to predict results for other types of deformation which are in good agreement with experiment.

For deformations which are not very large, λ_1 and λ_2 each approach the value unity and the calculation of $\partial W/\partial I_1$ and $\partial W/\partial I_2$ from (9.5.5) involves finding the quotients of quantities which are themselves not very large and which may, in fact, be of the same order of magnitude as the experimental errors. Other types of experiment therefore become necessary to examine the form of W in the region $I_1 = I_2 = 3$. In the remaining experiments carried out by Rivlin and Saunders $\partial W/\partial I_1$ was assumed to be constant, and the experimental results were used to determine a value of $\partial W/\partial I_2$ consistent with the assumed value of $\partial W/\partial I_1$.

9.6. Pure shear

From (9.5.3) and (9.5.4) the force per unit length of edge of the undeformed body in the x_1-direction required to produce the deformation (9.5.1) is

$$f_1 = 4h_0\left(\lambda_1 - \frac{1}{\lambda_1^3\lambda_2^2}\right)\left(\frac{\partial W}{\partial I_1} + \lambda_2^2\frac{\partial W}{\partial I_2}\right).$$

If the deformation is varied so that λ_2 remains constant, the curve of $\partial W/\partial I_1 + \lambda_2^2 \partial W/\partial I_2$ against I_2 gives a convenient means of evaluating $\partial W/\partial I_2$, provided $\partial W/\partial I_1$ can be assumed to have a known constant value.

When $\lambda_2 = 1$ and there is no volume change, the deformation is known as pure shear† and a convenient method of producing this type of deformation is to stretch a strip of rubber in such a way that its width remains unchanged. Fig. 9.4 illustrates the distribution of

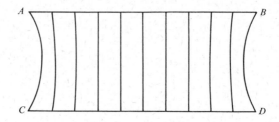

FIG. 9.4. Pure shear. Treloar.

strain in a wide sheet stretched between the clamped edges AB, CD. If the width of the sheet (i.e., AB or CD) is sufficiently large compared with its length, the non-uniformity of strain arising from the outer edges AC, BD being free may be neglected, and the extension ratio λ_2 in the centre in a direction parallel to the edges AB, CD is then substantially unaltered in any deformation produced by moving these edges farther apart.

Experimental results obtained by Rivlin and Saunders for values of $\lambda_2 = 1$ (curve I) and $\lambda_2 = 0.776$ (curve II) are shown in Fig. 9.5. To interpret the results for $\lambda_2 = 1$, the ratio $(\partial W/\partial I_2)/(\partial W/\partial I_1)$ was assumed to have the value found from the experiment of § 9.5 at a given value of I_2.‡ The measurement of $\partial W/\partial I_1 + \partial W/\partial I_2$ at this

---

† A. E. H. Love, loc. cit., p. 230.
‡ A value of $(\partial W/\partial I_2)/(\partial W/\partial I_1) = 1/8$ at $I_2 = 5$ was actually chosen.

point then yielded values for $\partial W/\partial I_1$ and $\partial W/\partial I_2$ separately. Assuming a strain energy of the form (9.5.9) and the value of $\partial W/\partial I_1$ ($= C_1$) thus determined, values of $\partial W/\partial I_2$ at other values of I_2 were found from the appropriate readings for $\partial W/\partial I_1 + \partial W/\partial I_2$. The values of

$$\partial W/\partial I_1 + \lambda_2^2 \partial W/\partial I_2$$

for $\lambda_2 = 0.776$ calculated using the values of $\partial W/\partial I_1$, $\partial W/\partial I_2$ thus obtained, agreed very closely with the experimental values derived directly from curve II.

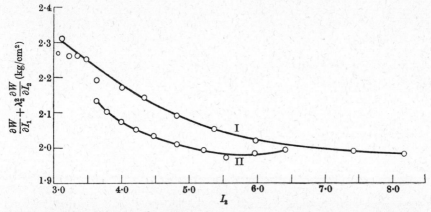

Fig. 9.5. Plot of $\partial W/\partial I_1 + \lambda_2^2 \partial W/\partial I_2$ against I_2 from experiments on pure shear (curve I; $\lambda_2 = 1$) and pure shear superposed on simple extension (curve II; $\lambda_2 = 0.776$).

9.7 Simple elongation

When $\sigma_{22} = \sigma_{33} = 0$ we have, from (9.5.3) and (9.5.2),

$$\frac{1}{\lambda_2^2} = \lambda_1 = \lambda \quad \text{(say)}, \tag{9.7.1}$$

$$\sigma_{11} = 2\left(\lambda^2 - \frac{1}{\lambda}\right)\left(\frac{\partial W}{\partial I_1} + \frac{1}{\lambda}\frac{\partial W}{\partial I_2}\right), \tag{9.7.2}$$

and
$$I_1 = \lambda^2 + \frac{2}{\lambda}, \quad I_2 = 2\lambda + \frac{1}{\lambda^2}. \tag{9.7.3}$$

The deformation then becomes a simple extension of ratio λ in the x_1-direction and the load N required to extend a test piece of initial uniform cross-sectional area A is

$$N = \frac{A}{\lambda}\sigma_{11} = 2A\left(\lambda - \frac{1}{\lambda^2}\right)\left(\frac{\partial W}{\partial I_1} + \frac{1}{\lambda}\frac{\partial W}{\partial I_2}\right). \tag{9.7.4}$$

The general shape of the load deformation curve obtained experimentally is shown in Fig. 9.6. A somewhat more interesting curve is obtained by plotting the quantity $\partial W/\partial I_1 + (1/\lambda)\partial W/\partial I_2$, calculated from experimental results by means of (9.7.4), against $1/\lambda$. From a typical curve obtained in this way by Rivlin and Saunders† and shown in Fig. 9.7, it is seen that a nearly linear relation is obtained over the range of values of $1/\lambda$ from 0·9 to 0·45, corresponding to values of I_2 lying between 3·03 and 4·65. It has been suggested that the departure

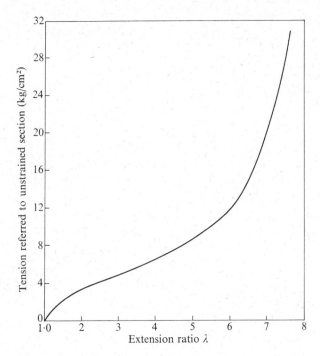

FIG. 9.6. Typical force extension curve for vulcanized rubber.

from linearity which occurs at values of $1/\lambda$ below about 0·45 is due to the finite extensibility of the long-chain molecules forming the molecular network. This is not taken into consideration in the simple statistical theory of rubber-like elasticity.‡

A possible interpretation of the linear portion of the curve of Fig. 9.7 would be obtained, in the absence of other evidence, by assuming that rubber has a strain energy function of the Mooney form (1.14.3),

† Loc. cit. p. 288.
‡ A fuller discussion is given by L. R. G. Treloar, loc. cit., p. 288.

in which $\partial W/\partial I_1$, $\partial W/\partial I_2$ have the constant values C_1, C_2 respectively.†
This, however, would lead to the prediction of constant values for
$\partial W/\partial I_1 + \lambda_2^2 \partial W/\partial I_2$ in the pure shear experiments described in § 9.6,
a result which is not borne out by experiment. Moreover, from the
linear portion of the curve a value of approximately 0·8 would be
obtained for the ratio C_2/C_1, a result inconsistent with the experiments
on two-dimensional deformation, where considerably smaller values of
$(\partial W/\partial I_2)/(\partial W/\partial I_1)$ are obtained throughout the entire range of values

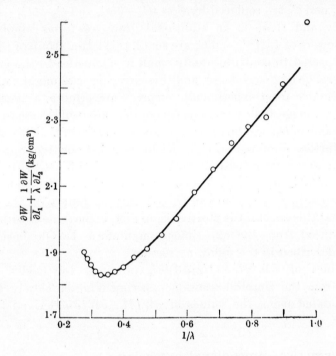

FIG. 9.7. Curve of $\partial W/\partial I_1 + (1/\lambda)\partial W/\partial I_2$ against $1/\lambda$ from experiment on simple extension.

of I_2 considered. A possible reason for the discrepancy becomes
apparent if in (9.7.4) we substitute a series expansion

$$W = C_1(I_1-3) + C_2(I_2-3) + C_3(I_2-3)^2 + C_4(I_2-3)^3 + \ldots$$

for W, consistent with (9.5.9). As far as the term in C_4 the expression

† The Mooney form of W would not, of course, give the non-linear part of the curve of Fig. 9.7.

for N becomes

$$N = 2A\left(\lambda-\frac{1}{\lambda^2}\right)\left\{\left[(C_1+4C_3-36C_4)+\frac{1}{\lambda}(C_2-6C_3+27C_4)\right]+\right.$$
$$\left.+\left[\left(12\lambda+\frac{12}{\lambda^2}-\frac{18}{\lambda^3}+\frac{3}{\lambda^5}\right)C_4+\frac{2}{\lambda^3}C_3\right]\right\}.$$

Comparing this with the expression for N which would be obtained by assuming the Mooney form of strain energy function to be valid, we see that C_1 and C_2 are replaced by $C_1+4C_3-36C_4$ and $C_2-6C_3+27C_4$ respectively and there is an additional group of terms involving different powers of λ. If C_3 and C_4 are small, this latter group of terms may be expected to contribute only small irregularities to the curve within the range of λ considered, and these irregularities might well be indistinguishable from experimental errors. Furthermore, a negative value of C_3 as suggested by the experiment of § 9.5 would decrease the apparent value of C_1, and increase the apparent value of C_2. This effect would be further reinforced if C_4 could be assumed positive. Such an assumption would be consistent with a curve of $(\partial W/\partial I_2)/(\partial W/\partial I_1)$ against I_2 of initially negative, but increasing slope. Although any inference regarding the nature of C_4 can only be tentative, the hypothesis that this constant is positive is, in fact, in agreement with the results deduced from the experiment on uniform two-dimensional extension described in the following section.

The values of $\partial W/\partial I_1+(1/\lambda)\partial W/\partial I_2$ obtained by Rivlin and Saunders from the simple extension experiment agree closely with those calculated using the values of $\partial W/\partial I_1$, $\partial W/\partial I_2$ derived from pure shear.

9.8. Uniform two-dimensional extension

If, in the pure homogeneous strain (9.5.1) the ratios λ_1 and λ_2 are equal and greater than unity, we obtain a uniform two-dimensional extension in the x_1-, x_2-plane. Putting

$$\lambda_1 = \lambda_2 = \lambda, \qquad \lambda_3 = 1/\lambda^2 \qquad (9.8.1)$$

in (9.5.2) we have

$$I_1 = 2\lambda^2+1/\lambda^4, \qquad I_2 = 2/\lambda^2+\lambda^4. \qquad (9.8.2)$$

Also, if p is again chosen so that $\sigma_{33} = 0$ we have, from (9.5.3),

$$\sigma_{11} = \sigma_{22} = 2\left(\lambda^2-\frac{1}{\lambda^4}\right)\left(\frac{\partial W}{\partial I_1}+\lambda^2\frac{\partial W}{\partial I_2}\right) = \sigma \quad \text{(say)}. \qquad (9.8.3)$$

The same state of strain could evidently be maintained by the stress system

$$\left.\begin{array}{c}\sigma_{11} = \sigma_{22} = 0 \\ \sigma_{33} = -2\left(\lambda^2 - \dfrac{1}{\lambda^4}\right)\left(\dfrac{\partial W}{\partial I_1} + \lambda^2 \dfrac{\partial W}{\partial I_2}\right)\end{array}\right\}, \qquad (9.8.4)$$

so that measurements of σ_{11} or σ_{22} in the case represented by (9.8.3) would also suffice to determine the stress associated with a unidirectional compression of the material of ratio $1/\lambda^2$ in the x_3-direction.

The conditions described by (9.8.1) can be reproduced approximately by inflating a thin circular sheet of rubber of uniform thickness, clamped around its edge, by applying an air pressure to one face as described in § 4.12. This type of experiment has been carried out by Treloar[†] and by Rivlin and Saunders.[‡] At the pole of the inflated sheet, the principal extension ratios in the plane of the sheet are equal, and the radial tension T at this point, measured per unit length on the major surface of the deformed sheet, is related to the inflating pressure P by the formula

$$P = 2T/r, \qquad (9.8.5)$$

r being the radius of curvature at the pole. If the initial thickness of the sheet is h and $h \ll r$ we may assume the components of stress at the pole to be very nearly uniform throughout the thickness of the sheet and to have the values (9.8.3) with $\sigma_{33} = 0$, the x_3-axis being assumed normal to the major surfaces of the sheet at its centre. The thickness of the deformed sheet at the centre is then h/λ^2 and, from (9.8.3) and (9.8.5),

$$P = \frac{2h\sigma}{r\,\lambda^2} = \frac{4h}{r}\left(1 - \frac{1}{\lambda^6}\right)\left(\frac{\partial W}{\partial I_1} + \lambda^2\frac{\partial W}{\partial I_2}\right). \qquad (9.8.6)$$

The foregoing results may be compared with (4.12.14)–(4.12.17).[§]

From measurements of P, r, and λ the quantity $\partial W/\partial I_1 + \lambda^2 \partial W/\partial I_2$ can be determined. Since for many types of rubber extension ratios of $\lambda = 6$ at the pole can be achieved without difficulty, and from (9.8.2) $I_2 > \lambda^4$, readings can be obtained for quite high values of I_2.

By writing $\lambda^2 = 1/\lambda'$ in (9.8.6), a comparison of the results for simple elongation and unidirectional compression may be obtained.

[†] L. R. G. Treloar, *Trans. Faraday Soc.* **39** (1943), 241.
[‡] Loc. cit., p. 288.
[§] $h = 2h_0$.

Thus in the graph of $\partial W/\partial I_1 + (1/\lambda')\partial W/\partial I_2$ against $1/\lambda'$ (Fig. 9.8) the simple extension portion of the curve ($\lambda' > 1$) is derived from the experiment of § 9.7 and in this region λ' is identical with the extension ratio λ of that section. In the compression portion of the curve, $\lambda' = 1/\lambda^2 (\lambda' < 1)$ where λ now has the significance ascribed to it in equations (9.8.1)–(9.8.6); in this region, therefore, λ' represents a compression ratio. The curve shows clearly the inadequacy of the Mooney form of W (1.14.3) for the treatment of unidirectional compression.

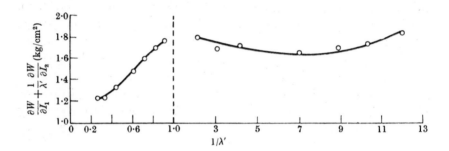

Fig. 9.8. Curve of $\partial W/\partial I_1 + (1/\lambda')\partial W/\partial I_2$ against $1/\lambda'$ from experiments on inflation (unidirectional compression) and simple extension. (Note change of scale at $1/\lambda' = 1$.)

Estimates of the value of $\partial W/\partial I_2$ throughout the range of deformation examined can be obtained by the procedure described in § 9.6 for pure shear. Thus a value of the ratio $(\partial W/\partial I_2)/(\partial W/\partial I_1)$ is assumed for a given value of I_2 (or λ'),† this value being based on previous experiments. From the measured value of $\partial W/\partial I_1 + (1/\lambda')\partial W/\partial I_2$ at this point separate values of the two derivatives can be determined and the assumption of the form (9.5.9) for W is then sufficient to allow the evaluation of $\partial W/\partial I_2$ at other values of I_2. The resulting variation

† In this case, Rivlin and Saunders used the value $(\partial W/\partial I_2)/(\partial W/\partial I_1) = 0.16$ at $I_2 = 4.25$ ($\lambda' = 2$) obtained from the experiment on pure shear. Since different specimens of rubber were used for the two experiments the actual value of this ratio for the inflation experiment might have been somewhat different from that assumed. Numerical calculations based on the actual readings and equation (9.8.6), however, are sufficient to show that values of $(\partial W/\partial I_2)/(\partial W/\partial I_1)$ calculated at high and low values of I_2 ($I_2 > 25$, $I_2 < 3.5$) are insensitive to changes in the value of this ratio assumed at $I_2 = 4.25$.

of $\partial W/\partial I_2$ with I_2 obtained by Rivlin and Saunders is shown in the accompanying table:

$1/\lambda'$	0·5	0·6	0·7	0·8	3	5	7	9	11
I_2	4·25	3·69	3·35	3·14	9·67	25·4	49·3	81·2	121
$\dfrac{(\partial W/\partial I_2)}{(\partial W/\partial I_1)}$	0·16	0·26	0·33	0·39	0·12	0·06	0·04	0·03	0·035

9.9. Experiments on torsion

The characteristic features of the form of the strain energy function for rubber inferred from the work described in §§ 9.5–9.8 are confirmed by experiments on torsion. Consider a tube of isotropic incompressible material of initial external and internal radii ρ_1, ρ_2 and length l_0. This is subjected to the simple torsion superposed on finite extension described in § 2.3, the dimensions ρ_1, ρ_2, l_0 being changed by the deformation into r_1, r_2, l respectively. The forces required to produce this deformation are derived from the analysis of §§ 2.1–2.3 by omitting terms involving Θ and Λ peculiar to the transversely isotropic case. Thus, remembering (2.3.2), when the outer surface of the tube is stress free, we obtain from (2.3.3) and (2.3.7), respectively, the expressions

$$\left.\begin{aligned}
M &= 4\pi\psi \int_{\rho_2}^{\rho_1} \mu\rho^3 \left(\lambda\mu \frac{\partial W}{\partial I_1} + \frac{1}{\lambda\mu}\frac{\partial W}{\partial I_2}\right) d\rho \\
N &= 2\pi \int_{\rho_2}^{\rho_1} \Big\{ \frac{2}{\lambda}\Big(\lambda^2 - \frac{1}{\lambda^2\mu^2}\Big)\Big(\frac{\partial W}{\partial I_1} + \mu^2\frac{\partial W}{\partial I_2}\Big) + \\
&\qquad + \frac{1}{\lambda^2\mu^2}\Big(\frac{\rho_2^2}{\rho^2}-1\Big)\Big(\mu^2 - \frac{1}{\lambda^2\mu^2}\Big)\Big(\frac{\partial W}{\partial I_1} + \lambda^2\frac{\partial W}{\partial I_2}\Big) + \\
&\qquad + \psi^2\Big[(\rho_2^2 - \rho^2)\frac{\partial W}{\partial I_1} - \frac{2\rho^2}{\lambda}\frac{\partial W}{\partial I_2}\Big]\Big\}\rho\,d\rho
\end{aligned}\right\}, \quad (9.9.1)$$

for the torsional couple M and the resultant longitudinal force N on the ends of the tube. Again, if the outer surface $\rho = \rho_1$ of the tube is free from surface traction, the inflating pressure P on the inner surface $\rho = \rho_2$ is given, from (2.3.6), by

$$P = -[\tau^{11}]_{\rho=\rho_2}$$

$$= 2\int_{\rho_2}^{\rho_1}\Big(\mu^2 - \frac{1}{\lambda^2\mu^2}\Big)\Big(\frac{\partial W}{\partial I_1} + \lambda^2\frac{\partial W}{\partial I_2}\Big)\frac{d\rho}{\lambda\mu^2\rho} + 2\psi^2\lambda\int_{\rho_2}^{\rho_1}\frac{\partial W}{\partial I_1}\rho\,d\rho. \quad (9.9.2)$$

With these expressions, from (2.1.13) and (2.2.1) we may couple the formulae

$$\left.\begin{aligned} I_1 &= \lambda^2+\mu^2+\frac{1}{\lambda^2\mu^2}+\psi^2\lambda^2\mu^2\rho^2 \\ I_2 &= \frac{1}{\lambda^2}+\frac{1}{\mu^2}+\lambda^2\mu^2+\psi^2\rho^2 \end{aligned}\right\} \qquad (9.9.3)$$

for the invariants.

If the tube is subjected to a simple extension only, then $\mu = \lambda^{-\frac{1}{2}}$. If, in addition, there is a small amount of torsion, so that $\rho_1\psi \ll 1$ then since, by symmetry, r must be an even function of ψ, (2.3.1) yields

$$\rho^2(1-\lambda\mu^2) = K = \beta\psi^2+O(\psi^4), \qquad (9.9.4)$$

where β is a constant. We then have

$$\mu = \frac{1}{\sqrt{\lambda}}\left(1-\frac{\beta\psi^2}{2\rho^2}\right)+O(\psi^4), \qquad (9.9.5)$$

and the invariants (9.9.3) become

$$\left.\begin{aligned} I_1 &= \left(\mu-\frac{1}{\lambda\mu}\right)^2+\lambda^2+\frac{2}{\lambda}+\psi^2\lambda^2\mu^2\rho^2 \\ &= \lambda^2+\frac{2}{\lambda}+\psi^2\lambda\rho^2+O(\psi^4) \\ I_2 &= \left(\lambda\mu-\frac{1}{\mu}\right)^2+2\lambda+\frac{1}{\lambda^2}+\psi^2\rho^2 \\ &= 2\lambda+\frac{1}{\lambda^2}+\psi^2\rho^2+O(\psi^4) \end{aligned}\right\} \qquad (9.9.6)$$

With these results equations (9.9.1) yield

$$\left.\begin{aligned} \left[\frac{M}{\psi}\right]_{\psi=0} &= \pi(\rho_1^4-\rho_2^4)\left(\frac{\partial W}{\partial I_1}+\frac{1}{\lambda}\frac{\partial W}{\partial I_2}\right)_{\psi=0} \\ [N]_{\psi=0} &= 2\pi(\rho_1^2-\rho_2^2)\left(\lambda-\frac{1}{\lambda^2}\right)\left(\frac{\partial W}{\partial I_1}+\frac{1}{\lambda}\frac{\partial W}{\partial I_2}\right)_{\psi=0} \end{aligned}\right\}, \qquad (9.9.7)$$

and from (9.9.2)

$$\begin{aligned} P &= 2\left(\frac{\partial W}{\partial I_1}+\lambda^2\frac{\partial W}{\partial I_2}\right)_{\psi=0}\int_{\rho_2}^{\rho_1}\left(-\frac{2\beta\psi^2}{\lambda\rho^3}\right)d\rho+2\psi^2\lambda\left(\frac{\partial W}{\partial I_1}\right)_{\psi=0}\int_{\rho_2}^{\rho_1}\rho\,d\rho+O(\psi^4) \\ &= \psi^2(\rho_1^2-\rho_2^2)\left[-2\left(\frac{\partial W}{\partial I_1}+\lambda^2\frac{\partial W}{\partial I_2}\right)_{\psi=0}\frac{\beta}{\lambda\rho_1^2\rho_2^2}+\lambda\left(\frac{\partial W}{\partial I_1}\right)_{\psi=0}\right]+O(\psi^4), \qquad (9.9.8) \end{aligned}$$

so that

$$\left[\frac{P}{\psi^2}\right]_{\psi=0} = (\rho_1^2 - \rho_2^2)\left[\lambda\left(\frac{\partial W}{\partial I_1}\right)_{\psi=0} - \frac{2\beta}{\lambda\rho_1^2\rho_2^2}\left(\frac{\partial W}{\partial I_1} + \lambda^2\frac{\partial W}{\partial I_2}\right)_{\psi=0}\right]. \quad (9.9.9)$$

In (9.9.7), (9.9.9) the invariants take the values

$$I_1 = \lambda^2 + \frac{2}{\lambda}, \qquad I_2 = 2\lambda + \frac{1}{\lambda^2}, \quad (9.9.10)$$

obtained from (9.9.6) by putting $\psi = 0$, and the derivatives $\partial W/\partial I_1$, $\partial W/\partial I_2$ are then constants. We observe that (9.9.7), (9.9.9) are consistent with the inference from symmetry considerations that M must be an odd function of ψ while N and P are even functions of this variable.

We observe that if after simple extension the value of P is chosen so that the internal radius of the tube remains constant at its value $\rho_2 \lambda^{-\frac{1}{2}}$ during torsion, then in (9.9.4) $K = 0$ and in (9.9.9) $\beta = 0$.

Alternatively, we may consider the case where the inner surface of the tube is force free, so that $P = 0$, and examine the variations of radial dimensions with small amounts of torsion ψ. Putting

$$\left.\begin{array}{ll} \mu_1 = r_1/\rho_1, & \mu_2 = r_2/\rho_2 \\ \epsilon_1 = \lambda\mu_1^2 - 1, & \epsilon_2 = \lambda\mu_2^2 - 1 \end{array}\right\}, \quad (9.9.11)$$

from (9.9.4) we have

$$\epsilon_1 = -\frac{\beta\psi^2}{\rho_1^2} + O(\psi^4), \qquad \epsilon_2 = -\frac{\beta\psi^2}{\rho_2^2} + O(\psi^4), \quad (9.9.12)$$

and the constant β is obtained by putting $P = 0$ in (9.9.8). Thus

$$\beta = \frac{\lambda^2 \rho_1^2 \rho_2^2 (\partial W/\partial I_1)_{\psi=0}}{2[(\partial W/\partial I_1) + \lambda^2(\partial W/\partial I_2)]_{\psi=0}}. \quad (9.9.13)$$

The second of equations (9.9.12) gives a measure of the decrease ΔV of the volume V contained by the inner surface of the tube consequent upon infinitesimally small amounts of torsion ψ. For, evidently,

$$\epsilon_2 = -\frac{\Delta V}{V}. \quad (9.9.14)$$

In the case of a solid circular cylinder, the conditions of § 2.2 apply and

$$\rho_2 = 0, \qquad \mu = \lambda^{-\frac{1}{2}}. \quad (9.9.15)$$

The formulae (9.9.1) for the torsional couple M and the resultant normal force N on the plane ends are then replaced by

$$\left.\begin{aligned} M &= 4\pi\psi \int_0^{\rho_1} \left(\frac{\partial W}{\partial I_1} + \frac{1}{\lambda}\frac{\partial W}{\partial I_2}\right)\rho^3\, d\rho \\ N &= 4\pi\left(\lambda - \frac{1}{\lambda^2}\right)\int_0^{\rho_1}\left(\frac{\partial W}{\partial I_1} + \frac{1}{\lambda}\frac{\partial W}{\partial I_2}\right)\rho\, d\rho - 2\pi\psi^2\int_0^{\rho_1}\left(\frac{\partial W}{\partial I_1} + \frac{2}{\lambda}\frac{\partial W}{\partial I_2}\right)\rho^3\, d\rho \end{aligned}\right\}, \quad (9.9.16)$$

where, from (9.9.3),

$$I_1 = \lambda^2 + \frac{2}{\lambda} + \psi^2\lambda\rho^2, \qquad I_2 = 2\lambda + \frac{1}{\lambda^2} + \psi^2\rho^2. \quad (9.9.17)$$

For small values of ψ, we have, in place of (9.9.7), the relations

$$\left.\begin{aligned} \left[\frac{M}{\psi}\right]_{\psi=0} &= \pi\left(\frac{\partial W}{\partial I_1} + \frac{1}{\lambda}\frac{\partial W}{\partial I_2}\right)_{\psi=0}\rho_1^4 \\ [N]_{\psi=0} &= 2\pi\left(\lambda - \frac{1}{\lambda^2}\right)\left(\frac{\partial W}{\partial I_1} + \frac{1}{\lambda}\frac{\partial W}{\partial I_2}\right)_{\psi=0}\rho_1^2 \end{aligned}\right\}, \quad (9.9.18)$$

and from these it follows that

$$\frac{[N]_{\psi=0}\rho_1^2}{[M/\psi]_{\psi=0}} = 2\left(\lambda - \frac{1}{\lambda^2}\right), \quad (9.9.19)$$

which implies that the ratio between the load N necessary to produce the simple extension of ratio λ and the torsional modulus for infinitesimal torsions superposed on that simple extension is independent of the form of W.

When the strain energy function has the Mooney form (1.14.3), equations (9.9.16) yield

$$\left.\begin{aligned} M &= \pi\psi\rho_1^4(C_1+C_2/\lambda) \\ N &= 2\pi\rho_1^2\left(C_1+\frac{1}{\lambda}C_2\right)\left(\lambda - \frac{1}{\lambda^2}\right) - \pi\psi^2\rho_1^4\left(\tfrac{1}{2}C_1+\frac{1}{\lambda}C_2\right) \end{aligned}\right\}. \quad (9.9.20)$$

If $M/(\pi\psi\rho_1^4)$ is then plotted against $1/\lambda$ the slope of the resulting straight line gives C_2 and its intercept on the line $1/\lambda = 1$ gives C_1+C_2. Again, if $\lambda = 1$ (9.9.20) become

$$\left.\begin{aligned} M &= \pi\psi\rho_1^4(C_1+C_2) \\ N &= -\pi\psi^2\rho_1^4(\tfrac{1}{2}C_1+C_2) \end{aligned}\right\}. \quad (9.9.21)$$

9.10   EXPERIMENTAL APPLICATIONS

If measurements of M and N are made for various values of ψ, the slopes of the curves of M against ψ and N against ψ^2 give values of C_1+C_2, $\tfrac{1}{2}C_1+C_2$ respectively and hence determine C_1 and C_2. Departures from linearity of either of these curves, or of the curve of $M/(\pi\psi\rho_1^4)$ against $1/\lambda$, indicate a departure of the strain energy function from the Mooney form.

For a material with the Mooney form of strain energy function, values of C_1 and C_2 may also be obtained from measurements of the normal traction at different points on the ends of the deformed cylinder. For a cylinder subjected to simple torsion without extension we obtain from (2.2.4) and (2.2.7) with $\lambda = 1$,

$$\tau^{33} = -\psi^2[2\rho_1^2 C_2 + (\rho_1^2-\rho^2)(C_1-2C_2)]. \qquad (9.9.22)$$

If, therefore, measurements of τ^{33} are made over one of the plane ends of the cylinder for a given value of ψ, and the resulting values of τ^{33}/ψ^2 plotted against $\rho_1^2-\rho^2$, the intercept of the resulting straight line on the line $\rho = \rho_1$ yields a value for C_2 and its slope gives a value for C_1-2C_2.

9.10. Torsion of a cylindrical rod

Assuming the Mooney form of strain energy function for rubber, an estimate of the ratio C_2/C_1 was obtained in 1947 by Rivlin† from measurements of the normal end traction on a twisted rubber cylinder. The ends of the cylinder were bonded to plane circular metal disks, one of which contained five circular holes spaced at different distances from its centre, that is, from the axis of the rod. The torsion was produced by rotating this disk in its plane about its centre relative to the other unperforated disk, and the rubber was observed to bulge out into the holes. Rivlin assumed the height of each bulge to be proportional to the pressure which would have been exerted on the rubber at that point by an unperforated plate, and found that it was approximately proportional to ψ^2 in agreement with the expression (9.9.22) for τ^{33}. The experimentally obtained pressure distribution yielded a ratio $C_1/C_2 = 7\cdot 1$ in good agreement with the results of § 9.5.

In subsequent experiments on the torsion of a rod Rivlin and Saunders† concentrated attention on measurements of the resultant forces M and N on the plane ends. The curves obtained for M against ψ (Fig. 9.9) for a large range of variation of ψ (about 2 radians) show

† Loc. cit., p. 288.

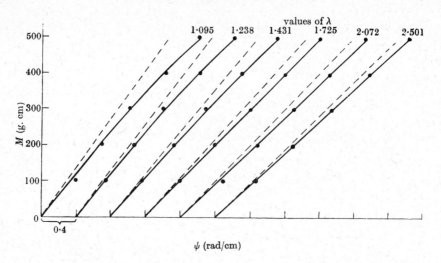

FIG. 9.9. Plot of torsional couple M against amount of torsion ψ for various values of the extension ratio λ (large range of variation λ; the origin is shifted by 0·4 units parallel to the abscissa for each increment in the value of λ).

departures from linearity which are to be expected from the results of § 9.5. The broken lines in Fig. 9.9 are the straight lines

$$M = [M/\psi]_{\psi=0}\psi.$$

These lines are tangential to the corresponding curves of M against ψ at $\psi = 0$, and are calculated by evaluating $\partial W/\partial I_1 + (1/\lambda)\partial W/\partial I_2$ from measurements of $[N]_{\psi=0}$ at various values of λ by means of the second of (9.9.18) and then using the first of these equations to determine $[M/\psi]_{\psi=0}$. The manner in which $\partial W/\partial I_2$ varied with I_2 was found from the known values of $\partial W/\partial I_1 + (1/\lambda)\partial W/\partial I_2$ by the method of § 9.6, that is, by assuming the functional form (9.5.9) for W and a given value (0·16) for the ratio $(\partial W/\partial I_2)/(\partial W/\partial I_1)$ at a fixed reference point ($I_2 = 4.25$). The dependence of $\partial W/\partial I_2$ upon I_2 thus obtained is shown in Fig. 9.10.

From measurements of M, N, and ψ at various values of λ Rivlin and Saunders also found the values of the ratio

$$\{2(\lambda - 1/\lambda^2)[M/\psi]_{\psi=0}\}/\{\rho_1^2[N]_{\psi=0}\}$$

to lie in the range 0·98 to 1·00, in good agreement with (9.9.19).

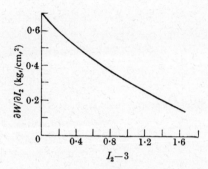

Fig. 9.10. Plot of $\partial W/\partial I_2$ against I_2-3.

9.11. Inflation, extension, and torsion of a cylindrical tube

The results of §§ 9.5–9.10 have been confirmed by experiments reported by Gent and Rivlin.† In these, measurements of M, N, and P were made on tubes subjected to small amounts of torsion ψ superposed on an initial finite extension λ, the inflating pressure P being adjusted so that the superposed torsion produced no further change in radial dimensions. The formulae (9.9.7), (9.9.9) could thus be applied with $\beta = 0$. In each experiment the (M, ψ) and (P, ψ^2) curves were found to be linear, in agreement with (9.9.7) and (9.9.9), the slopes of these curves giving values for $[M/\psi]_{\psi=0}$ and $[P/\psi^2]_{\psi=0}$. From (9.9.9) the derivative $\partial W/\partial I_1$ was evaluated and in confirmation of earlier work found to remain approximately constant over the limited range $(1 \cdot 0 < \lambda < 1 \cdot 35)$ of extension ratios examined. This constant value with the second of (9.9.7) and the measurements of $[N]_{\psi=0}$, then yielded values of $\partial W/\partial I_2$ throughout the given range of values of λ. Assuming a strain energy of the form (9.5.9), these results were sufficient, with (9.9.10), to determine the variation of $\partial W/\partial I_2$ with I_2 in the region investigated. Measurements of $[M/\psi]_{\psi=0}$, with the first of (9.9.7), provided confirmation of the values of $\partial W/\partial I_1 + (1/\lambda)(\partial W/\partial I_2)$ (and hence $\partial W/\partial I_2$) obtained from the values of $[N]_{\psi=0}$. The method, which is particularly suitable for low values of I_2 (< 4), gave results in general agreement with the experiments of §§ 9.5–9.10.

Further confirmation of the form of W was obtained by Gent and Rivlin from measurements of the change during torsion of the volume inside a rubber tube, these measurements yielding, by virtue of (9.9.12)–(9.9.14), values for the ratio $(\partial W/\partial I_2)/(\partial W/\partial I_1)$.

† Loc. cit., p. 288.

For a discussion of the bearing of these experiments upon the molecular theory of elasticity for rubber, and a tentative explanation of the small amounts of hysteresis which occur in the measurements of M and N, reference may be made to the original paper.

9.12. Further experiments

For other experiments related to the theory of large elastic deformations reference may be made to original papers. Gent and Rivlin† have considered the information about the form of W for rubber obtainable from measurements on rubber tubes turned inside out. The same workers‡ have confirmed experimentally theoretical predictions for the small torsion of stretched prisms. From the theoretical results for the symmetrical expansion of a thick spherical shell,§ Gent and Lindley‖ have been able to suggest a possible explanation for the internal rupture, at comparatively small deformations, of rubber test pieces when bonded to metal end pieces.

Further confirmation of the theoretical and experimental work on the form of W for rubber†† is provided by the work on rubber membranes described in Chapter IV.

† A. N. Gent and R. S. Rivlin, *Proc. phys. Soc.* B, **65** (1952), 118. A theoretical treatment of this problem is given by R. S. Rivlin, *Phil. Trans. R. Soc.* A, **242** (1949), 173, and *T.E.*, § 3.6.

‡ A. N. Gent and R. S. Rivlin, *Proc. phys. Soc.* B, **65** (1952), 645. The theory is given by A. E. Green and R. T. Shield, *Phil. Trans. R. Soc.* A, **244** (1951), 47, and also in *T.E.*, § 4.3.

§ *T.E.*, § 3.10.

‖ A. N. Gent and P. B. Lindley, *Proc. R. Soc.* A, **249** (1959), 195.

†† See also L. J. Hart-Smith, loc. cit., p. 288.

APPENDIX

REDUCTION OF A MATRIX POLYNOMIAL OF TWO MATRICES

Let $\mathbf{A}$, $\mathbf{B}$ be two 3×3 matrices defined by

$$\mathbf{A} = \|A_{ij}\|, \qquad \mathbf{B} = \|B_{ij}\|. \tag{A1}$$

Consider any product $\mathbf{Q}$ of these defined by

$$\mathbf{Q} = \mathbf{A}^{\alpha_1}\mathbf{B}^{\beta_1}\mathbf{A}^{\alpha_2}\mathbf{B}^{\beta_2}\ldots\mathbf{A}^{\alpha_r}\mathbf{B}^{\beta_r} \tag{A2}$$

and formed by a finite number of successive multiplications of the matrices in any order. The indices $\alpha_1, \beta_1, \ldots, \alpha_r, \beta_r$ are positive integers or zero. A polynomial $\mathbf{P}$ in the matrices $\mathbf{A}$, $\mathbf{B}$ is defined as any sum of polynomials of the type (A2), with scalar coefficients which are expressible as polynomials in the elements of the matrices $\mathbf{A}$, $\mathbf{B}$.

We now reduce $\mathbf{P}$ to a canonical form† by the help of the Cayley–Hamilton theorem for a matrix $\mathbf{A}$, namely

$$\mathbf{A}^3 - \mathbf{A}^2 \operatorname{tr} \mathbf{A} + \tfrac{1}{2}\mathbf{A}[(\operatorname{tr} \mathbf{A})^2 - \operatorname{tr} \mathbf{A}^2] - \mathbf{I} \det \mathbf{A} = 0, \tag{A3}$$

where

$$6 \det \mathbf{A} = (\operatorname{tr} \mathbf{A})^3 - 3 \operatorname{tr} \mathbf{A} \operatorname{tr} \mathbf{A}^2 + 2 \operatorname{tr} \mathbf{A}^3, \tag{A4}$$

together with results which can be deduced from (A3). If in (A3) we replace $\mathbf{A}$ by $\mathbf{A} + \lambda \mathbf{B}$, where λ is a scalar, and compare coefficients of λ, we obtain

$$\mathbf{ABA} = -\mathbf{A}^2\mathbf{B} - \mathbf{BA}^2 + \mathbf{G}(\mathbf{A}, \mathbf{B}), \tag{A5}$$

where

$$\mathbf{G}(\mathbf{A}, \mathbf{B}) = \mathbf{A}^2 \operatorname{tr} \mathbf{B} + (\mathbf{AB} + \mathbf{BA}) \operatorname{tr} \mathbf{A} + \tfrac{1}{2}\mathbf{B}[\operatorname{tr} \mathbf{A}^2 - (\operatorname{tr} \mathbf{A})^2] +$$
$$+ \mathbf{A}[\operatorname{tr} \mathbf{AB} - \operatorname{tr} \mathbf{A} \operatorname{tr} \mathbf{B}] + \mathbf{I}[\operatorname{tr} \mathbf{A}^2\mathbf{B} - \operatorname{tr} \mathbf{A} \operatorname{tr} \mathbf{AB} + \tfrac{1}{2} \operatorname{tr} \mathbf{B}\{(\operatorname{tr} \mathbf{A})^2 - \operatorname{tr} \mathbf{A}^2\}]. \tag{A6}$$

Also, replacing $\mathbf{B}$ by $\mathbf{B}^2$ in (A5) we have

$$\mathbf{AB}^2\mathbf{A} = -\mathbf{A}^2\mathbf{B}^2 - \mathbf{B}^2\mathbf{A}^2 + \mathbf{G}(\mathbf{A}, \mathbf{B}^2). \tag{A7}$$

If we multiply (A5) on the left by $\mathbf{A}$ and then on the right by $\mathbf{A}$, add the results and use (A6) and (A3), we obtain

$$\mathbf{A}^2\mathbf{BA} = -\mathbf{ABA}^2 + \mathbf{ABA} \operatorname{tr} \mathbf{A} + \mathbf{A}^2 \operatorname{tr} \mathbf{AB} +$$
$$+ \mathbf{A}(\operatorname{tr} \mathbf{A}^2\mathbf{B} - \operatorname{tr} \mathbf{A} \operatorname{tr} \mathbf{AB}) - \mathbf{B} \det \mathbf{A} + \mathbf{I} \det \mathbf{A} \operatorname{tr} \mathbf{B}. \tag{A8}$$

Again, multiplying (A8) on the right by $\mathbf{A}$ and using (A3) gives

$$\mathbf{A}^2\mathbf{BA}^2 = \tfrac{1}{2}\mathbf{ABA}[(\operatorname{tr} \mathbf{A})^2 - \operatorname{tr} \mathbf{A}^2] - (\mathbf{AB} + \mathbf{BA}) \det \mathbf{A} + \mathbf{A}^2 \operatorname{tr} \mathbf{A}^2\mathbf{B} +$$
$$+ \mathbf{A}[\det \mathbf{A} \operatorname{tr} \mathbf{B} - \tfrac{1}{2}\{(\operatorname{tr} \mathbf{A})^2 - \operatorname{tr} \mathbf{A}^2\} \operatorname{tr} \mathbf{AB}] + \mathbf{I} \det \mathbf{A} \operatorname{tr} \mathbf{AB}. \tag{A9}$$

† A. J. M. Spencer and R. S. Rivlin, *Arch ration. Mech. Analysis* **2** (1958), 309, 435; **4** (1960), 214. These authors have considered the general case of a polynomial in R matrices. See also R. S. Rivlin, *J. rat. Mech. Analysis* **4** (1955), 681.

Using the definition (A5) for $\mathbf{G}(\mathbf{A}, \mathbf{B})$ we can verify that, if $\mathbf{X}$ is any 3×3 matrix,

$$3\mathbf{A}^2\mathbf{X}\mathbf{B}^2 = \mathbf{A}^2\mathbf{G}(\mathbf{B}, \mathbf{X}) - \mathbf{G}(\mathbf{A}, \mathbf{B})\mathbf{X}\mathbf{B} + \mathbf{A}\mathbf{G}(\mathbf{B}, \mathbf{A}\mathbf{X}) - \mathbf{G}(\mathbf{A}, \mathbf{B}^2)\mathbf{X} + \mathbf{G}(\mathbf{B}, \mathbf{A}^2\mathbf{X}). \tag{A10}$$

In particular,

$$3\mathbf{A}^2\mathbf{B}\mathbf{A}\mathbf{B}^2 = \mathbf{A}^2\mathbf{G}(\mathbf{B}, \mathbf{B}\mathbf{A}) - \mathbf{G}(\mathbf{A}, \mathbf{B})\mathbf{B}\mathbf{A}\mathbf{B} + \mathbf{A}\mathbf{G}(\mathbf{B}, \mathbf{A}\mathbf{B}\mathbf{A}) -$$
$$- \mathbf{G}(\mathbf{A}, \mathbf{B}^2)\mathbf{B}\mathbf{A} + \mathbf{G}(\mathbf{B}, \mathbf{A}^2\mathbf{B}\mathbf{A}). \tag{A11}$$

Also, multiplying (A8) on the right by $\mathbf{B}^2$, we have

$$\mathbf{A}\mathbf{B}\mathbf{A}^2\mathbf{B}^2 = -\mathbf{A}^2\mathbf{B}\mathbf{A}\mathbf{B}^2 + \mathbf{A}\mathbf{B}\mathbf{A}\mathbf{B}^2 \operatorname{tr} \mathbf{A} + \mathbf{A}^2\mathbf{B}^2 \operatorname{tr} \mathbf{A}\mathbf{B} +$$
$$+ \mathbf{A}\mathbf{B}^2(\operatorname{tr} \mathbf{A}^2\mathbf{B} - \operatorname{tr} \mathbf{A} \operatorname{tr} \mathbf{A}\mathbf{B}) - \mathbf{B}^3 \det \mathbf{A} + \mathbf{B}^2 \det \mathbf{A} \operatorname{tr} \mathbf{B},$$

and with the help of (A3) and (A5) this becomes

$$\mathbf{A}\mathbf{B}\mathbf{A}^2\mathbf{B}^2 =$$
$$-\mathbf{A}^2\mathbf{B}\mathbf{A}\mathbf{B}^2 - (\mathbf{A}^2\mathbf{B}^3 + \mathbf{B}\mathbf{A}^2\mathbf{B}^2) \operatorname{tr} \mathbf{A} + \mathbf{G}(\mathbf{A}, \mathbf{B})\mathbf{B}^2 \operatorname{tr} \mathbf{A} + \mathbf{A}^2\mathbf{B}^2 \operatorname{tr} \mathbf{A}\mathbf{B} +$$
$$+ \mathbf{A}\mathbf{B}^2(\operatorname{tr} \mathbf{A}^2\mathbf{B} - \operatorname{tr} \mathbf{A} \operatorname{tr} \mathbf{A}\mathbf{B}) + [\tfrac{1}{2}\mathbf{B}\{(\operatorname{tr} \mathbf{B})^2 - \operatorname{tr} \mathbf{B}^2\} - \mathbf{I} \det \mathbf{B}] \det \mathbf{A}. \tag{A12}$$

By using (A3)–(A12) it follows that any matrix product $\mathbf{Q}$ of $\mathbf{A}$ and $\mathbf{B}$ can be expressed in the form

$$\mathbf{Q} = \phi_0\mathbf{I} + \phi_1\mathbf{A} + \phi_2\mathbf{A}^2 + \phi_3\mathbf{B} + \phi_4\mathbf{B}^2 +$$
$$+ \phi_5\mathbf{A}\mathbf{B} + \phi_6\mathbf{B}\mathbf{A} + \phi_7\mathbf{A}^2\mathbf{B} + \phi_8\mathbf{B}\mathbf{A}^2 + \phi_9\mathbf{A}\mathbf{B}^2 +$$
$$+ \phi_{10}\mathbf{B}^2\mathbf{A} + \phi_{11}\mathbf{A}^2\mathbf{B}^2 + \phi_{12}\mathbf{B}^2\mathbf{A}^2 + \phi_{13}\mathbf{A}\mathbf{B}\mathbf{A}^2 +$$
$$+ \phi_{14}\mathbf{B}\mathbf{A}\mathbf{B}^2 + \phi_{15}\mathbf{A}\mathbf{B}^2\mathbf{A}^2 + \phi_{16}\mathbf{B}\mathbf{A}^2\mathbf{B}^2, \tag{A13}$$

where the ϕ's are polynomials of scalar invariants of $\mathbf{A}$ and $\mathbf{B}$ expressed as polynomials in their elements.

Now suppose that any product $\mathbf{\Pi}$ formed from matrices $\mathbf{A}$, $\mathbf{B}$ is of degree m in $\mathbf{A}$ and n in $\mathbf{B}$. Then $\mathbf{\Pi}$ may be expressed either as $\mathbf{AQ}$ or $\mathbf{BQ}$ where $\mathbf{Q}$ is a product of the form (A13) formed from $\mathbf{A}$ and $\mathbf{B}$. Suppose $\mathbf{\Pi} = \mathbf{AQ}$, then $\mathbf{Q}$ is of degree $m-1$ in $\mathbf{A}$ and n in $\mathbf{B}$. Also, from (A13),

$$\operatorname{tr} \mathbf{\Pi} = \phi_0 \operatorname{tr} \mathbf{A} + \phi_1 \operatorname{tr} \mathbf{A}^2 + \phi_2 \operatorname{tr} \mathbf{A}^3 + \phi_3 \operatorname{tr} \mathbf{A}\mathbf{B} +$$
$$+ \phi_4 \operatorname{tr} \mathbf{A}\mathbf{B}^2 + \phi_5 \operatorname{tr} \mathbf{A}^2\mathbf{B} + \phi_6 \operatorname{tr} \mathbf{A}\mathbf{B}\mathbf{A} +$$
$$+ \phi_7 \operatorname{tr} \mathbf{A}^3\mathbf{B} + \phi_8 \operatorname{tr} \mathbf{A}\mathbf{B}\mathbf{A}^2 + \phi_9 \operatorname{tr} \mathbf{A}^2\mathbf{B}^2 +$$
$$+ \phi_{10} \operatorname{tr} \mathbf{A}\mathbf{B}^2\mathbf{A} + \phi_{11} \operatorname{tr} \mathbf{A}^3\mathbf{B}^2 + \phi_{12} \operatorname{tr} \mathbf{A}\mathbf{B}^2\mathbf{A}^2 +$$
$$+ \phi_{13} \operatorname{tr} \mathbf{A}^2\mathbf{B}\mathbf{A}^2 + \phi_{14} \operatorname{tr} \mathbf{A}\mathbf{B}\mathbf{A}\mathbf{B}^2 + \phi_{15} \operatorname{tr} \mathbf{A}^2\mathbf{B}^2\mathbf{A}^2 +$$
$$+ \phi_{16} \operatorname{tr} \mathbf{A}\mathbf{B}\mathbf{A}^2\mathbf{B}^2. \tag{A14}$$

If $\mathbf{X}$, $\mathbf{Y}$ are any two 3×3 matrices, then

$$\operatorname{tr} \mathbf{X}\mathbf{Y} = \operatorname{tr} \mathbf{Y}\mathbf{X}.$$

Thus, using results of this type and equations (A3) and (A5), we see that $\operatorname{tr} \mathbf{\Pi}$

may be expressed as a polynomial in

$$\left.\begin{array}{llll} \text{tr } \mathbf{A}, & \text{tr } \mathbf{A}^2, & \text{tr } \mathbf{A}^3 \\ \text{tr } \mathbf{B}, & \text{tr } \mathbf{B}^2, & \text{tr } \mathbf{B}^3 \\ \text{tr } \mathbf{AB}, & \text{tr } \mathbf{A}^2\mathbf{B}, & \text{tr } \mathbf{AB}^2, & \text{tr } \mathbf{A}^2\mathbf{B}^2 \end{array}\right\}, \tag{A15}$$

and

$$\text{tr } \mathbf{ABA}^2\mathbf{B}^2, \tag{A16}$$

and also traces of products of $\mathbf{A}$ and $\mathbf{B}$ of degree $m-1$ or less in $\mathbf{A}$ and n or less in $\mathbf{B}$. Similarly, if $\mathbf{\Pi}$ can be expressed as $\mathbf{BQ}$ then tr $\mathbf{\Pi}$ can be expressed as a polynomial in the traces (A15) and

$$\text{tr } \mathbf{BAB}^2\mathbf{A}^2, \tag{A17}$$

and also traces of products of $\mathbf{A}$ and $\mathbf{B}$ of degree m in $\mathbf{A}$ and $n-1$ in $\mathbf{B}$. It follows by repeated application of these results that tr $\mathbf{\Pi}$, and hence any scalar invariant, can be expressed as a polynomial in the scalar invariants (A15), (A16), and (A17).

Finally, any matrix polynomial in the matrices $\mathbf{A}$, $\mathbf{B}$ can be expressed in the form

$$\begin{aligned} \mathbf{P} = {} & \alpha_0 \mathbf{I} + \alpha_1 \mathbf{A} + \alpha_2 \mathbf{A}^2 + \alpha_3 \mathbf{B} + \alpha_4 \mathbf{B}^2 + \\ & + \alpha_5 \mathbf{AB} + \alpha_6 \mathbf{BA} + \alpha_7 \mathbf{A}^2\mathbf{B} + \alpha_8 \mathbf{BA}^2 + \alpha_9 \mathbf{AB}^2 + \\ & + \alpha_{10} \mathbf{B}^2\mathbf{A} + \alpha_{11} \mathbf{A}^2\mathbf{B}^2 + \alpha_{12} \mathbf{B}^2\mathbf{A}^2 + \alpha_{13} \mathbf{ABA}^2 + \\ & + \alpha_{14} \mathbf{BAB}^2 + \alpha_{15} \mathbf{AB}^2\mathbf{A}^2 + \alpha_{16} \mathbf{BA}^2\mathbf{B}^2, \end{aligned} \tag{A18}$$

where the α's are polynomials in the invariants (A15)–(A17).

Suppose now that $\mathbf{A}$ and $\mathbf{B}$ are *symmetric* matrices and $\mathbf{P}$ is a *symmetric* matrix polynomial in $\mathbf{A}$ and $\mathbf{B}$. Then, in view of (A8), (A7), and (A5), $\mathbf{P}$ reduces to the form

$$\begin{aligned} \mathbf{P} = {} & \alpha_0 \mathbf{I} + \alpha_1 \mathbf{A} + \alpha_2 \mathbf{A}^2 + \alpha_3 \mathbf{B} + \alpha_4 \mathbf{B}^2 + \\ & + \alpha_5 (\mathbf{AB} + \mathbf{BA}) + \alpha_6 (\mathbf{A}^2\mathbf{B} + \mathbf{BA}^2) + \\ & + \alpha_7 (\mathbf{AB}^2 + \mathbf{B}^2\mathbf{A}) + \alpha_8 (\mathbf{A}^2\mathbf{B}^2 + \mathbf{B}^2\mathbf{A}^2), \end{aligned} \tag{A19}$$

where the α's are polynomials in the invariants (A15)–(A17). Since $\mathbf{A}$ and $\mathbf{B}$ are symmetric

$$\text{tr } \mathbf{ABA}^2\mathbf{B}^2 = \text{tr } \mathbf{B}^2\mathbf{A}^2\mathbf{BA} = \text{tr } \mathbf{A}^2\mathbf{BAB}^2.$$

Hence, using (A12), tr $\mathbf{ABA}^2\mathbf{B}^2$ can be expressed in terms of the invariants (A15). Similarly, tr $\mathbf{BAB}^2\mathbf{A}^2$ can be expressed in terms of the invariants (A15). The polynomials $\alpha_0, \ldots, \alpha_8$ in (A19) may therefore be expressed in terms of the invariants (A15) only.

Again, suppose that $\mathbf{A}$ is *antisymmetric*, $\mathbf{B}$ is *symmetric*, and $\mathbf{P}$ is an *antisymmetric* polynomial of $\mathbf{A}$ and $\mathbf{B}$. Then, using (A8), $\mathbf{P}$ reduces to

$$\begin{aligned} \mathbf{P} = {} & \beta_1 \mathbf{A} + \beta_2 (\mathbf{AB} + \mathbf{BA}) + \beta_3 (\mathbf{AB}^2 + \mathbf{B}^2\mathbf{A}) + \beta_4 (\mathbf{A}^2\mathbf{B} - \mathbf{BA}^2) + \\ & + \beta_5 (\mathbf{A}^2\mathbf{B}^2 - \mathbf{B}^2\mathbf{A}^2) + \beta_6 (\mathbf{BA}^2\mathbf{B}^2 - \mathbf{B}^2\mathbf{A}^2\mathbf{B}), \end{aligned} \tag{A20}$$

where $\beta_1, \ldots, \beta_6$ are polynomials in

$$\begin{matrix} \operatorname{tr} \mathbf{A}^2, & \operatorname{tr} \mathbf{B}, & \operatorname{tr} \mathbf{B}^2, & \operatorname{tr} \mathbf{B}^3 \\ \operatorname{tr} \mathbf{A}^2\mathbf{B}, & \operatorname{tr} \mathbf{A}^2\mathbf{B}^2, & \operatorname{tr} \mathbf{A}\mathbf{B}\mathbf{A}^2\mathbf{B}^2 & \end{matrix} \quad\quad (A21)$$

if we note that $\operatorname{tr} \mathbf{A}$, $\operatorname{tr} \mathbf{A}^3$, $\operatorname{tr} \mathbf{A}\mathbf{B}$, $\operatorname{tr} \mathbf{A}\mathbf{B}^2$ are all zero and

$$\operatorname{tr} \mathbf{B}\mathbf{A}\mathbf{B}^2\mathbf{A}^2 = -\operatorname{tr} \mathbf{A}\mathbf{B}\mathbf{A}^2\mathbf{B}^2. \quad\quad (A22)$$

AUTHOR INDEX

Adkins, J. E., 26, 33, 37, 38, 59, 75, 87, 89, 101, 105, 123, 127, 133, 143, 165, 166, 168, 200, 218, 225, 227, 228, 229, 239, 253, 272.

Baker, M., 27.
Biot, M. A., 273.
Blackburn, W. S., 185, 188.

Carlson, D. E., 200.
Chadwick, P., 273.
Coleman, B. D., 273.
Corneliussen, A. H., 133.
Crisp, J. D. C., 134.

Dana, E. S., 11.
Dana, J. D., 11.
Doyle, T. C., 1.
Duhamel, J. M. C., 273.

Ericksen, J. L., 1, 24, 27, 37, 84, 133.

Ford, W. E., 11.

Gee, G., 290.
Gent, A. N., 288, 313, 314.
Green, A. E., 1, 8, 37, 59, 81, 101, 105, 127, 133, 143, 174, 178, 182, 185, 188, 198, 200, 225, 227, 314.

Hart-Smith, L. J., 134, 288, 314.
Hurlbut, C. S., 11.

Jeffreys, H., 273.

Klingbeil, W. W., 101, 134, 161, 168.
Kydoniefs, A. D., 133, 185.

Lindley, P. B., 314.
Love, A. E. H., 230, 296, 300.

Misicu, M., 174.
Mooney, M., 27, 101, 302, 303, 304, 311.
Murnagham, F. D., 185, 199.

Muskhelishvili, N. I., 198, 219.

Nicholas, G. C., 133, 143, 200.
Noll, W., 1, 273, 290.

Pipkin, A. C., 10, 273, 280.

Radok, J. R. M., 219.
Rivlin, R. S., 1, 9, 11, 24, 37, 133, 156, 165, 166, 168, 174, 185, 197, 198, 228, 243, 269, 273, 280, 288, 289, 290, 297, 298, 299, 300, 302, 305, 306, 311, 312, 313, 314, 315.

Saunders, D. W., 269, 288, 289, 297, 298, 299, 300, 302, 305, 306, 312.
Sheng, P. L., 185.
Shield, R. T., 101, 105, 127, 133, 134, 161, 168, 185, 198, 200, 225, 314.
Signorini, A., 174, 175.
Smith, G. F., 1, 9, 11, 290.
Sokolnikoff, I. S., 159, 191.
Southwell, R. V., 96.
Spencer, A. J. M., 133, 174, 185, 315.
Spratt, E. B., 174, 178, 182, 185, 199.
Stern, J., 290.
Stoppelli, F., 175.

Thomas, A. G., 133, 156, 288.
Topakoglu, C., 174.
Toupin, R. A., 1.
Treloar, L. R. G., 168, 288, 289, 290, 296, 300, 302, 305.
Truesdell, C., 1, 133, 273.

Voigt, W., 273.

Weyl, H., 7.
Wilkes, E. W., 8, 37, 101.
Wineman, A. S., 10, 280.

Zerna, W., 1.

SUBJECT INDEX

acceleration, 5; *see also* gradient.
aeolotropic bodies, 10, 292; *see also* crystal class, curvilinear aeolotropy, cylindrical aeolotropy, flexure, transversely isotropic bodies.
Airy's stress function, 101, 104, 136, 238.
anisotropy, *see* aeolotropic bodies.
annulus: concentric circular, 97; cylindrical, 96; eccentric circular, 100; elliptic, 100.
axially symmetric problems, 133, 156.

base vectors, 2, 30, 102, 178, 242.
boundary conditions, 96, 107, 118, 138, 198, 207.
— value problems: for circular boundary, 220; for network of cords, 249; for reinforced sheet, 243; simultaneous solution of, 216; *see also* circular hole, inclusion.

catenary of revolution, *see* shell.
Cayley–Hamilton equation, 282, 315; generalizations of, 315.
central inversion, 13.
centre of symmetry, 27, 283, 285.
circular cylinder; compressible, extension and torsion of, 67; incompressible, extension and torsion of, 43.
— cylindrical tube, *see* tube.
— hole, 152, 222.
— inclusion, 152, 226; *see also* boundary value problems.
classical theory, 8, 19, 101, 112, 133, 177, 191, 200, 202, 212, 285.
complex displacements, 114, 202; potential, 92, 195, 200, 203, 209, 215.
— strain tensor, 115; stress tensor, 92, 116, 197.
— variable, 91, 112, 188, 200; *see also* coordinates.
compressible bodies, 5, 65, 180, 212.
concentric circular cylinders, *see* annulus.
conduction, *see* heat.

conformal transformation, *see* transformation.
constraints, *see* geometrical.
coordinates: complex, 91, 101, 112, 188, 201; convected, 2; curvilinear, 1, 27, 30, 34, 37, 102, 134, 234; in deformed body, 101, 113, 200; in undeformed body, 101, 113, 201, 205; material, *see* convected.
cords, *see* inextensible.
crystal class, 1, 8, 10, 280; cubic, 17, 107; hexagonal, 19, 56, 59, 107; monoclinic, 14, 56, 59, 107, 123; rhombic, 15, 28, 31, 32, 107, 108, 118, 292; tetragonal, 15, 56, 59, 61, 107; triclinic, 13, 27, 32, 56, 59, 118, 292.
cube: under hydrostatic pressure, 185.
cubic system, *see* crystal class.
curvilinear aeolotropy, 29, 38, 102, 292; *see also* cylindrical aeolotropy.
— coordinates, *see* coordinates.
cylinder: deformation of, 188; in generalized shear, 92; *see also* circular cylinder.
cylindrical aeolotropy, 59, 75.
cylindrically symmetrical problems, 39; for compressible bodies, 65, 72, 75, 254; for cylindrically aeolotropic bodies, 59, 72, 75; for incompressible bodies, 39, 59, 77, 255; for isotropic bodies, 39, 84; parametric forms, 75; for reinforced bodies, 228, 252; for transversely isotropic bodies, 39, 65, 82; *see also* circular cylinder, geometrical constraints, torsion, tube.

deformation, 2.

elastic body, 5, 273, 275, 290.
— potential, 1, 5, 37, 273; *see also* energy.
elongation, 292, 301; *see also* extension.
energy: equation, 273; Helmholtz free, 273, 275; internal, 273; kinetic, 274; strain, 1, 5, 8, 29, 35, 56, 58, 59, 61, 64, 70, 72, 76, 229, 272, 273, 288, 293, 298, 303.

entropy, 9, 273, 275, 290; production inequality, 275.
equations of equilibrium, 7, 38, 41, 86, 90, 101, 104, 108, 117, 136, 141, 147, 153, 159, 184; of motion, 4.
existence of series solution, 175.
expansion of spherical shell, 84, 169, 314.
experimental applications of theory, 288.
— determination of strain energy, 293, 297.
experiments on rubber, 167, 289, 296.
extension: combined with flexure, 267; combined with inflation, 44, 68, 261, 307, 313; combined with plane strain, 102; combined with torsion, 43, 67, 307, 313; of reinforced sheet, 232, 267; simple, 186, 232, 262, 267, 297, 301; uniform two-dimensional, 229, 304.

flexure: of aeolotropic cuboid, 57, 69, 74, 82, 123; combined with homogeneous strain, 62, 74, 82, 264; of compressible cuboid, 69, 74, 82; of curved cuboid, 47, 49, 62, 73, 79, 127, 259; of incompressible cuboid, 50, 57, 62, 82, 125, 265; of isotropic cuboid, 84; parametric forms, 82, 265; of reinforced cuboid, 264; of transversely isotropic cuboid, 50, 84.
form invariance, 7, 11; see also crystal class, isotropic bodies, symmetry properties, transversely isotropic bodies.

generalized shear, 87, 127.
geometrical constraints, 33, 228, 271, 293; in cylindrically symmetrical problems, 252; geometrical conditions, 34, 229, 236, 252, 260, 261, 265; produced by cords, 33, 229, 272; in reinforced sheet, 230, 237; see also stress–strain relations.
— relations for deformed shell, 143.
gradient: displacement, 9, 276; temperature, 276.

heat conduction vector, 273, 275, 280, 283.
— flux, 273.

Helmholtz free energy, see energy.
hexagonal system, see crystal class.
Hilbert problem, 201, 219.
hole, see circular.
homogeneous body, 8, 29, 290.
— strain, 55, 230, 294, 297; combined with flexure, 62, 74, 82, 264.
hydrostatic pressure, 185.

inclusion, see circular.
incompressible body, 6, 29, 32, 35, 39, 87, 107, 116, 120, 127, 140, 142, 150, 152, 182, 201, 232, 253, 278, 293.
inextensible cords: layer of, 33, 228, 234, 252, 271; network of, 248; reinforcement by, 33, 228.
infinitesimal deformation, 285; see also classical theory.
inflation: of circular sheet, 161, 305; of spherical shell, 169; see also tube.
internal energy, see energy.
invariants: scalar, 281; strain, 4, 40, 110, 115, 116, 139, 149, 178, 295; theorems on, 7; see also crystal class, transversely isotropic bodies.
isentropic deformation, 290.
isothermal deformation, 290.
isotropic body, 8, 26, 29, 39, 84, 112, 119, 120, 141, 149, 180, 200, 201, 232, 267, 273, 280, 292.
iterative procedure, 155.

kinetic energy, see energy.

loads, 134.

material coordinates, see coordinates.
matrix: anti-symmetric, 281; symmetric, 281.
membrane, circular, 161.
— theory of shells, 133.
molecular theory of rubberlike elasticity, 289, 302, 314.
monoclinic system, see crystal class.
Mooney material, 27, 101, 122, 157, 213, 271, 302, 306, 311; special solutions for, 90, 129, 161, 170.

network of inextensible cords, see inextensible cords.

orthorhombic, *see* crystal class, rhombic.
orthotropic bodies, 28, 59, 82, 118; *see also* crystal class, rhombic.

parametric forms, *see* cylindrically-symmetrical problems, flexure.
physical components of stress, 4, 29, 56, 89, 97, 116, 128, 231, 294.
— stress resultants, 135, 153, 159, 231, 238.
plane deformation: of network of inextensible cords, 248; of reinforced sheet, 234.
— strain, 101; approximation methods, 200; complex variable formulation, 112, 201; for compressible bodies, 107; for incompressible bodies, 107, 120, 201; for isotropic bodies, 112, 119, 201; for rhombic symmetry, 108, 118; for transversely isotropic bodies, 109, 119.
— stress, 133; for isotropic body, 141; for rhombic system, 139; for transversely isotropic bodies, 140.
polynomial basis, 7.
— form of strain energy, 10.
proper orthogonal transformation, *see* transformation.

reciprocal relations, 101, 120, 236.
reinforced body, 33, 228, 271.
— cuboid, 264; curved block, 259; cylindrical tube, 252, 261; isotropic sheet, 232, 267; plane sheet, 229, 267.
reinforcement, *see* inextensible cords, reinforced body.
rhombic system, *see* crystal class.
rigid-body rotation, 9, 24, 276; translation, 9, 276.
rod, cylindrical, *see* circular cylinder, torsion.
rubber: balloon, 170; bush mounting, 96; experiments on, 156, 168, 296; rupture of, 314; strain energy function for, 27, 39, 156, 289, 298; *see also* molecular theory, Mooney material.

scalar invariant, *see* invariants.
second-order theory, 178, 192, 200.

series expansion: for displacement, 178, 202; for invariants, 178, 202; for strains, 178; for stress, 180.
— solution, 175, 200.
shear; of cuboid, 57, 187; generalized, 87, 127; pure, 296, 300; simple, 57, 87, 128, 187, 296.
shell: equations of equilibrium for, 147; in form of catenary of revolution, 172; general membrane theory of, 133, 143; of revolution, 156; spherical, 84, 169.
simple elongation, *see* extension.
simple shear, *see* shear.
statistical theory, *see* molecular theory.
strain: complex components, 115; dimensionless components, 31; expansion for, 178; tensor, 3.
— energy, *see* energy.
— invariants, *see* invariants.
stress: resultants and couples, 133, 135, 141, 144, 147; *see also* physical stress resultants.
— tensor, *see* tensor.
stress–strain relations: for aeolotropic bodies, 27; in complex coordinates, 118; for compressible bodies, 5, 27, 32, 107, 118, 278; for curvilinear aeolotropy, 32; for incompressible bodies, 5, 27, 32, 35, 107, 118, 279; for isotropic bodies, 29, 112, 119; for materials subject to constraints, 34; for plane strain, 107, 118; for rhombic system, 28, 108, 118; for transversely isotropic materials, 28, 109, 119; for triclinic system, 27.
successive approximations, 134, 174, 200.
surface conditions, 145.
— of revolution, 156, 171.
symmetry properties, 8, 10, 31, 36, 56, 292.
— transformations, 11; *see also* crystal class.

tensor: cartesian, 3; metric, 2, 30, 33, 37, 102, 114, 138, 144, 179, 235; strain, 3, 33, 285; stress, 4, 29, 33, 104, 116, 180; surface, 135; symmetric, 5.

torsion: of circular cylinder, 43, 67, 309, 310; of circular tube, 44, 68, 307, 313; classical theory, 191; experiments on, 307; second-order theory, 192; of stretched prisms, 314.

transformation: conformal, 93, 99; orthogonal, 26; proper orthogonal, 7, 9, 24, 26, 276, 277, 284; tensor, 30, 33, 113, 116, 197.

transversely isotropic bodies, 19, 24, 28, 31, 39, 59, 65, 82, 84, 107, 109, 119, 126, 140, 261, 263, 293.

triclinic system, *see* crystal class.

two-dimensional problems, 101, 200, 228.

— extension, 229, 297, 304.

tube: reinforced with cords, 261; under finite extension, 44, 68, 80, 261, 313; under finite inflation, 44, 68, 80, 261, 313; under finite torsion, 44, 68, 80, 313; turned inside out, 47, 68.

uniqueness of solution, 175.

vector: acceleration, 5; body force, 5; displacement, 3, 174, 178; heat conduction, 273, 276; stress, 4.